M000288152

STATISTICS FOR EARTH AND ENVIRONMENTAL SCIENTISTS

STATISTICS FOR EARTH AND ENVIRONMENTAL SCIENTISTS

John H. Schuenemeyer
Southwest Statistical Consulting, LLC
Cortez, Colorado

Lawrence J. Drew
United States Geological Survey
Reston, Virginia

WILEY

A JOHN WILEY & SONS, INC., PUBLICATION

Published by John Wiley & Sons, Inc., Hoboken, New Jersey.
Published simultaneously in Canada.

For general information on our other products and services or for technical support, please contact
our Customer Care Department within the United States at (800) 762-2974, outside the United States
at (317) 572-3993 or fax (317) 572-4002.

Wiley also publishes its books in a variety of electronic formats. Some content that appears in print
may not be available in electronic formats. For more information about Wiley products, visit our
web site at www.wiley.com.

Library of Congress Cataloging-in-Publication Data:

Schuenemeyer, J. H.
   Statistics for earth and environmental scientists / John H. Schuenemeyer, Lawrence J. Drew.
      p. cm.
   Includes index.
   ISBN 978-0-470-58469-9 (cloth)
   1.   Geology–Statistical methods.   2.   Environmental sciences–Statistical methods.   I. Drew,
Lawrence J. II. Title.
   QE33.2.S82S38 2010
   550.72′7–dc22   2010006819

Printed in Singapore

10 9 8 7 6 5 4 3 2 1

CONTENTS

PREFACE

This book is intended for students and practitioners of the earth and environmental sciences who want to use statistical tools to solve real problems. It provides a range of tools that are used across earth science disciplines. Statistical methods need to be understood because today's interesting problems are complex and *involve uncertainty*. These complex problems include energy resources, climate change, and geologic hazards. Through the use of statistical tools, an understanding of process can be obtained and proper inferences made. In addition, through design of field trials and experiments, these inferences can be made efficiently.

We stress data analysis, modeling, model evaluation, and an understanding of concepts through the use of real data from many earth science disciplines. We also encourage the reader to supplement exercises with data from his or her discipline. The reader, especially the student, is encouraged to collect his or her own data. This may be as simple as the recording of temperature and precipitation or the travel time to work or school. The downside to using real data is that the resulting analysis may not always be as clean as when artificial data are used. In the real world, however, important structure often is not readily apparent. The goal of this book is to engage you, the reader, in the application of statistics to assist in the solution of important problems. We use statistics to explore, model, and forecast.

Statistics is a blend of science and art. Statistics cannot be learned or practiced by rote application of a method. Every problem is different and requires careful examination. The reader needs to gain an understanding of when and why methods work. Sometimes, different methods perform equally well, and at times none of the standard methods are suitable and a new method must be developed. Most often, model assumptions do not hold exactly. A challenge is to determine when they are "close enough." Simulation is a useful tool to evaluate assumptions.

Most of the statistical models in this book are introduced by concept and application, given as equations and then heuristic justification provided rather than a formal proof. Some of the mathematics, especially in the chapters on spatial statistics (Chapter 6) and multivariate analysis (Chapter 7), may be challenging and can be omitted without loss of basic understanding. Those with the necessary background will benefit from having them available.

The use of graphs to illustrate concepts, to identify unusual observations, and to assist in model evaluation is strongly encouraged. Graphs combined with statistics lead to more informative results than those for either taken separately.

There are a variety of paradigms in statistics. We introduce models using the frequentist approach; however, we also discuss Bayesian, nonparametric, and computer-intensive methods. There is no single approach that works best in all circumstances, and we tend to be pragmatic and use whatever method seems appropriate to solve a given problem.

It is assumed that the reader has had at least a one-semester undergraduate course in statistics or equivalent experience and is familiar with basic probability and statistical distributions, including the normal, binomial, and uniform. However, these concepts, with the exception of basic probability, are covered in the first four chapters. Further, we have assumed a general ability to recognize basic matrix computations. The book may be used for a one-semester course for students who have a minimal background in statistics. A more advanced reader or student may begin with concepts from multiple regression, time series, spatial statistics, multivariate analysis, discrete data analysis, and design. During many years of university teaching, presenting workshops, and working with practitioners, we have discovered that the mathematical and statistical background of earth scientists is diverse. At the expense of an occasional uneven level of technical presentation, we have attempted to provide information that will be useful to students and practitioners of varied backgrounds.

The Web site for this book is www.EarthStatBook.com. Appendixes I through V can be downloaded from this Web site. This site also contains other selected data sets, answers to some exercises, R-code for selected exercises and examples, a blog, and an errata page.

Some of the exercises we present are conceptual. Many require the use of a computer. Our expectation is that students will develop insight in solving problems using statistics rather than a rote application of methods and computer programs. We expect that the reader has access to and is familiar with a standard statistical computing package. Most standard statistical packages will do all of the computations required of students to complete the assignments. A major exception may be spatial statistics. Spatial statistical modeling and analysis and most other computations have been done in R, an open-source code statistics and graphics language.

Acknowledgments

We appreciate discussions with many earth scientists. Some have shared their data, and credit is given where used. We especially acknowledge the help of Anne Schuenemeyer, BSN, RN. Without her invaluable assistance, this book would not have come to fruition.

JOHN H. SCHUENEMEYER

LAWRENCE J. DREW

1 Role of Statistics and Data Analysis

1.1 INTRODUCTION

The purpose of this chapter is to provide an overview of important concepts in data analysis and statistics. Types of data, data evaluation, and an introduction to modeling and estimation are presented. Random variation, sampling, and different statistical paradigms are also introduced. These concepts are investigated in detail in subsequent chapters. An important distinguishing feature in many earth and environmental science analyses is the need for spatial sampling. Problems are described in the context of case studies, which use real data from earth science applications.

1.2 CASE STUDIES

Wherever possible, case studies are used to illustrate methods. Two studies that are used extensively in this and subsequent chapters are water-well yield data and observations from an ice core.

1.2.1 Water-Well Yield Case Study

A concern in many parts of the world is the availability of an adequate supply of fresh water. Planners and managers want to know how much water is available. Scientists want to gain a greater understanding of transport systems and the relationship of water to other geologic phenomena. Homeowners who do not have access to municipal water want to know where to drill for water on their property. A subset of 754 water-well yield observations (water-well yield case study, Appendix I; see the book's Web site) from the Blue Ridge Geological Province, Loudoun County, Virginia (Sutphin et al., 2001) is used to illustrate graphical procedures. The variables are water-well yield in gallons per minute (gpm) for rock type Yg (Yg is a Middle Proterozoic Leucocratic Metagranite) and corresponding coordinates called *easting* (x-axis) and *northing* (y-axis). In Chapter 6 spatial applications are discussed.

Statistics for Earth and Environmental Scientists, By John H. Schuenemeyer and Lawrence J. Drew
Copyright © 2011 John Wiley & Sons, Inc.

1.2.2   Ice Core Case Study

Ice core data help scientists understand how Earth's climate works. The U.S. Geological Survey National Ice Core Laboratory (2004) states that "Over the past decade, research on the climate record frozen in ice cores from the Polar Regions has changed our basic understanding of how the climate system works. Changes in temperature and precipitation, which previously we believed, would require many thousands of years to happen were revealed, through the study of ice cores, to have happened in fewer than twenty years. These discoveries have challenged our beliefs about how the climate system works."

A record that can extend back many thousands of years may include temperature, precipitation, and chemical composition. An example of ice core data (ice core case study, Appendix II; see the book's Web site) submitted to the National Geophysical Data Center (2004) by Arkhipov et al. (1987) has been chosen. Data submitted by Arkhipov are from 1987 in the Austfonna Ice Cap of the Svalbard Archipelago and go to a depth of 566 m. Melting of ice masses is thought to be contributing to sea-level rise. Only data in the first 50 m are presented. In addition to depth, the variables are pH, HCO_3^- (hydrogen carbonate), and Cl (chlorine), all in milligrams per liter of water.

1.3   DATA

Sir Arthur Conan Doyle, physician and writer (1859–1930), noted: "It is a capital mistake to theorize before one has data. Insensibly one begins to twist facts to suit theories, instead of theories to suit facts." Data are fundamental to statistics. Most data are obtained from measurements. Increasingly, these measurements are obtained from automated processes such as ground weather stations and satellites. However, field studies are still an important way to collect data. Another important source of data is expert judgment. In areas where few hard data (measurements) are available, such as in the Arctic, experts are called upon to express their opinions.

Data may be rock type, wind speed, orientation of a fault, temperature, and a host of other variables. There are several ways to classify data. Two of the most useful classifications are continuous versus discrete and ratio–interval–ordinal–nominal (Table 1.1). A continuous process generates *continuous data. Discrete data* typically result from counting. Continuous data can be ratio or interval. Discrete data are nominal data. Data classification systems help to select appropriate data analytic techniques and models.

To distinguish between ratio and interval data, consider the following example. With a *ratio scale*, zero means an absence of something, such as rainfall. With an *interval scale*, zero is arbitrary, such as zero degrees Celsius, which is not an absence of temperature and has a different meaning than zero degrees Fahrenheit. The terms *quantitative* and *qualitative* are also used. Sometimes *qualitative data* is considered synonymous with *nominal data*; and sometimes it just refers to something subjective or not precisely defined. *Categorical data* are data classified into categories. The terms *categorical* and *nominal* are sometimes used interchangeably.

TABLE 1.1   Data Classification Systems

	Examples
Continuous vs. Discrete Data	
Continuous: measurements can be made as fine as needed	Temperature, depth, sulfur content, well water yield
Discrete: data that can be categorized into a classification where only a finite number of values are possible, typically count data	Number of days above freezing, number of water wells producing among a sample of 50 holes
Ratio, Interval, Ordinal, and Nominal Data	
Ratio: continuous data where an interval and ratio are meaningful	Depth, sulfur content
Interval: continuous data with no natural zero	Temperature measured in degrees Celsius
Ordinal: data that are rank ordered	Survey responses such as good, fair, poor; water yields as high, medium, low
Nominal: Data that fit into categories; cannot be rank ordered	Location name, rock type

Another way to view data is as primary or secondary. *Primary data* are collected to answer questions related to a particular study, such as sampling a site to ascertain the level of coal bed methane seepage. *Secondary data* are collected for some other purpose and may be used as supportive data. Typically, secondary data are *historical data*. Numerous government agencies routinely collect and publish both types of data on the earth sciences.

In the beginning chapters of this book, properties of a single variable are discussed. This variable may be temperature, water-well yield, or mercury level in fish. A single variable may change over time or space. In later chapters, multivariate data are examined, that is, data where multiple attributes are recorded at each sample point. Most data are multivariate. For example, in a study of climate, the relationships among temperature, atmospheric pressure, and precipitation can be analyzed. Geochemical data often contain dozens of variables.

1.4   SAMPLES VERSUS THE POPULATION: SOME NOTATION

A critical distinction for the analyst to make is sample versus population. A *population* comprises all the data of interest in a study. In most earth science applications, the population is large to infinite. In air quality studies, it may be the troposphere. A *sample* is a subset of a population. A *statistic* is a number derived from a sample. The method used to obtain a sample (the sampling plan) determines the type of inferences that can be made. Generally, in earth science applications, the sample size will be small with respect to the population size. The notations that are used in this book to represent populations and samples are those commonly used in the statistics literature. Statistics involves the use of random variables. A *random variable* is a function, that maps events into numbers. Each number or range of numbers is assigned a probability. There are two types of random variables, continuous and discrete. For example, a discrete random variable Y may be defined as mapping the event of tossing a fair coin into the numbers 0 and 1, corresponding to tail and head, respectively, where the outcome of 0 is assigned a probability of 1/2 and 1 is assigned a probability of 1/2.

- An uppercase italic letter denotes a general reference to a data element: more specifically, a random variable. For example, Y may denote water-well yield in an aquifer.
- A lowercase italic letter refers to a specific element of a population: for example, y. A sample of size n yields from this aquifer is $y_1, y_2, \ldots, y_n$. The distinction between the use of upper- and lowercase italic letters is not always obvious and is of minimal importance for this applied treatment of material. Generally, in this book we refer to specific samples and use lowercase letters.
- Population attributes are generally unknown and are usually denoted by Greek letters. For example, the population mean and *standard deviation* (a measure of variability) of a yield are typically denoted by μ and σ, respectively. When working with several types of random variables, such as temperature and pressure, the authors may use subscripts for clarification, as, for example, μ_Y to indicate the mean of the variable Y.
- Statistics are typically designated by a putting a "hat" over the parameter, as in $\hat{\mu}$ and $\hat{\sigma}$ for the sample mean and standard deviation, respectively, or with upper- or lowercase italic letters. For example, $\overline{Y}$ is the mean of a sample of Y's and S may be used to represent the sample standard deviation; $\overline{y}$ and s represent specific values. Both the hat and italic letter notations are used in this book.

1.5   VECTOR AND MATRIX NOTATION

Vector and matrix notation provide a shorthand way to express columns of numbers. In subsequent chapters, vector and matrix notation are used to express model relationships. Vector and matrix notation also make manipulation of equations easier. A *vector* is a column of numbers or symbols. A sample $y_1, \ldots, y_n$ written in column vector notation is

$$\mathbf{y} = \begin{pmatrix} y_1 \\ \vdots \\ y_n \end{pmatrix}$$

In the text line it is more convenient to denote this as a row vector $\mathbf{y}' = (y_1, \ldots, y_n)$. The prime symbol represents a transpose; some books use a superscript T. A *transpose* of a column vector moves the element in the ith column to the ith row. A *matrix* is a collection of elements whose position is denoted by a row and a column. For example, the matrix **A** with m rows and n columns is

$$\mathbf{A} = \begin{pmatrix} a_{11} & \cdots & a_{1n} \\ \vdots & \ddots & \vdots \\ a_{m1} & \cdots & a_{mn} \end{pmatrix}$$

A bold uppercase letter typically denotes a matrix. A matrix for which $m = n$ is called a *square matrix*. Matrices and vectors may be added, multiplied, and inverted, subject to certain rules and restrictions. Readers wishing to learn more about matrix computation are referred to works by Gentle (2007) and Golub and Van Loan (1996).

1.6   FREQUENCY DISTRIBUTIONS AND HISTOGRAMS

The importance of graphing data is stressed repeatedly because its application is fundamental to understanding data, including unusual and possibly erroneous values. One way to describe univariate data (a single variable) is to construct a *frequency distribution*, which is a tabulation of data into classes, and then graph it. The graph, called a *histogram*, provides general information about the form of a sample and may be useful in constructing a theoretical model. Sometimes the terms frequency distribution and histogram are used interchangeably. For a small data set, a line plot often suffices.

In Figure 1.1a, the first seven water-well yield observations for rock type Yg are graphed. A concentration of points at smaller yields and two large values are observed, which may warrant further investigation. For larger data sets, a line plot is not useful. Figure 1.1b is a histogram of the 81 samples in the water-well yield case study for rock type Yg. The vertical axis is frequency or counts. (An alternative is to display relative frequency, which is the percentage or fraction of the counts in each class.) The histogram indicates, for example, that slightly over 50 of the 81 observations are between 0 and 10, slightly less than 20 are between 10 and 20, and so on. The important fact is that most of the yields tend to be small; only a relatively few are large. A frequency distribution that has this general form is called a *right-* or *positively skewed distribution*. Properties of a frequency distribution will be discussed shortly. The data used in Figure 1.1 are assumed to be generated from a continuous process.

Most statistical packages select a default bin width using some combination of the sample size and spread. In Figure 1.1b it is 5; however, the user has the option of changing it. There is no best bin width. Clearly, a very narrow bin results in histogram bars that do not summarize the data, and a very wide bin lumps all the data in a few classes.

Discrete data can also be represented graphically. An example is the frequency of occurrence of toxic waste sites by state on the Final National Priority List (Figure 1.2) (U.S. Environment Protection Agency, 2004). Of the 50 states plus the District of

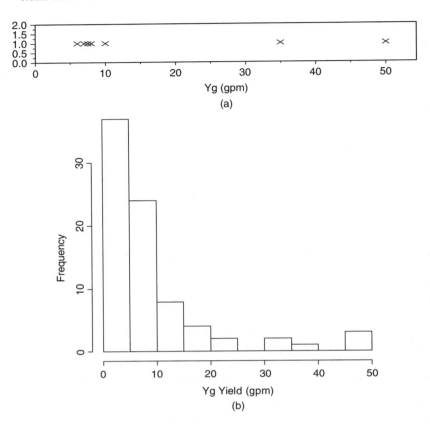

FIGURE 1.1   (a) Line plot of the first seven water-well yields (rock type Yg) from the water-well yield case study. (b) Histogram of water-well yield case study, rock type Yg.

Columbia, this graph shows that only one had no toxic waste sites (North Dakota) and one had 112 toxic waste sites (New Jersey). The most frequently occurring number of toxic waste sites is 14. This is the mode of the distribution. Five states have 14 toxic waste sites. This distribution also appears to be right-skewed since many states have 14 or fewer toxic waste sites, and a few states contain many more sites.

There are numerous other ways to display data. For small data sets, dotplots and stem-and-leaf plots, which resemble histograms except that values are actually displayed, may be appropriate (Cleveland, 1993).

1.7   DISTRIBUTION AS A MODEL

In addition to serving as a graphical device to display data, a histogram may suggest a theoretical model or distribution. The reason for these models is to connect

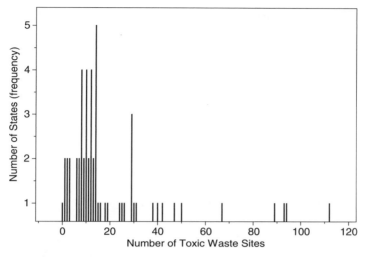

FIGURE 1.2   Histogram of the number of states with toxic waste sites.

observation with theory. For example, the number of occurrences of toxic waste sites by state, the proportion of successful wells in a drilling project, or the intensity of earthquakes can be observed and the question becomes: Can these be represented by well-studied theoretical distributions? Often, the answer is yes. In subsequent chapters we discuss discrete and continuous distributions, which often effectively represent a population for that which is observed in nature.

A *probability density function for a continuous random variable* can be represented as the pair $(f(Y), Y)$ where Y may be a variable such as temperature, parts per million of arsenic, or percent porosity. Probability density can be viewed as an area under a curve. Specifically, the probability that a random variable Y will be between a and b inclusive is

$$\Pr(a \leq Y \leq b) = \int_a^b f(y)\, dy \quad f(y) > 0$$

Further, the total area under the curve described by a probability density function is 1. The domain of Y may assume finite or infinite values, depending on the specific distributional form. Most distributions (continuous and discrete), both theoretical (expressed as frequency curves) and observed (empirical), fall into four general forms (shapes):

1. A symmetric, bell-shaped distribution (Figure 1.3a)
2. A right (positively)-skewed distribution (Figure 1.3b)
3. A uniform (equally likely) distribution (Figure 1.3c)
4. A left (negatively)-skewed distribution (Figure 1.3d)

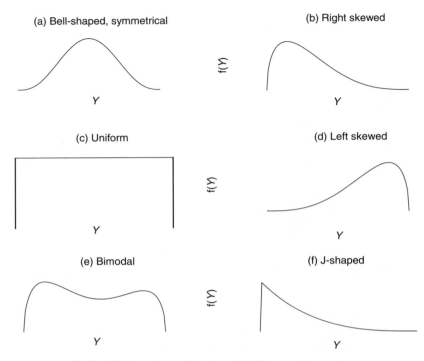

FIGURE 1.3 General shapes of continuous distributions.

Several probability density functions can be used to describe each of these general shapes. Occasionally, a bimodal distribution (Figure 1.3e) will be observed; however, a *bimodal distribution* usually results from the mixture of two or more distributions. An example of a bimodal distribution is heights of adults in the U.S. population since men are, on average, taller than women. When possible, a mixed distribution should be separated into homogeneous populations. Should this not be possible, computational procedures are available to fit mixed distributions (Titterington et al., 1986). It is also useful to distinguish skewed distributions that have a mode of zero versus those that have a nonzero mode. Figure 1.3f shows a right-skewed distribution with a zero mode. This is often referred to as a *J-shaped distribution*.

A *probability mass function* is the analog of the probability density function for a discrete random variable. The form of this function is

$$\Pr(Y = y) = \Pr(\{s \in S : Y(s) = y\})$$

where S is the sample space. So in the toss of a single fair coin, $S = \{\text{head, tail}\}$ and $\Pr(\{\text{head} \in S : Y(s) = y\}) = 0.5$, since a head and a tail are equally likely. The sum over all y is $\sum_y \Pr(Y = y) = 1$. A major difference is that the probability that Y, say,

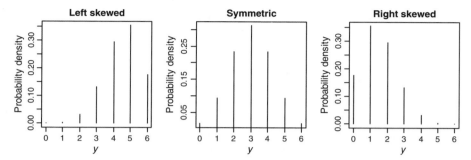

FIGURE 1.4   General shapes of a discrete distribution.

is equal to 2, is exact. Three forms of the binomial distribution, a common discrete distribution, are shown in Figure 1.4. Discrete distributions are discussed in detail in Chapter 8.

Most distributions that are encountered in the earth and environmental sciences are symmetric, typically bell-shaped or positively (right) skewed. Earthquake intensity is an example of a right-skewed distribution because there are many small tremors but relatively few episodes of large seismic activity.

1.8   SAMPLE MOMENTS

In addition to viewing data, it is useful to compute statistics to describe properties of the sample data. Some basic statistics are illustrated using the water-well yield case study data (Appendix I). In later chapters, parameters are introduced that describe population attributes. The distributions associated with many sample data sets can be characterized by their first few moments. The term *moment* comes from physics to describe a quantity that represents an amount of force applied to a rotational system at a distance from the axis of rotation, as in a seesaw. In statistics, moments describe properties of a distribution. The first moment is the mean, the second moment is the variance, and the third moment is the skewness.

For the following formulas and computations, *sample statistics* are displayed on the left, where the sample of size n is $y_1, y_2, \ldots, y_n$; on the right, the results from a sample of water-well yields of rock type Yg are displayed.

1.8.1   Measures of Location

For every sample, it is necessary to determine location. Three commonly used measures of location are the mean, the median, and the mode. Each measure of location describes a different attribute of the data. Frequently, all of these measures are computed.

Mean   The *mean* is the arithmetic average of data. It is a part of any set of summary statistics and is used in many statistical procedures.

$$\frac{\overline{Y} = \sum_{i=1}^{n} Y_i}{n} \qquad \overline{y} = \frac{y_1 + \cdots + y_{81}}{81}$$

$$= 9.64$$

A disadvantage of the mean is that it may be strongly influenced by outliers, especially when the data set is small. An *outlier* is "an observation that deviates so much from other observations as to arouse suspicions that it was generated by a different mechanism" (Hawkins, 1980). Suppose, for example, that observation 1 in Appendix I (see the book's Web site) is recorded as 750 instead of 7.50. Since 750 is far from the body of the data, it is considered to be an outlier. The mean with the outlier present is $\overline{y}_o = 18.8$. This is a significant change in the value of the sample mean $\overline{y} = 9.64$ without the outlier, and thus is highly influential. Outliers may be the result of a mistake or they may contain important information. They are discussed in depth in subsequent chapters.

Median   The *median* is the middle observation or average of the two observations closest to the middle when the data is sorted in ascending order. Only the rank of the data and the middle observation(s) affect its value. The median is defined as

$$\widetilde{Y} = \begin{cases} Y_{[(n+1)/2]} & n \text{ odd} \\ \dfrac{Y_{[n/2]} + Y_{[(n+2)/2]}}{2} & n \text{ even} \end{cases} \qquad \begin{aligned} \widetilde{y} &= y_{[(81+1)/2]} \\ &= 6 \end{aligned}$$

The median is significantly less sensitive to an outlier than is the mean. Note that a change in observation 1 (Appendix I) from 7.50 to 750 does not change the median. A disadvantage of the median is that it is sensitive only to the values of the one (*n* odd) or two (*n* even) middle observations.

Mode   The *mode* is the most frequently occurring observation, or the value associated with the maximum probability for a continuous distribution. When the data display is a histogram, the mode can only be identified as a value within the domain of the tallest bar. In the water-well yield data (Figure 1.1b), the mode is between 0 and 5. The distributions shown in Figure a, b, d, and f have unique modes. Figure 1.3e has two modes, which are usually the result of mixing of two or more populations. The uniform distribution (Figure 1.3c) often is used in the generation of random numbers.

For a right (positively)-skewed distribution, the mean > median > mode. For the water-well yield data, the sample mean is 9.64, the median is 6, and the mode is in the range 0 to 5. For a left (negatively)-skewed distribution, the mean < median < mode. This relationship is always true for the population. For a sample, especially a small sample, it may not hold. In a symmetric population, the mean = median = mode. In a sample from a symmetric distribution, all three should be approximately equal.

Trimmed Mean   The (10%) *trimmed mean* is defined as

$$\widehat{\mu}_{T10} = \sum_{M90} Y_{[i]}$$

where $y_{[i]}$ refers to the y_i's in ascending order and $M90$ is the middle 90% of the data. A 10% trimmed mean excludes the lower and upper 5% of the observations. This has the advantage of being less sensitive to outliers than the mean is but has the disadvantage that it does not use all the data. However, it does use more of the data than are used by the median. Other variations on this statistic down-weight the lower and upper observations rather than discounting them totally. Clearly, any other percentage value may be trimmed.

1.8.2   Measures of Spread or Variability

Two data sets can have the same mean and very different spread or variability. There are a number of useful measures of variability, including the sample variance, standard deviation, interquartile range, and range.

Variance   The *sample variance* is defined as

$$s_Y^2 = \sum_{i=1}^{n} \frac{(Y_i - \overline{Y})^2}{n-1} \qquad s_y^2 = \sum_{i=1}^{81} \frac{(y_i - 9.64)^2}{81-1}$$
$$= 130.0$$

where the sum of squares of the observations about the sample mean is divided by $n - 1$. It is commonly used and appropriate for a well-behaved set of data. A disadvantage is that the variance is influenced by outliers more strongly than is the mean. Another equivalent notation in common use in this book and elsewhere is the abbreviation $\widehat{\text{V}}\text{ar}(Y)$ to represent the sample variance of the random variable Y.

Standard Deviation   The sample standard deviation is the positive square root of the sample variance and is in the same units as the data.

$$s_Y = \sqrt{s_Y^2} \qquad s_y = \sqrt{130}$$
$$= 11.4$$

Interquartile Range (IQR)   First, three quartiles, Q_1, Q_2, and Q_3, are defined. Assume that the y_i's are sorted in ascending order. Then $Q_1 = y_{[0.25n]}$ (also known as the 25th percentile); $Q_2 = y_{[0.50n]}$ (also known as the 50th percentile, or median); and $Q_3 = y_{[0.75n]}$ (also known as the 75th percentile).

$$\text{IQR} = Q_3 - Q_1 \qquad \text{IQR} = 10.0 - 2.5$$
$$= 7.5$$

The *interquartile range* measures the spread of the middle 50% of the data and is therefore less sensitive to outliers than is the variance.

Range   The *range* is the maximum value minus the minimum value:

$$\text{range} = Y_{\max} - Y_{\min} \qquad \begin{aligned} \text{range} &= 50.00 - 0.09 \\ &= 49.91 \end{aligned}$$

The range is strongly influenced by outliers.

Mean Absolute Deviation (MAD)   The *mean absolute deviation* is

$$\text{MAD} = \frac{1}{n} \sum_{i=1}^{n} |Y_i - \overline{Y}| \qquad \text{MAD} = \frac{1}{81} \sum_{i=1}^{81} |y_i - \overline{y}| = 7.74$$

This measure is used in time series analysis when the interest is in the absolute difference between observed and forecasted values. In the related measure called the *median absolute deviation*, the mean is replaced by the median.

1.8.3   Skewness

Two examples of right (positively)-skewed distributions (Figure 1.3b and f) and one of a left (negatively)-skewed distribution (Figure 1.3d) have been seen. A measure Sk_y of the degree of skewness is

$$\text{Sk}_y = \frac{\sqrt{n(n-1)}}{n-2} \frac{\sum_{i=1}^{n} (Y_i - \overline{Y})^3 / n}{\left[\sum_{i=1}^{n} (Y_i - \overline{Y})^2 / n \right]^{3/2}}$$

$$\text{Sk}_y = \frac{\sqrt{(81)(81-1)}}{81-2} \frac{\sum_{i=1}^{81} (y_i - 9.64)^3 / 81}{\left[\sum_{i=1}^{81} (y_i - 9.64)^2 / 81 \right]^{3/2}}$$

$$= 2.19$$

A symmetric distribution (e.g., Figure 1.3a and c) has a skewness of zero. A left-skewed distribution will have a skewness of less than zero. Skewness provides information on the form of the distribution.

1.9   NORMAL (GAUSSIAN) DISTRIBUTION

In general, distributions will be introduced in context; however, one form of the bell-shaped curve (Figure 1.3a) has a special place in statistics. That form is the *normal* or *Gaussian distribution*. The terms *normal* and *Gaussian* are equivalent and are used

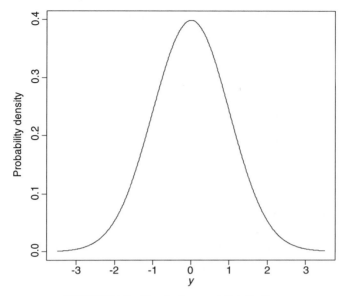

FIGURE 1.5   Standard normal distribution.

interchangeably. The probability density of a Gaussian distribution with mean 0 and variance 1 is shown in Figure 1.5. This distribution was first described by French mathematician de Moivre in 1733, but popularized by Carl Friedrich Gauss (Stigler, 1986).

The assumption of normality is basic to many statistical methods. The equation of this curve, called the *normal density function* is

$$f(y \mid \mu, \sigma) = \frac{1}{\sigma(2\pi)^{1/2}} \exp\left[\frac{-(y-\mu)^2}{2\sigma^2}\right]$$

where μ is the population mean, σ is the population standard deviation, and $-\infty < y < \infty$. Other properties of the normal distribution will be described as needed.

1.10   EXPLORATORY DATA ANALYSIS

Exploratory data analysis (EDA) consists of tools and procedures to help reveal structure and problems that may exist in data. It represents a disciplined approach to examining data. The seminal work in EDA was done by Tukey (1977). A more current treatment is that of Cleveland (1993). Results of EDA often serve as a basis for model development.

TABLE 1.2   Statistics of Water-Well Yield Case Study, Rock Type Yg

Statistic	Value
Minimum	0.09
Q_1	2.50
Q_2 (median)	6.00
Q_3	10.00
Maximum	50.00
IQR $= Q_3 - Q_1$	7.50

Numerous tools comprise EDA. Many of these are explored in the context of specific case studies, which appear throughout this book. Basic tools include the histogram, the boxplot, the scatter plot, and the time series plot. It is assumed that the reader is familiar with these tools; they are reviewed briefly here.

1.10.1   Boxplot

A *boxplot* is a graphical device for displaying data and is an alternative to the histogram (Figure 1.1b). A boxplot presents a distribution using a few quantiles. Although the information displayed in boxplots varies somewhat, a boxplot typically displays a minimum value, quartiles Q_1, Q_2, and Q_3, a maximum, and possibly outliers. These values from the analysis of water-well yield of rock type Yg are summarized in Table 1.2.

The simplest form of the boxplot is shown in Figure 1.6. From bottom to top (minimum to maximum value), the boxplot is described as follows:

- The horizontal line at 0.09 is the minimum.
- The bottom of the box is Q_1.
- The middle line is Q_2.
- The top line of the box is Q_3.
- The next horizontal line is $Q_3 + 1.5$IQR $= 21.25$. The reason for drawing this line is that the 4 points above it may be outliers; however, an alternative explanation is that the distribution is right-skewed, which is believed to be true in this example. The maximum value would be displayed if it were less than 21.25.

The rectangular box captures the middle 50% of the data, the IQR. The box width in this example is arbitrary, however, in the case of multiple data sets displayed on the same graph, the width may be set proportional to the number of observations. This boxplot is generated by the R-project command boxplot; those generated by other packages may differ.

The real power of a boxplot is its ability to assist in comparing several distributions. In Figure 1.7, water-well yields from rock types Yg, Ygt, Ymb, and Zc are compared.

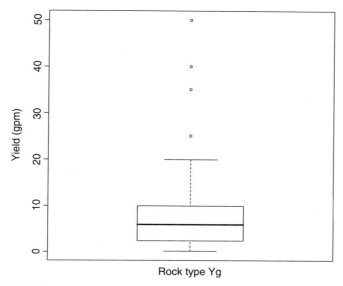

FIGURE 1.6   Boxplot of water-well yield case study data, rock type Yg.

Two new options are used. One is to create notches around the median. The notches are designed to give roughly a 95% confidence interval for the difference between two medians. Lack of overlap of the notches, assuming a representative sample, suggests that the population medians may differ. The second new option is to make the width of

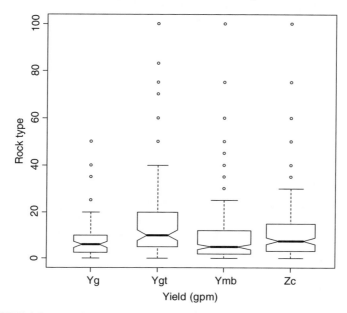

FIGURE 1.7   Boxplots of water-well yield case study data for four rock types.

the boxplots proportional to the square root of the number of observations for a given rock type. All distributions are highly right-skewed. Rock type Ygt has the largest sample median, and viewing the notches suggests that its population median may be larger than the rest. The boxplot widths imply that rock type Ymb has the most observations, and thus confidence in the form of this distribution will be higher than that for rock type Yg, which has the fewest observations. The number of observations for rock types Yg, Ygt, Ymb, and Zc are, respectively, 81, 115, 204, and 171.

1.10.2   Time Series Plot

A *time series plot* is defined as a plot with time on the horizontal axis and the attribute or variable on the vertical axis. It is a valuable tool for detecting trends, cyclical behavior, and shifts over time. A time series plot (Figure 1.8) is illustrated using a subset of northern hemisphere temperature data (Mann et al., 1999). Among the interesting features shown in Figure 1.8 is a long-term decline in temperature from the year 1000 to approximately 1900. Some of this decline occurs in what is called the "little ice age." Experts disagree on the duration of this period (Cutler, 1997), with some stating that it began around 1200 and lasted until almost 1900. Others define the end more narrowly at around the year 1445. Unprecedented warming over a short time span begins with the start of the industrial revolution in 1900. A time series plot may also be constructed by using distance, say along a transect, in place of time. Sometimes only the order of occurrence is available; this plot must be interpreted more cautiously but is still valuable.

Time is usually not a causal variable; however, changes over time in a response variable, such as global temperature, can indicate an important process (i.e., the increased burning of fossil fuel). Thus, time can be a lurking variable. We strongly

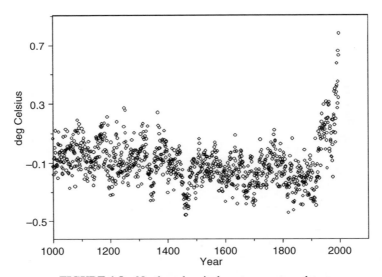

FIGURE 1.8   Northern hemisphere temperature data.

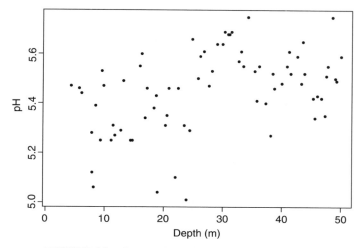

FIGURE 1.9   Scatter plot of ice core depth versus pH.

advocate time-stamping all data and plotting data versus time. Additional examples
are presented in Chapter 5.

1.10.3   Scatter Plot

A *scatter plot*, the plot of one variable against another, is an important tool in EDA
because it allows an investigation of the relationship between variables and may help
identify possible outliers. An example from the ice core case study is depth versus pH
(Figure 1.9). A possible increase in pH as a function of increasing depth is observed.
A next step, which is addressed in Chapter 2, may be to fit a model (an equation) that
describes this relationship.

1.11   ESTIMATION

Occasionally, interest in a study may be solely in understanding relationships within
the sample. A good example is "The Best and Worst Used Cars" report presented in
Consumer Reports' annual auto issue. They indicate that their car reliability histories
are based on "almost 480,000 responses to our annual subscriber survey" (*Consumer
Reports*, 2003). There is no suggestion that these results hold for the general
population of used cars.

Most often, the interest is in what information the sample can give about some
characteristics of the population. For example, the mean water-well yield from rock
type Yg is 9.64 based on 81 observations. The primary interest of the director of a
water conservation district is: What does this tell me about the yield from rock type Yg
in my district? Assuming that these 81 observations constituted a representative
sample, the 9.64 is a statistically based estimate of the population mean, which of
course in this and most instances is impossible to know with certainty. The process is

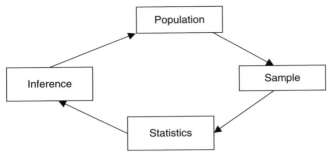

FIGURE 1.10   Sampling.

to take a representative sample from the population, compute an appropriate statistic, which will serve as an estimate, and then make some inference about a population attribute (Figure 1.10).

A key question that needs to be asked after an estimate has been proposed is: How good is it? To answer this question, properties of the estimate are investigated. An estimate has many properties; bias and variance are among the most important.

1.12   BIAS

Bias in everyday language is defined as "inclined to one side." It is sometimes used interchangeably with *prejudice*. In statistics it can be good or bad but most often is undesirable and difficult to estimate. The three broad categories are sampling bias, estimation bias, and model bias.

1.12.1   Sampling Bias

Sampling bias results from an inability to obtain a representative sample. To make a proper inference, a sample needs to be representative of the population, which typically is achieved by introducing some random element in the way that data are collected. As an example, in the case study of water-well yield from rock type Yg, there is geologic evidence that because of faulting, the yield may be lower in the southern part of the district. If the samples come mainly from the northern region, the estimate of yield will probably be biased high. Not every sample will yield a sample mean that is higher than the population mean, but on average this will be the case. In some types of demographic sampling, young adults are often underrepresented because they are more difficult to contact. Failure to obtain a representative sample may occur if an instrument is miscalibrated such that the readings are on average too high or too low. Another source of bias is an improperly designed questionnaire. A properly designed questionnaire should elicit unbiased responses from the target audience. Sometimes, deliberate bias is introduced in questionnaires to achieve a result desired for political or economic reasons. Experts can also be biased because of management pressure or other factors. Most often, sampling bias is inadvertent, usually caused by an improper sampling design. Occasionally, it is deliberate and results from the desire to obtain a specific answer.

1.12.2   Expected Value and Estimation Bias

The expected value of a random variable is its population mean or first moment. It is typically represented by the letter E. If the random variable, say Y, is discrete,

$$E[Y] = \sum_{i=1}^{N} Y_i p_i$$

where the summation is over the entire population and p_i is the probability or weight assigned to the occurrence of Y_i. Note that $\sum_{i=1}^{N} p_i = 1$. Consider the following simple example. Let Y be a random variable corresponding to the toss of a six-sided die. Then Y can assume values $\{1, 2, 3, 4, 5, 6\}$. If the die is fair, then $p_i = 1/6$ for $i = 1, \ldots, 6$. Then $E[Y] = (1 + 2 + 3 + 4 + 5 + 6)/6 = 3.5$. If the random variable Y is continuous, its expected value is found by integration over its probability density function $f(y)$, defined as

$$E[Y] = \int y f(y)\, dy$$

An *unbiased statistic* is one that is correct on average. The sample mean $\hat{\mu}$ is an unbiased estimator of μ if $E[\hat{\mu}] = \mu$. The letter E implies summation for a discrete distribution or integration for a continuous distribution over all Y, the variable of interest. If $E[\hat{\mu}] - \mu = \theta$, $\theta \neq 0$, the estimate is biased by an amount equal to θ. Note that μ and therefore θ are parameters and thus are usually unknown and unknowable. However, there are ways to gain insight into possible causes and amount of biases. The classical example of estimation bias is in the estimate of population variance. By analogy with the sample mean, one might expect the unbiased estimate of the population variance to be $\sum_{i=1}^{n} (Y_i - \hat{\mu})^2/n$: in other words, the average of squared deviations. However, the expected value of this quantity is $\sigma^2(n-1)/n$. To obtain an unbiased estimate of the population variance, the statistic $\hat{\sigma}^2 = \sum_{i=1}^{n} (Y_i - \hat{\mu})^2/(n-1)$ is used. In the water-well yield variance computation, if the previous incorrect formula is used with the n in the denominator (as opposed to $n-1$), the estimate of variance is 128.4, as opposed to the correct answer, 130.0. For large n, this bias is negligible for most applications. It is important to note that in addition to minimum variance and unbiasedness, there are other criteria on which to choose an estimator. For example, the biased variance estimator $\sum_{i=1}^{n} (Y_i - \hat{\mu})^2/n$ is by one criterion superior to the unbiased estimator.

1.12.3   Model Selection Bias

Choosing an inappropriate model can significantly bias model estimates. Often, the choice of a model is not obvious, and many possibilities exist. A model that omits an important explanatory variable, or one that misspecifies it, can be an important source of selection bias. Misspecification can occur, for example, when, say, a model to predict ice melt specifies temperature squared when the log of temperature is the appropriate formulation.

Another issue that relates to model selection is the need to recognize the sampling procedure used to obtain the data. If the investigator assumes that the data are obtained via simple random sampling (i.e., every element in the population has an equal chance of being selected), but in fact, the sampling is proportional to the size of the element being selected, any inference (estimate) may be seriously biased.

1.12.4   Bias Versus Variance

In future chapters, various types of estimators and properties are discussed. Consider two important properties of estimators, minimum variance and unbiasedness. Clearly, a minimum variance unbiased estimator is the first choice. However, this is not always possible, as Figure 1.11 shows. In this illustration, the × marks the target and the circles are the result of samples of projectiles fired at the target. Figure 1.11a shows a pattern that appears to lack bias and has a small variance. Figure 1.11b shows the same scatter but with an apparent bias that may be the result of miscalibration. Figure 1.11c shows a pattern with a larger variance but minimum bias. Among the three, the estimator used to obtain Figure 1.11a is an obvious choice. However, if it is not possible to obtain this result, the choice between the estimator used to achieve parts b and c of Figure 1.11 is less clear. If it is possible to obtain a reasonable estimate of bias, the choice may be Figure 1.11b. Of course, given the small sample, it is not possible to

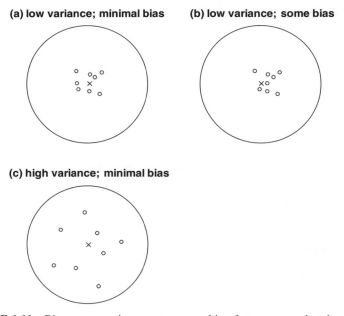

FIGURE 1.11   Bias versus variance patterns resulting from a target-shooting sample.

conclude definitively that any of the patterns are biased or unbiased, but in a larger sample the evidence may be convincing.

1.13   CAUSES OF VARIANCE

Understanding, and whenever possible, accounting for sources of variation are critically important. In most studies in the earth and environmental sciences, some inherent natural irreducible (random) variability exists. In addition, there is measurement error. Measurement error can be a function of humans reading measuring devices such as rulers and compasses, fluctuations in electronic monitoring equipment, or misjudgments of experts. In explanatory models, it is often possible to account for some sources of variability. For example, in a model to predict climate temperature change, ocean salinity and CO_2 generation may explain some of total variability associated with climate change.

1.14   ABOUT DATA

Data come in all shapes, sizes, and quality. Ensuring that the data quality is appropriate for an application is essential for success in modeling, estimation, and forecasting. Over a century ago, biologist and writer Thomas Huxley noted: "What you get out depends on what you put in; and as the grandest mill in the world will not extract wheat-flour from peascods, so pages of formulae will not get a definite result out of loose data." To put it more colloquially, "garbage in–garbage out." The best analytic methods and most powerful computation will not save a poorly designed experiment that uses or generates bad data. We strongly encourage all investigators to plan, design, and in all ways think carefully about the need for data consistent with the problem being studied.

In scientific investigation, sometimes there is the luxury of designing an experiment and then collecting the data. These are called primary data. Clearly, it is preferred to have data collected to address a specific question. When this occurs, the sampling plan is known along with measurement error and other factors. Unfortunately, at least in part, secondary data must be relied on, that is, data collected for another purpose or for some general purpose by other investigators or agencies. In the next few sections, issues of concern when using secondary data are discussed; however, these same issues need to be considered in the context of a designed experiment where data are collected. These data concerns include the trustworthiness of the data, the availability of information on the accuracy and precision of data elements, the sampling plan used to collect data, and the information on missing data.

1.14.1   Trustworthiness

Trustworthiness describes the comfort level of the scientist with the accuracy of the data. It can be enhanced by a description of data verification procedures, references to persons or agencies collecting the data, a general description of the possible uses and limitations of the data, and providing the name of a contact person. The issue is especially important now that more data are retrieved from Internet sites.

1.14.2   Missing Data

At the conclusion of an experiment, data may be missing. The reasons are many and varied. For example, a planned soil sample may be inaccessible because it is in the middle of a busy road, in the water, or on private property. When conducting a household survey, a large mean-looking dog may block the way. In addition, data that are collected may be missing due to a lost shipment or improper handling in a laboratory. It is important to try to ascertain which data elements are missing and *why* they are missing. Sometimes missing data are coded. Entries such as 9999 or -9999 are often used. However, it is not always easy to identify missing data. Sometimes missing data are coded as zeros and it is necessary to know when zeros are real: that is, indicating an absence of something and when they represent missing data. When data are missing according to a systematic pattern as opposed to being missing at random, serious bias can result. For example, if weather stations are missing in an area of high or low precipitation, biased estimates of mean rainfall will result. How are missing data to be treated? Sometimes they can be "filled in" by a statistical procedure. At other times, models handle data sets with missing data. For a complete discussion of missing data, see the book by Little and Rubin (2002).

1.14.3   Truncated and Censored Data

Truncated and censored data represent special types of missing data. A *truncated sample* consists of a set of observations where an observation or event X is detected under one of the following circumstances:

- Only if it is at least as large as some value a, called *left truncation*
- When it is less than or equal to some value b, called *right truncation*
- When it is between a and b (i.e., $a \leq X \leq b$), called *double truncation*

Table 1.3 shows a left-truncated titanium dioxide sample. It is known that the measurement device could not detect any value below 0.100. Thus, the number of observations less than 0.100 is unknown. In geochemical studies, the term *trace element data* usually refers to the smallest value of data that can be measured or analyzed. An added complication occurs when the truncation point is unknown.

A *censored sample* is similar to a truncated sample except that the observation is counted even though its value is unknown. Table 1.3 also shows an example of censored data. Observation 4 has been detected, but all that is known is that it is less than 0.1. Right and double censoring occur occasionally. Censoring is common in

TABLE 1.3   Titanium Dioxide Levels in Soil

Observation	TiO_2 (%)
Left-Truncated	
1	0.334
2	0.117
3	0.834
4	0.334
Left-Censored	
1	0.235
2	0.107
3	0.634
4	<0.1
5	0.509

medical studies, where participants may drop out of a long-term longitudinal study because they move, die, or for other reasons choose not to continue.

Censoring, particularly in reliability studies and clinical trials, is of two major types, type I and type II. To illustrate the difference, consider an example of a test of five (n) portable emergency power generators in preparation for an Arctic expedition, where time to failure is important. One way to conduct this experiment is to let the generators run until all fail (no censoring); however, this can be time consuming and costly. Another approach, called *type I censoring*, is to let the generators run a certain number of hours (i.e., fix the time t in advance). This provides a control; however, a disadvantage is that if there are no failures in time t, no information is available on estimating time to failure. An alternative approach, called *type II censoring* is to stop the experiment after r (say, three failures). This ensures a known number of failures but is open-ended timewise.

Truncated and censored samples occur in earth science applications. Recognizing truncated and censored observations in historical data is not always easy; however, failure to recognize truncation or censoring can lead to serious bias. Whenever possible it is best to go back to the source of the data. When not possible, the data should be sorted and/or plotted. For example, if all TiO_2 values are at least 0.100 and theory shows that values much closer to zero are possible, it may be reasonable to suspect that the data have been truncated. If the data column contains "less than" signs (e.g., < 0.100), left censoring may be a reasonable assumption. Another clue may be the presence of numerous minimum values equal to 0.100.

When truncation or censoring are recognized, approaches to address this problem include:

- Fitting the data to a truncated or censored distribution to specific distributional forms
- Using nonparametric estimators
- Filling in the missing data

If a significant part of the data is censored or truncated, results may be dependent on the technique used to model such data; if only a small portion of the data is censored or truncated, the course of action may not matter. Specific techniques used to model such data are beyond our scope here but may be found in Cohen (1991) and Helsel (2005).

1.14.4 Time

In an article, "Lurking Variables," the author (Joiner, 1981) notes potential problems for modeling and forecasting when the investigator fails to recognize that data may not be consistent across time. Changes in measurement methodology, effects of aging, changes in atmospheric conditions, new power plants, and the addition or removal of recording stations can all influence data values over time. An investigator should always attempt to see if changes have occurred over time that may influence the results of a study.

1.14.5 Data Validation

Data validation procedures may include:

- Establishing an allowable range for each continuous data element. Values outside this range should be examined.
- Verifying that the new data element is consistent with existing data elements. For continuous data elements, one may check to see that the new element is within ± 3 standard deviations of the mean.
- Examining histograms or bar charts to see if they are consistent with similar known distributions. Observe any unusual gaps in the data, which might indicate missing or transformed data. Note values that occur too frequently, which might indicate rounding, guesses, or truncation.
- Examining correlation patterns among data elements to see if results are consistent with theory or analogs.
- Plotting of spatial data. Check to see if state codes or other geopolitical indicators are consistent with latitude and longitude or other spatial measures. See if spatial trends occur.

There are many sources of quality data, including government, professional societies, and not-for-profit agencies. However, we suggest strongly that users evaluate all data before use.

1.15 REASONS TO CONDUCT STATISTICALLY BASED STUDIES

In the simplest form, there are two reasons to conduct statistically based studies: to assist in decision making, and to gain greater insight into a system or process. Even in the latter case, the ultimate goal is usually to aid decision makers. In most challenges faced by the earth science community, significant uncertainty exists. So how can statistically based studies aid in this process? Consider the following benefits of using statistical methodology:

- Statistical tools provide a basis for quantifying and estimating uncertainty, thus allowing the investigator to work on important problems.
- Statistical tools provide a foundation for scientific methods. A properly documented statistical study can be reproduced by anyone.
- Statistically based studies provide a solid foundation for future work and improvements.
- Statistical methods, including graphics, allow patterns to be seen and thus help us to find information in large, complex data sets.
- Properly designed statistical studies allow for appropriate inference and use data efficiently, thus frequently resulting in reduced cost and more useful results.

1.16   DATA MINING

One of the challenges that today's statisticians and data analysts face in many applications is a plethora of data. Even a local grocery store may collect gigabytes of data daily. Intelligence agencies, climate models, and geneticists are drowning in data. A major problem for an analyst is that useful information may reside in only a small subset of these mammoth data sets. It is clearly impossible to investigate these data sets in the same manner as one would a data set with five variables and 1000 observations. Thus, there is a need for automated learning and other analytic tools. Data mining uses a variety of automated statistical and pattern recognition techniques. Many of the multivariate statistical methods introduced in Chapter 7 are used in data mining to find patterns in large data sets. See the book by Han and Kamber (2000) for an overview of data mining and that of Hastie et al. (2003) for an introduction to data mining from a statistical perspective.

1.17   MODELING

A model is an abstraction that is used to gain insight into a problem. A statistical model is an equation or system of equations that represent some process.

The ice core case study scatter plot (Figure 1.9) suggests a possible linear relationship between pH and core depth (Figure 1.12). A goal of this analysis may be to determine if depth can account for some of the variability in pH. In other words, does knowledge of depth provide additional information about the pH value? For example, at greater depths, is the pH likely to be less than, more than, or about the same as the pH at a lesser core depth? One way to investigate this problem is by formulating a model, estimating model parameters, and evaluating the results of the

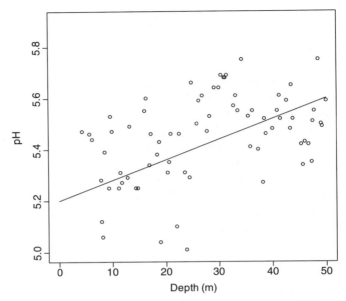

FIGURE 1.12   Ice core case study data with a graphically fit model indicated by the line suggesting a possible linear relationship between pH and depth.

model. Viewing the data in the domain 0 to 50 m suggests that there may be a linear relationship between depth and pH. Consider the equation (model) $Y = \alpha + \beta X$ as a candidate model. Of course, considerable variability in pH at every depth can be seen. Thus, the term ε, which represents the error or vertical deviation from the line, is added. Now there is a statistical model of the form $Y = \alpha + \beta X + \varepsilon$, where α and β are the unknown intersection and slope (parameters). This is a candidate model because no evaluation has occurred to help us determine if this is a reasonable model for this problem. In Chapter 2 there is further discussion of the use of the error term.

How are values assigned to α and β (i.e., how are α and β estimated)? One way is to lay a straightedge on the plot to determine the intersection and slope (Figure 1.12). Of course, this may give a reasonable first approximation but is generally considered to be unsatisfactory. This graphical fit yields the estimates $\widehat{\alpha}_G = 5.2$ and $\widehat{\beta}_G = 0.008$. Others may differ. There are however, statistically valid ways to fit a model to data, which are discussed in subsequent chapters. Assuming a valid model, there are numerous questions to be asked of it. For example, the desire may be to know exactly how much the pH changes with depth. It may be to predict pH at some point not observed: for example, at 25 m. What if the task is to estimate pH at a greater depth: for example, at 60 m? These and other questions and answers give insight into the relationship between depth and pH. Modeling concepts are discussed in detail in Chapter 2.

1.18   TRANSFORMATIONS

Transformations play a useful role in statistics. Often, a transformation is made to achieve symmetry of the data about the mean to satisfy model assumptions, such as a requirement that the data be normally distributed. Sometimes standardization is used to ensure that variables in different units are comparable. For example, suppose that temperature is measured in degrees Celsius and pressure in pounds per square inch. Their relative impact in a forecasting model of precipitation cannot be determined unless the variables are transformed to standardized units. Data can be put into standardized units using the transformation

$$Y = \frac{X - b}{a}$$

where b is often the sample mean and a is the sample standard deviation.

Another reason for making a transformation is to spread out data and make them interpretable from a graphic viewpoint. It is often easier to spot outliers in symmetric data. However, we *strongly caution against the indiscriminant use of transformations* because it is difficult to think in transformed units (i.e., the log of temperature). In addition, transforming results back into original units can be tricky, and if not done carefully, can result in biased estimates.

As noted previously, many earth science data sets are right-skewed and may need to be transformed to permit the use of methods that require a normal distribution assumption. Three ways to make a skewed distribution more symmetric are examined.

First, if the investigator believes that the data are lognormally distributed, the correct choice is $Y = \log_b(X)$, where common choices for the base b are 2, e, and 10. Often, the choice is arbitrary. Frequently in science, the base e is used. Two forms of notation to express the log base e of a variable X are $\log_e(X)$ and $\ln(X)$. Both are used. Figure 1.13 shows a histogram of $\log_e$(yield). This distribution is more symmetric than that of Figure 1.1b. The observations are centered approximately around the most frequently occurring interval, the interval between 1.5 and 2.5. Of course, for the log transformation to be applied, the data elements must be positive.

Second, an alternative to the log transformation is a root transformation: square root, cube root, and so on. This is typically a trial-and-error procedure but may be useful when the data set contains zeros.

A third alternative is the *Box–Cox transformation*. It is defined as

$$W = \begin{cases} \dfrac{Y^\lambda - 1}{\lambda} & \lambda \neq 0 \\[2mm] \ln(Y) & \lambda = 0 \end{cases}$$

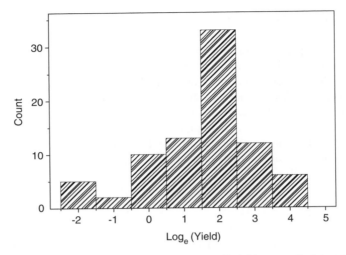

FIGURE 1.13 Histogram of $\log_e$ transformed water-well yield case study data, rock type Yg.

and must be solved iteratively for $\lambda \neq 0$. For the water-well yield case study data for rock type Yg, $\hat{\lambda} = 0.225$. The histogram of the Box–Cox transformed Y (not shown) and the histogram of the fifth root of Y (also not shown) are quite similar to that of the log-transformed data (Figure 1.13).

1.19   STATISTICAL CONCEPTS

In the following section, important statistical concepts that are utilized throughout the book are described. These concepts include the need for a disciplined approach to investigation, some general principles governing estimation of parameters, biased estimates, the need for randomization, determining sample size, statistical versus practical significance, the use of simulation, and different statistical paradigms.

1.19.1   Disciplined Approach to Investigation

Scientific investigation requires discipline in organization and documentation. Not only is a detailed plan of data collection, estimation, and analysis required, but it is also essential that the investigator keep detailed notes of actions taken. If not, at the study's conclusion, the investigator will have scores of graphs and other output and little insight into the problem.

1.19.2   Random Component

The concept of a random sample for polling results that appear in newspapers is familiar. For example, 60% of respondents in a poll favor more federal funding for teacher training in the sciences, and the margin of error based on a sample of 1000 respondents is ±3.2%. What is seen in newspapers is almost always sampling error. Sometimes this error may be partitioned by ancillary information on the respondents, such as education, job type, or age. However, even people of similar demographic characteristics hold differing opinions. This is a source of random variation or unexplainable error. Another source of variation in a survey is measurement error caused by improper wording of questions or failure to record an answer accurately. The statistical modeler must attempt to identify and partition sources of error to better highlight the important relationships in the systems being studied and ultimately, reduce error.

1.19.3 Sample Size (Large Versus Small)

A question that always occurs in a conversation concerning the collection of data is: What sample size do I need? The answer is: It depends. It depends on the precision required and the amount of variability in the process or region of interest. If the mean value of uranium in a region is to be estimated to within ±0.001 ppm at a 5% error, more samples are needed than if the answer is to be within ±0.1 ppm at the same error level. Similarly, if uranium values vary considerably, more samples are needed to achieve the same level of precision than if the amount of variability is small.

A large sample size does not guarantee success. It is more important to ensure that the data are representative of the population of interest. For example, if uranium samples are collected in a part of the study area known to contain unusually high values, the estimate of the mean is biased high. Studies that depend on volunteer responses should be suspect because often those who respond are not representative of the population. The best way to obtain a representative sample is to introduce randomness into the data selection process. Sampling designs to ensure a representative sample can be as simple as allowing each element in the population an equal chance of selection or as complex as a multistage hierarchical design, such as first choosing a city, next a neighborhood, and then a household. To achieve spatial coverage, a sampling grid is often chosen.

With a large representative sample, it may be possible to select a known distribution from among well-studied models. When a sample is small, it is difficult to estimate a population distribution. Thus, for a small sample, the best choice may be a nonparametric procedure, which does not require a distributional assumption.

The question is: What is a large data set? The answer is: It depends. Some elementary books refer to sample sizes of 30 or more as large and to fewer than 30 as small. Unfortunately, this is too simplistic because small versus large depends on the type of data, variability, and application. A sample of size 10 can be quite large if taken from a homogeneous population. A sample of size 100 can be quite small if taken from a heterogeneous population.

1.19.4   Sampling

In subsequent chapters, the importance of a representative sample and ways to obtain it are discussed. Many of the examples of sampling presented in the media involve a random sample of people or businesses, usually obtained from some type of list. Data requirements in many earth and environmental science problems involve spatial sampling. Examples include assessing water quality in a bay or locating a mineral deposit. Sampling may be on a grid or by selecting geographic coordinate randomly or by some other process. Complications include possible correlation between nearby measurements and trends over space and time.

1.19.5   Statistical Significance Versus Practical Significance

One issue that practitioners should always be concerned about is statistical significance versus practical significance. In theory and sometimes in practice, given sufficient resources it is possible to detect very small differences or shifts in population means or other parameters. Consider the uranium example. To estimate uranium content to within ± 0.001 as opposed to estimating to within ± 0.1 requires an order of 10^4 more samples. It is foolish to waste resources collecting these additional data unless there is a need for this higher level of precision. This is discussed in Chapter 3 under the subject of hypothesis testing.

1.19.6   Simulation

Many complex problems cannot be solved analytically, so computer simulation is used. A frequent candidate is Monte Carlo simulation, which means that some component of randomness is incorporated into the system under study. A major application of simulation is the study of the effects of global warming. By constructing simulation models, investigators can study the effects of changing energy mixes, ocean current, polar melt, and a myriad of other factors. It is important to recognize that a simulation result does not constitute a formal proof. However, even when a closed-form solution (a mathematical solution) exists, simulation may provide insight and verification.

1.20   STATISTICS PARADIGMS

In subsequent chapters, three approaches to problem solving are mentioned:

1. The frequentist approach developed by Jerzy Neyman, R. A. Fisher, and Egon Pearson
2. The maximum likelihood approach of R. A. Fisher
3. The Bayesian approach developed by Harold Jeffreys and others based on the work of the Reverend Thomas Bayes, an eighteenth-century English mathematician.

A historical perspective on these approaches has been described by Stigler (1986). For additional details, see the article by Berger (2003). The approaches are described briefly.

1.20.1   Frequentist Approach

Frequentists define *probability* as the limiting result of observing outcomes in a large number of trials. For example, a frequentist may conjecture that 60% of the time that a well is drilled in rock type Yg, the water-well yield is at least 9 gpm. A measure of probability of that event is obtained if, for example, in a designed experiment, the investigator decides to drill 50 holes in rock type Yg. If 27 of these wells have a yield of 9 gpm or more, the relative frequency is $0.54 = 27/50$. This result is consistent with the conjecture that the population probability will be $0.54 = 27/50$. One may ask if this empirical result is consistent with the conjecture that the population probability is 0.6. Hypothesis testing is examined in Chapter 3.

1.20.2   Maximum Likelihood Approach

The maximum likelihood approach, pioneered by Sir Ronald Fisher, is a tool used to estimate model parameters. It is based on fitting a model after data are collected. In the method the parameter estimates are selected that are most likely to occur given the data. This typically involves setting up an equation of data and parameters (unknowns, such as the population mean) and taking the derivative of this function with respect to the parameter. As in the frequentist approach, the maximum likelihood approach is based on observed data.

1.20.3   Bayesian Approach

Bayesian statistics flow from a theorem stated by Thomas Bayes published post-humously in 1763, which provides a systematic approach to combining data resulting from an experiment with the use of prior information to improve the quality of estimates. *Bayes' theorem*, stated in terms of probabilities, is

$$\Pr(A \mid B) = \frac{\Pr(B \mid A)\Pr(A)}{\Pr(B)} \qquad \Pr(B) > 0$$

where $\Pr(A \mid B)$ is the conditional probability of event A given B. It is also called the *posterior probability*. $\Pr(B \mid A)$ is the conditional probability of B given A and $\Pr(A)$ is called the prior or marginal probability of event A. $\Pr(A)$ represents a degree of belief about event A before conducting an experiment. It does not take into account information about event B. $\Pr(B)$ is a normalizing constant.

There are two types of priors, noninformative and informative. A classical Bayesian will use a *noninformative prior*, often a uniform distribution, which conveys very little information and lets the data speak for themselves. *Informative priors* come in two general categories. One is the *conjugate prior*, chosen largely for computation convenience. For example, members of the exponential family often serve as conjugate priors. The other is a distribution that reflects prior knowledge. A subjective Bayesian will typically pick this form of distribution and not worry about computational issues. A strong prior, typically one with little variance, overwhelms the data, especially for a small sample, and may lead to biased results if incorrect. A weak prior, giving more weight to the data, may lead to biased estimates for small samples.

Consider a simple example based on the previously mentioned water-well yield counts tabulated by rock type and yield category (a yield ≤ 40 gpm is a low yield; a yield >40 gpm is a high yield). The problem is to determine the probability of drilling a high-yield well (event A) given that the drilling is in rock type Ygt (event B). Prior knowledge indicates that 6% of wells drilled in a similar geologic area (consisting of rock type Ygt and rock types not Ygt) are high yield [$\Pr(A) = 0.06$]. A sampling pattern is developed, wells are drilled in the area of interest, and results are obtained. An analysis of rock types for high-yield wells indicates that $\Pr(B \mid A) = 0.50$; that is, 50% of the high-yield wells are drilled in rock type Ygt. A geologic map shows that 20% of the rock is type Ygt. Since $\Pr(A) = 0.06$, $\Pr(B) = 0.20$, and $\Pr(B \mid A) = 0.50$,

$$
\Pr(A \mid B) = \frac{(0.50)(0.06)}{0.20}
$$

$$
= 0.15
$$

Thus, the probability of drilling a high-yield water well given that the drilling is in rock type Ygt, is greater than finding a high-yield water well by random drilling, which will occur only 6% of the time. The prior probability estimate of $\Pr(A) = 0.06$ typically will be obtained by some combination of historical data and/or expert judgment.

Another example is from mineralogy (Donald Singer, U.S. Geological Survey, personal communication, June 5, 2005). The problem is to find the (posterior) probability of a deposit D given some evidence E. The probability of a deposit $\Pr(D)$ is based on information obtained prior to conducting the current experiment. The probability of evidence given a deposit $\Pr(E \mid D)$ may be derived from observation, drilling, or other methods of investigation conducted as part of the current exploration study. The problem is to find the probability of a deposit given evidence, which by Bayes' theorem is

$$
\Pr(D \mid E) = \frac{\Pr(E \mid D)\Pr(D)}{\Pr(E)} \qquad \Pr(E) > 0
$$

There are other ways of expressing Bayes' theorem, which are presented in subsequent chapters. For an overview of Bayesian approaches, see the article by Berger (2000). Numerous Bayesian textbooks have appeared in recent years that emphasize data analysis (e.g., Carlin and Louis, 2000; Gelman et al., 2004).

Historically, most hypothesis testing and modeling have been undertaken using a frequentist and/or maximum likelihood approach. This was due, in part, to a lack of computational power and available software to implement Bayesian methods. Now faster computers and software for Bayesian computation permit such implementation (see, e.g., WinBugs Project, 2005).

1.20.4   Frequentist Versus Bayesian Approaches

Differences in estimates resulting from the frequentist and Bayesian approaches are of significant practical interest particularly, but not exclusively, in hypothesis testing. In the frequentist approach to hypothesis testing, the parameters are considered fixed, and a conjecture is either accepted or rejected at some level of probability. The Bayesian methods conditioned on the observed data and posterior probabilities can be assigned to any number of conjectures. Unfortunately, these two approaches can lead the investigator to different conclusions. The choice and consequence of using the frequentist or Bayesian approach is discussed in Chapter 3 in the context of hypothesis testing.

For a discussion of the frequentist versus Bayesian controversy, see the article by Efron (1986). Both approaches have strengths and weakness (Little, 2005). The frequentist theory strength, because it is based on repeated sampling, is in some sense well calibrated but it lacks a unifying theory. Bayesian statistics have a more coherent unifying theory; however, a more complete specification of a model is required than in the frequentist approach and can yield multiple answers, one for each prior distribution.

Many statisticians take a pragmatic approach and use whatever approach seems best for the problem at hand. For most of the model-based inferences discussed in this book, we believe that frequentist and Bayesian approaches will yield similar results. However, it is possible to find even simple examples for which the two approaches yield very different results. Results are more likely to diverge as models become more complex.

1.21   SUMMARY

Statistical methods help gain insight into data and the processes that generate them. They are an important vehicle for studying and quantifying uncertainty. Statistical methods facilitate scientifically based decision making under uncertainty. Statistical methods require data, which come in various forms: nominal, ordinal, interval, and ratio. Data can be continuous or discrete. Sometimes, expert judgment is relied upon for data. In statistical analysis, the interest is most often in what a statistic conveys about some characteristic of a larger population. Basic statistics include the mean, median, mode, variance, standard deviation, and range, which are measures of the location and scale (variability) of data. Various graphical techniques, including histograms, boxplots, scatter plots, and time series plots, illustrate features of data.

Statistics and graphical tools give information about the form of the distribution and can help identify trends and possible outliers. Many of the standard statistical methods require an assumption of normality. Transformations may make it possible to achieve normality. The idea of a random sample is critical to statistical inference. Statistics based on a random sample convey information about attributes of a population. Two key properties of an estimator are bias and variance. Estimators that are unbiased and have minimum variance are desired. A frequentist approach is based on repeated sampling. The Bayesian approach implies a prior distribution, which combined with data, yields a posterior distribution.

EXERCISES

1.1 An oil company has leased land in northern Alaska that has resource potential. The company conducts an internal study to determine the likelihood of commercial oil deposits. Identify possible sources of bias that might occur in this study. What might the company do to avoid potential bias?

1.2 Find an example that illustrates an appropriate earth science use of graphics, and explain why.

1.3 Find an example of an inappropriate earth science use of graphics, and explain why.

1.4 The United States is divided into 10 cold-hardiness climate zones (1 to 10, with 1 being the coldest). What type (classification) of data are these? Does it make sense to find the mean of the climate zones in your state by averaging the climate zones? Comment.

1.5 Find an example of earth science data that are symmetrically distributed and comment.

1.6 Suppose that one observes earthquakes of a magnitude of 4, 5, and 6 on the Richter scale. Is the change in intensity from 4 to 5 the same as the intensity change from 5 to 6? Explain.

1.7 Has precipitation changed in your community over time? Data are available on the Internet. Use graphical techniques, including a time series plot and a boxplot.

1.8 For this exercise we use the ice core case study data in Appendix II.
  (a) Calculate the IQR, mean, median, mode, skewness, variance, and standard deviation of HCO_3.
  (b) Draw a boxplot.
  (c) Draw a scatter plot between HCO_3 and Cl.
      Comment on the findings.

1.9 Of the measures of variability presented in this chapter (the IQR, variance, standard deviation, and range), which is best? Explain.

1.10 Some scientists say that the sea level is rising about 1.5 mm per year. Is it reasonable to conclude that in the next decade a 15-mm rise in sea level will be experienced? Comment.

1.11 How does one design a study to investigate global warming? Describe possible sources of bias.

1.12 Give examples of type I censoring and of truncation.

2 Modeling Concepts

2.1 INTRODUCTION

A *model* is an abstract or synthetic characterization of a process. Statistical models are expressed as equations. The structure of a model can be as simple as a mean or as complex as a set of nonlinear equations. Models are constructed to:

- Study systems
- Make estimates
- Test conjectures
- Understand and partition sources of error
- Make forecasts

Models are fundamental to the scientific method. Results from properly documented models can be understood and replicated. Models can be evaluated and tested using generally accepted scientific procedures. They become building blocks for future studies.

Statistical models are characterized by one or more error components. The ice core depth (variable X) versus pH (variable Y) model $Y = \alpha + \beta X + \varepsilon$ that is formulated in Chapter 1 has an error component ε that expresses unexplained variation or deviation from the mean response $\mu_Y = \alpha + \beta X$ given X and known parameters α and β. The word *error* in the context of statistics indicates an unexplained random deviation, not a mistake as it does in popular use of the term.

The goal in scientific investigation is to formulate useful models. A useful model is one that will provide estimates to within the level of accuracy and precision needed for the problem under consideration. In formulating a model, decisions are made about the structure and variables to be included.

2.2 WHY CONSTRUCT A MODEL?

The reasons are many and varied and include increasing understanding of relationships among variables, making an inference, testing a conjecture, and/or making a forecast.

Statistics for Earth and Environmental Scientists, By John H. Schuenemeyer and Lawrence J. Drew
Copyright © 2011 John Wiley & Sons, Inc.

2.2.1   Understanding

Models permit an understanding of relationships, such as the association between depth and pH. For example, if pH changes in a significant way as a function of depth, how does it change: linearly, exponentially, or in some other manner? Insight is gained into a system by varying input data, examining sensitivity to assumptions, and changing the model structure.

2.2.2   Inference

Statistical inference provides a mechanism to say something about a characteristic of a population based on a sample. In the ice core case study, an investigator may wish to estimate the pH at a given depth where no data are recorded: for example, at 5.5 m. Not only will the investigator be able to make a point estimate of pH at 5.5 m but will also be able to provide a measure of its uncertainty at this depth.

2.2.3   Testing a Conjecture

Most decisions in science, business, and personal lives are made under uncertainty. Rarely is there the luxury of knowing all necessary information and assimilating it to obtain an unambiguous answer. The systems discussed in this book contain inherent uncertainty, and therefore a mathematical model should not be formulated.

A conjecture or hypothesis is formulated to address a question of interest. It is tested using a statistical model and data. For example, earlier work, current data, and/ or theory may suggest a linear relationship between pH and depth. This conjecture is tested by seeing if a linear model adequately describes the relationship between pH and depth observed in the data. An outline of a method for doing so will be discussed shortly.

2.2.4   Forecasting

Forecasting is one of the more challenging aspects of modeling. Estimates are made corresponding to observations that extend beyond the range of data, usually in time or space. For example, in the context of the ice core case study, an investigator wants to estimate the pH at 60 m. Measurements are taken only to 50.04 m (Figure 1.9). The forecaster must make the strong assumption that relationships formulated within the domain of the data also hold outside it.

2.3   WHAT DOES A STATISTICAL MODEL DO?

A statistical model partitions variability. For examples variability in pH can often be accounted for by depth (Figure 1.12). Depth and pH appear to be related, although there is considerable variability. One expression of this relationship is the linear model $Y = \alpha + \beta X + \varepsilon$. The expression $\alpha + \beta X + \varepsilon$ represents the total variability in a system, the expression $\alpha + \beta X$ represents the variability accounted for by the model, and ε represents the unexplained variability:

total variability $=$ variability accounted for by model
$+$ unexplained variability

In an ideal world, it may be desirable to have the model account for the total variability (i.e., 100%), but in the real world, most processes are inherently noisy. This is especially true in the earth sciences. It is not unusual in earth science processes for a good model to explain less than 50% of the variability in the response, as in the ice core case study. By contrast, it is often possible to explain a much higher percentage of the variability in chemical and engineering processes.

2.4 STEPS IN MODELING

A statistical approach to modeling involves model formulation, estimation of model parameters, initial model evaluation, model use, and feedback. These steps are valid across a variety of applications and methods. The procedure is outlined in the following sections and is discussed in detail in the context of specific models, such as regression. Regression-type models are not the only models involving uncertainty; however, they are among the most common. Many of the formulation and evaluation procedures described in this chapter are applicable to a wider class of models.

2.4.1 Formulating a Model

In the ice core case study, pH is plotted against depth. A candidate model $Y = \alpha + \beta X + \varepsilon$ is chosen based on viewing the data and parameters estimated graphically (Figure 1.12). Is there a better way to approach this problem? When possible, the authors suggest an approach based on *first principles*, which is the use of available theory to guide model formulation. Unfortunately, in some earth science problems, unambiguous theories are not available. Often, the investigator must formulate a model based on data and combine it with whatever theory is available to serve as a check, especially when data are sparse. Sometimes there is no theory, or conflicting theories exist and models are formulated based solely on empirical observations (data). Occasionally, hard data (data related directly to the study) do not exist and a model must be based solely on theory or analogy, often elicited via expert judgment.

Consider an example of model formulation using northern hemisphere temperature data (Figure 1.8). This example illustrates some of the hazards associated with forecasting. A scientist living in the early part of the nineteenth century with access to the northern hemisphere temperature data (Figure 1.8) might reasonably have conjectured that temperature declines linearly over time. He might formulate and fit a model of the relationship between time X and temperature Y to be $\widehat{Y} = 0.154 - 1.89 \times 10^{-4} X$, where $\widehat{Y}$ is the estimated or fitted value of Y. Suppose that this scientist predicts temperature in the year 1900 using data collected from the years 1000 through 1800. He obtains a value $\widehat{Y} = -0.20\,°C$, which is a reasonable value based on his "advanced knowledge." However, he is now asked to forecast to the year 2000, which he does reluctantly and obtains a value $\widehat{Y} = -0.22\,°C$, a data value hardly consistent with what is now known to have occurred. Indeed, since 1978, all temperature measurements have been greater than $0°C$.

Why is the estimate for 1900 reasonable and that for 2000 unreasonable? There are two plausible reasons. One is that no theory is built into the model. The predictor variable is time. The other is that the prediction at 1900 is 100 years into the future. The forecast at the year 2000 is 200 years into the future. The further the predictor variable is outside the domain of the data, the more inaccurate the predicted value $(\widehat{Y})$ is likely to be. However, return to the first point: namely, that no theory is used in formulating the model.

In studies of global warming, scientists investigate relationships among clouds, ice, oceans, and, of course, heating mechanisms. Further, they transfer this understanding into equations. By incorporating principles of physics, researchers increase the chance that the model will respond to changing conditions in a reasonable manner. Mahlman (1998) of the Geophysical Fluid Dynamics Laboratory/NOAA, Princeton University has given an excellent presentation on the nature of climate change models. Of course, predictive models must be checked continually against data. Model predictions that deviate significantly from data may indicate an improperly formulated model.

An important first step in model formulation is specifying anticipated results. Sometimes the results surprise, and this is part of the fun of scientific investigation. However, if the results differ significantly from expectations, one should be skeptical and ensure that such a difference is not due to a misspecified model, bad data, or some fluke of nature.

A second important step in model formulation is the *principle of parsimony*, attributed to mediaeval philosopher William of Occam: "One should not increase, beyond what is necessary, the number of entities required to explain anything." The importance of this statement lies in the fact that for any given set of data and any process, there are an infinite number of possible explanatory models. For example, in the ice core case study a linear relationship between pH and depth (Figure 1.12) is observed. Now it is conceivable that the relationship can be quadratic, cubic, or some higher-order relationship. However, in the absence of evidence pointing to a more complex model, the linear relationship should be chosen.

2.4.2   Estimating Model Parameters

The linear model postulated for the relationship between depth (X) and pH (Y) is of the form $Y = \alpha + \beta X + \varepsilon$. The model parameters α and β and the error ε are estimated from data and/or prior beliefs. In Chapter 1 the model is fit to the data visually. Although this procedure may provide a reasonable initial guess, it is an unsatisfactory way to obtain a final result. A reproducible and scientifically valid procedure is desired. One such procedure to estimate the parameters is least squares, which is

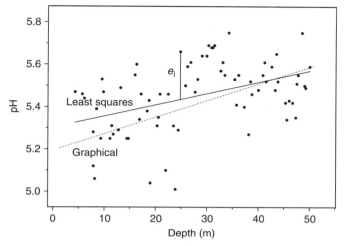

FIGURE 2.1 Ice core data showing least squares and graphical fits.

presented in Chapter 4. The least-squares estimates of model parameters are $\widehat{\alpha} = 5.30$ and $\widehat{\beta} = 0.0053$. The fitted model is

$$\widehat{Y} = \widehat{\alpha} + \widehat{\beta} X$$

It is the solid line in Figure 2.1. A graphical fit (parameters $\widehat{\alpha}_G = 5.20$ and $\widehat{\beta}_G = 0.0090$) is shown by the dashed line in Figure 2.1. The sample model (in this case the ice core data fit by least squares) may be expressed as

$$\widehat{y}_i = \widehat{\alpha} + \widehat{\beta} x_i \qquad i = 1, \ldots, n$$

where n, the sample size, is 74. The slope $\widehat{\beta} = 0.0053$ is an example of a point estimate. However, statistically based estimation procedures allow computation of interval estimates to quantify the uncertainty associated with point estimates. The $\widehat{\alpha}$ and $\widehat{\beta}$ estimators introduced above are statistics. They are functions of the y_i's and thus have variances, which are important to an understanding of their uncertainty. They are described in Chapter 4.

In addition to least-squares estimation, high-speed computing permits the implementation of other estimation procedures, which carry different and often less restrictive assumptions. These are discussed in the context of specific methods.

2.4.3 Model Evaluation

Anyone can formulate a model describing almost anything. Consider the conversation between Hotspur and Glendower in Shakespeare's *Henry IV*. Glendower says: "I can call spirits from the vasty deep!" To which Hotspur replies: "Why so can I, or so can any man; but will they come when you do call for them?" The challenge, akin to calling for the spirits, is to formulate a sensible model that addresses the problem under consideration and meets acceptable research standards.

A key question that needs to be answered is: How good is the model? This question must be addressed in context; namely, does the model do the job for which it is designed? Some of the questions posed are:

- Does the model account for a statistically significant percentage of the variability in the response, which is pH in the ice core study?
- Does the model capture the important explanatory variables? In the ice core study, are there variables in addition to depth that perhaps should be included in the model to help explain variation in pH?
- Are the model assumptions satisfied?
- Is the model overly sensitivity to small changes in explanatory variables?
- Does the model perform satisfactorily when tested on a set of data not used to estimate parameters?

As noted, statistical models partition variability into that accounted for by the model and that left over, which is called error. A good model accounts for the important systematic variability, and that left over is error or random variability. An example is the variability in pH accounted for by depth in the ice core data. A key component of model evaluation is studying the error to see if it is in fact random. Of course, the true error is unknown and unknowable. However, in regression the error is estimated by a *residual*, defined as

$$e_i = y_i - \widehat{y}_i \qquad i = 1, \ldots, n$$

where y_i represents the observed data and $\widehat{y}_i$ the fitted values. The magnitude of a single residual e_i is shown as the vertical line in Figure 2.1. An important tool to allow examination of the error is a *residual plot* (Figure 2.2). This graph shows

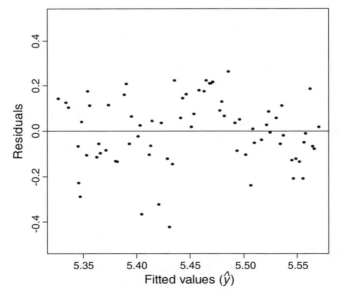

FIGURE 2.2   Residuals plotted against fitted values for the ice core data.

residuals plotted against the fitted values $(e_i, \widehat{y}_i)$, $i = 1, \ldots, n$. If the model is satisfactory, the residual should appear random and thus not contain a trend or other evidence of systematic variability. The general approach to evaluation is common to a variety of models, including regression, time series analysis, and spatial statistics; however, some diagnostic tools are model specific.

2.4.4 Evolutionary Nature of Modeling

Rarely can an investigator view a complex problem and formulate an appropriate model on the first attempt. *Modeling* is an iterative process of model formulation, fitting, evaluation, revision, and reevaluation, as shown in Figure 2.3.

An investigator begins with a candidate model using the best information available. Consider the regression model $Y = \alpha + \beta X + \varepsilon$. Once the model has been formulated, model parameters α and β are fitted (estimated) using data, and then the response variable $\widehat{y}_i = \widehat{\alpha} + \widehat{\beta} x_i$, $i = 1, \ldots, n$ is estimated. The $\widehat{y}_i$'s are compared to existing and, hopefully, new data. New data are data not used to estimate model

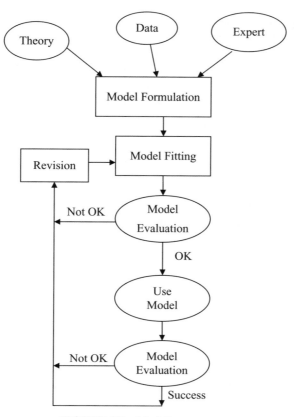

FIGURE 2.3   Modeling process.

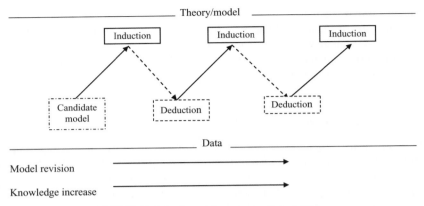

FIGURE 2.4   Iterative process of modeling.

parameters. Revisions to the structure of the model are made as needed. A decision characterizing a model as "OK" or "not OK" is a judgment call. Model evaluation tools provide guidance. The decision to characterize a model as OK means that it is useful for the problem at hand.

Deductive reasoning is learning about data from theory and the model. *Inductive reasoning* is learning about the theory and the model from a study of the data. The investigator iterates between deductive and inductive reasoning, hopefully improving the model at each stage. A diagram of the evolutionary nature of modeling is shown in Figure 2.4.

2.5   IS A MODEL A UNIQUE SOLUTION TO A PROBLEM?

In statistics, models are rarely unique. One need not venture too far in the direction of complexity before the existence of a unique solution is difficult or impossible to guarantee. Suppose, for example, that the desire is to model the relationship between temperature and years 1000 through 1800 in the northern hemisphere data (Figure 1.8). Here are just a few models that may reasonably describe the data (Y is temperature and X is year):

- A linear model of the form $Y = \alpha + \beta X + \varepsilon$
- A model that is quadratic in X: $Y = \alpha + \beta X + \gamma X^2 + \varepsilon$
- An exponential model of the form $Y = \alpha e^{\beta(X - 1000)} \varepsilon$

The quadratic and exponential models capture a curvilinear effect that may exist in the data. There is an added complexity in these data, periodicity, which is discussed in Chapter 5.

When a theory is available to guide the development of a model, the number of possible candidate models may be reduced. However, there are often conflicting theories in poorly understood processes.

2.6 MODEL ASSUMPTIONS

A model carries with it a certain set of assumptions that must hold to a greater or lesser degree for the model to be useful. Many of the models studied are of the form $Y = f(X_1, X_2, \ldots, X_k) + \varepsilon$, where Y is a function of the explanatory variables $X_1, X_2, \ldots, X_k$ plus the error ε. Often, an assumption is made about the distribution of the error term. In the real world, no assumption is ever satisfied completely; however, large departures from assumptions can invalidate model results. The art and science of modeling is to know how stringently model assumptions must hold and what happens if they do not.

2.6.1 Normally Distributed Errors

A common assumption required in least-squares procedures is that the errors, the ε's, are normally distributed. Testing of significance of model parameters may depend on a normality assumption, but not all errors are normally distributed. Sometimes they are right-skewed. Graphical and statistical procedures are available to test normality assumptions. Alternative methods not requiring normality assumptions are available.

2.6.2 Independence of Errors

Typically, it is assumed that the ε's are independent of each other. This is important because lack of independence implies structure, which will need to be modeled. Failure to recognize *dependence*, or lack of independence, in data can weaken or invalidate tests of significance on model parameters. Correlated errors often occur in time series, as in precipitation and spatial data, because observations taken near each other are often similar.

> Two errors (events) are *independent* if the magnitude and direction of one has no influence on the outcome of the other.

2.6.3 Equal-Variance Assumption

Some models are used to compare means from multiple populations. A standard assumption is that population variances are equal, which is the *homogeneity assumption*. Other models are used to investigate trends over time or distance. For example, Figure 1.8 shows temperature trends over time. Standard time series and regression models relating time or distance to a response variable such as temperature require that the variance remain constant. If the equal-variance assumption is violated, it may be possible to use a weighting scheme to equalize the variance among populations, along time, or along distance. Sometimes homogeneity can be attained by making a log, square root or other transformation of the Y's. Failure to recognize nonhomogeneous variance (inhomogeneity) may lead to biased estimates of model forecasts.

2.6.4   Additive Error Assumption

In the model $Y = \beta_0 + \beta_1 X + \varepsilon$ that is used to describe the relationship between pH and depth, the error is additive, which means that a Y at a given X can be represented by the expression $\beta_0 + \beta_1 X$ plus the error ε. In an additive model, the error is in the same unit as the Y (pH in the ice core example). In a model of the form $Y = e^{\beta_0 + \beta_1 X} \varepsilon$, the error ε is multiplied by the model expression $e^{\beta_0 + \beta_1 X}$. An error expressed as a percent is a multiplicative error. It is important in modeling and testing to understand the form of the error structure.

2.6.5   Measurement Error

As noted previously, a statistical model partitions variability into systematic and random components. A component of this random variability may be measurement error. All recording devices (mechanical, electrical, and human) are subject to some error. Often, the measurement error is negligible, and the variability is just the natural variability inherit in a processes. When measurement error is recognized to be large, it is important to model and thereby account for its effect. For example, an issue of concern is the amount of thinning of the Greenland and Antarctic ice sheets. If thinning is occurring at the rate of 2 cm per year but the measurement error is ±2.5 cm, can one be convinced that thinning is actually occurring? Conversely, suppose that the measurement error is ±2 mm. Will the conclusion differ? Clearly, it is important to attempt to determine the measurement error, which is often done by reviewing specifications of the measuring device, examining data records, talking to on-site investigators, and better still, designing an experiment to minimize error.

2.6.6   Comments on Assumptions

Some assumptions are critical—if violations occur, an inference can be severely flawed. Conversely, some assumptions can be violated to a degree and the model results still hold. One of the challenging things to learn is which assumptions are required to hold to a high degree and which can be relaxed. In part, the answer comes through experience working with data. In addition, simulation is a useful tool to use to investigate the effect of changing assumptions.

When the assumptions associated with least-squares procedures do not hold or cannot be validated, other estimation procedures are available, including robust and nonparametric estimation. These alternative methods are discussed in Chapter 4 in the context of regression. Very briefly, robust procedures provide important protection against the violation of certain assumptions, including protection against an outlying point exerting undo influence. Bayesian procedures facilitate the efficient use of prior information, such as the form of the distribution of errors. Nonparametric procedures provide protection against the violation of assumptions, including the assumption that errors are normally distributed. Least-squares procedures are sensitive to skewed error distributions. None of the alternative procedures is employed without cost. Bayesian procedures can yield biased estimates when a strong, incorrect prior assumption is used. When an alternative to least squares is used, model results may be less powerful if normal assumptions hold but more powerful than if a least-squares procedure is misapplied.

2.7   DESIGNED EXPERIMENTS

Capturing data in the earth and environmental sciences is often time consuming and expensive. A single sample obtained from an oil well, satellite, or undersea exploration can cost $1 million or more. Therefore, data must be collected as efficiently as possible. It is equally important to ensure that a statistically based study is capable of answering the question of interest. For example, it is a waste of resources to drill too few holes to detect mineralization, if present. These goals often can be achieved by applying principles of statistical design. Designed experiments take two forms. One, called sampling design, typically is implemented in field studies to collect data according to some prescribed method. The other, design of experiments, is usually executed under controlled conditions, such as in a lab where factors such as temperature, humidity, and dosage can be carefully controlled.

2.7.1   Sampling Designs

A *sampling design* is a set of procedures used to collect data in the field in a way that best achieves the goal or goals of the study. It can be defined in many ways, including meeting a specified level of precision and minimizing sampling cost. Often, compromises are required. Pollsters and market researchers construct sampling designs to ensure a representative sample. Suppose, for example, that a human geographer wants to assess water use in Phoenix, Arizona. Since this is a large city, it is impossible to interview all users. The geographer may begin by stratifying water users according to government, business, and individuals. A random sample within each of these strata may be selected from water company records for interviews. Random sampling involves chance in the selection of observations.

Consider a second example in which data are obtained via quadrat sampling (a *quadrat* is a two-dimensional shape such as a square or circle). Suppose that an investigator wants to measure the diversity of grass. It is not feasible to count the number of different types of grass that occur in a large region. However, the investigator can randomly "toss" a number of small rings in the region of interest and count the number of different grasses in each quadrat.

Sampling in spatial statistics often occurs on a grid or at points located randomly. Sampling design is discussed in Chapter 9. See Levy and Lemeshow's (1991) book for additional details.

2.7.2   Design of Experiments

Field studies play a primary role in earth science research. In the field, one learns how things work in the real world. However, there are concerns about field studies. Factors that the investigator fails to recognize may have a significant effect on the results. In the real world, there are so many factors in play that it may be difficult to recognize the effect of possible interactions. Another concern is that the levels of factors in a field study cannot be changed. For example, the amount of sunlight, rainfall, or humidity that a section of prairie grass in a Nebraska plain receives usually cannot be changed. Sometimes it is possible to design an experiment in a laboratory where factors (e.g., rainfall, sunlight, humidity) can be changed under controlled conditions and thus gain insights that would not be possible in the field. The methodology that permits one to do this in a systematic way is *design of experiments* (DOE). It is widely used in biology, engineering, and many industrial applications.

In the prairie grass example, a laboratory can control the soil type, moisture, sunlight, and other factors to see how changes in these factors and interactions between factors affect the growth and health of the grass. The importance of DOE is that there is a plan as to how to change levels of factors (such as the amount of moisture that a grass receives) to obtain the maximum amount of information from a given number of observations. Thus with DOE, efficiency is gained and results that can be generalized are obtained.

A more complex and important application of design of experiment is the study of tsunami behavior. In December 2004, a tsunami off the western coast of northern Sumatra resulted in the death of hundreds of thousands of people. One cause of a tsunami is an underwater earthquake, which displaces huge amounts of water. The relationships among earthquake location, distance from population centers, shoreline, and other factors are complex. While scientists gain information monitoring actual events, additional information can be gained by studying phenomena such as a tsunami under controlled conditions. Clearly, investigators do not pretend that an actual tsunami could be replicated in a wave basin constructed at a research facility. However, replication allows scientists to examine and monitor changes in factors such as ocean depth, earthquake magnitude, and shoreline slope under controlled conditions. For more information on tsunami wave research, see the Oregon State University (2004) Web site. We discuss experimental design further in Chapter 9. For more information on design of experiments, see the books by Box et al. (1978) and Montgomery (2000).

2.8 REPLICATION

Replication is used in several contexts in statistics. One is to repeat all or part of an experiment multiple times. Another is to take multiple measurements at a single location or sampling unit. For example, agencies of the federal government often evaluate the reintroduction of plant species. Prior to field testing, a laboratory experiment may be conducted. In such a study, characteristics of prairie clover are observed under various soil characteristics and environmental variables. An entire growing experiment may be repeated several times in order to obtain more precise estimates of model parameters. A second type of replication is done by repeating measurements on the same data or at the same location. Taking several measurements on the same data at a given point in time can reduce measurement error. For example, in a geological field study, three observers may take independent measurements on the direction of a fault. Sometimes measurements are taken at the same place over time, such as at a monitoring station, to determine the rate of ice melt. In addition to the reasons above, replication protects the experiment against failure due to the loss of a single observation.

2.9 SUMMARY

In this chapter we focused on general principles of modeling: formulation, estimation, and evaluation. Constructing useful models is a key to scientific understanding. It is important, when possible, to formulate models using theory. In addition, it is necessary to recognize that modeling is an interactive process between theory and data and inductive and deductive reasoning. In statistics, there can be a variety of models that solve a given problem. An important function of a model is to partition variability. A good model accounts for the important systematic variability in a system or process. Many techniques are available to fit models. The choice typically depends on available data and the types of assumptions the investigator is willing to make. A model needs to undergo careful evaluation prior to use.

EXERCISES

2.1 Identify factors that may cause landslides. Design a laboratory experiment that provides insight into the cause of landslides. Comment on how well the experiment matches actual landslide conditions.

2.2 What is the value of visually (graphically) fitting a model? What are the disadvantages?

2.3 What are some general principals in scientific investigation to which any reputable scientist should adhere?

2.4 Give an earth science example in which the error may be multiplicative as opposed to being additive. Explain.

2.5   What potential problems may surface when a statistical model is formulated solely on data? What if the model is developed solely on theoretical considerations?

2.6   Paleontology is defined as the study of life in past geologic time based on fossil plants and animals. Identify a major source of bias that a paleontologist must address in her studies.

2.7   Why is random sampling important?

2.8   Geologists conducting field studies often examine outcrops (visible exposure). Comment on outcrop data as being representative of unseen geologic phenomena.

2.9   Antarctica is a prolific source of meteorites. Are the location and spatial distribution of meteorites in Antarctica representative? Explain.

2.10   If one is asked to construct a global climate model, what factors need to be incorporated into the model?

2.11   If the global climate change model of Exercise 2.10 were used to forecast global mean surface air temperature for 40 years ahead, what would be the concerns?

2.12   Suppose that one is given the following set of data, which is a measure of heat for the southeastern United States taken in July for the years 1965–1970 (measurements are in degrees Celsius): 30.0, 30.2, 30.4, 30.6, 30.8, and 31.0. What model can be formulated? What percent of the variability in temperature can be accounted for by time? Are there any concerns about this data set?

2.13   Suppose that the following data represents measurements of dissolved oxygen from Lake Erie: 2.2, 3.3, 2.5, 3.4, 4.0, 3.9, 5.1, and 5.3. What additional information does one want to know?

2.14   Why are model assumptions important?

2.15   A goal of archaeology is to understand the behavior of ancient peoples by studying material remains. A typical excavation site may be a few meters square. A principal source of records in the southwestern United States is pottery. In an attempt to piece together a historical record, what biases should an archaeologist consider?

3 Estimation and Hypothesis Testing on Means and Other Statistics

3.1 INTRODUCTION

Data are collected and analyzed to help gain insight into a problem and/or make a decision. To assist in decision making, a measure of uncertainty of the sample mean or other statistic is needed. An interval estimate provides a measure of uncertainty about a population parameter. Often, interest is in changes in a mean over time or differences between means from two or more populations. Classical estimation and testing on means typically require a normal (Gaussian) distribution assumption, which for moderate sample sizes is usually satisfied. However, due to advances in statistical methodology and computing power, many procedures need less restrictive assumptions. These include Bayesian, nonparametric, and bootstrap methods.

Before interval estimates and hypothesis testing are presented, some building blocks are required. One is the issue of the relationship between elements (observations) in a sample. Many estimation procedures require an assumption of independence between observations. Another is the distributional assumption about the test statistic. The central limit theorem shows that under certain assumptions the distribution of means is normally distributed. Finally, the distribution of a test statistic is required.

> *Statistical independence* implies that the value of one of the observations does not influence the values of any other observation.

3.2 INDEPENDENCE OF OBSERVATIONS

The assumption of statistical independence between observations is required in many statistical procedures; however, it needs to be investigated, especially for spatial and time series data. For spatial data sets, nearby observations are often more alike than

Statistics for Earth and Environmental Scientists, By John H. Schuenemeyer and Lawrence J. Drew
Copyright © 2011 John Wiley & Sons, Inc.

those separated by a greater distance. This is true for physical phenomena such as water and soil samples or samples of people in the same household (Chapter 6). For time series, nearby observations taken over a short interval are more likely to be similar than those taken over a long interval (Chapter 5).

Why is the independence assumption important? Dependence (a lack of independence) reduces the effective number of observations (samples). Why does this happen? Suppose that one is asked to estimate dissolved oxygen content in a lake and decide that a sample of size 20 is appropriate. A boat is not available, so instead of taking samples throughout the lake, they are taken at one spot near shore. The results show that all 20 samples are 3.2 mg/L to one decimal place. What has been learned other than that in the small sampling area the values are homogeneous? Because of nearly perfect agreement of the samples in this small sampling area, there is, in effect, not 20 but one observation. Thus, a test conducted on water quality using 20 identical observations will not be valid because the observations are not independent. Another example is to take a soil sample from one location, mix it, and send a packet to each of 10 laboratories. Even though 10 independent analyses are obtained, which is useful for determining measurement error, there is only one sample. Yet another example is tree leaf data collected by a biologist. Multiple samples from a single tree can be used to estimate within-tree variability, but they do not provide the same information as does sampling multiple trees. Samples need not be identical, simply correlated for a reduction in the apparent number of observations. The estimate of the mean, given highly correlated samples, may be unbiased, but the estimate of the standard error of the mean is not.

3.3   CENTRAL LIMIT THEOREM

The sample mean describes an important characteristic of a data set and is used to estimate the population mean. However, to quantify uncertainty and make inferences using classical statistics, the mean is assumed to be normally distributed. Fortunately, this is not a harsh requirement because the *central limit theorem* (CLT) guarantees normality under certain assumptions. It states that the distribution of a sample mean approaches that of a normal (Gaussian) distribution as the sample size increases, even when the distribution of the original sample is severely nonnormal. Let $Y_1, Y_2, \ldots, Y_n$ be an independent random sample from a distribution with mean μ and variance $\sigma^2 > 0$. Let $\widehat{\mu} = \overline{Y} = \sum_{i=1}^{n} Y_i/n$. As n becomes large, the statistic

$$Z = \frac{\widehat{\mu} - \mu}{\sigma/\sqrt{n}}$$

approaches a standard normal distribution. A precise definition of the CLT has been given by Arnold (1990). Further properties of the normal distribution have been summarized by Evans et al. (2000). As noted previously, $\widehat{\mu}$ and $\overline{Y}$ are used interchangeably; however, when there are multiple random variables, a subscript often is used (i.e., $\hat{\mu}_Y$) to represent the estimated mean of the random variable Y.

A z-ratio or statistic has a mean 0 and variance 1, denoted as $N(0,1)$. This is also called the *standard normal distribution*.

The quantity $\sigma/\sqrt{n}$ is the scale factor in the Z-statistic. It combines the standard deviation and sample size into one quantity, which is the standard deviation of $\overline{Y}$ and is referred to as the *standard error of the mean* or $\sigma_{\hat{\mu}}$. Sometimes $\sigma_{\hat{\mu}}$ is denoted as $SE(\hat{\mu})$ or $SE(\overline{Y})$. When σ is unknown, which is the usually situation, the estimate is $\hat{\sigma}_{\hat{\mu}} = \hat{\sigma}/\sqrt{n}$. The standard error needs to accompany $\hat{\mu}$ because the estimated mean by itself provides incomplete information. In subsequent sections, the CLT is employed to justify the use of the normal distribution in interval estimation and hypothesis testing.

How large must the sample size n be for the distribution of $\hat{\mu}$ to be approximately normally distributed? The answer depends on the form of the distribution of the Y's, but unless the Y's are highly skewed, $n = 30$ is sufficient. Often, a sample size of 10 or less will suffice if the distribution being sampled is nearly symmetric. The water-well yield for rock type Yg (Figure 3.1a) is highly right-skewed. Repeated samples with a replacement of size $n = 5$ are taken and a histogram of the means are shown in Figure 3.1b. This histogram is less skewed than the original (Figure 3.1a), and the variability is less. Note that the spread on the horizontal axis is less. Figure 3.1c shows the distribution of means resulting from taking samples of $n = 10$. There is less skewness and the variance has again declined. Finally, for samples of $n = 30$ (Figure 3.1d), the distribution is close to normal and the variability has once again shrunk. Thus, for population distributions similar in shape to that of Figure 3.1a, a sample of size 30 is sufficient for most applications. The authors leave it to the reader to ponder why the variance of the sample mean decreases as n increases. If the random variables $Y_1, Y_2, \ldots, Y_n$ are each normally distributed with mean μ and variance σ^2 [denoted $N(\mu, \sigma^2)$], $\overline{Y}$ is normally distributed for any size n.

3.4 SAMPLING DISTRIBUTIONS

Three distributions commonly used in interval estimation and hypothesis testing are the t-distribution, the chi-square distribution, and the f or Fisher's distribution. They are called *sampling distributions* because they are functions of statistics derived from samples. They provide the distributions needed to quantify uncertainty and test means, variances, and ratios of variances. They are described and illustrated briefly in examples throughout this and subsequent chapters. Evans et al. (2000) have provided additional properties.

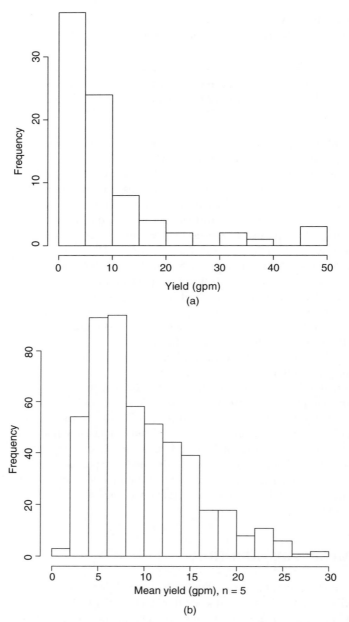

FIGURE 3.1   (a) Central limit theorem: histogram of water-well yield, rock type Yg. (b) Central limit theorem: histogram of water-well yield, means of size $n = 5$, rock type Yg.

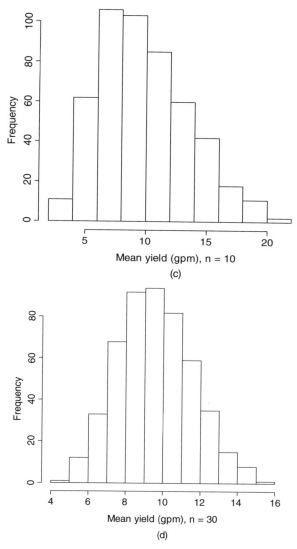

FIGURE 3.1 (*Continued*)   (c) Central limit theorem: histogram of water-well yield, means of size $n = 10$, rock type Yg. (d) Central limit theorem: histogram of water-well yield, means of size $n = 30$, rock type Yg.

3.4.1   *t*-Distribution

Student's *t*-distribution is used in classical statistics for interval estimation and hypothesis testing. A *t-distribution* (Figure 3.2) is symmetric about zero, bell-shaped, and normal looking, except that its tails are more spread out than the normal. This means that there is more probability in the tails and thus more variability than in a

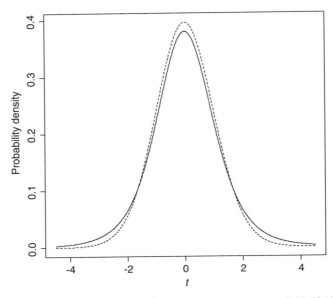

FIGURE 3.2 t-Probability density functions for 5 df (solid line) and 30 df (dashed line).

standard normal distribution. The t-distribution is a family of distributions where the specific functional form of each member of this family is determined by its degrees of freedom (df). Observe the difference in the tails between t-distributions at 5 and 30 df. The probability density function is defined as

$$f(t; v) = \frac{\Gamma((v+1)/2)}{\sqrt{v\pi}\ \Gamma(v/2)} \left(1 + \frac{t^2}{v}\right)^{-(v+1)/2} \qquad -\infty < t < \infty$$

where v is the df and Γ is the gamma function.

To illustrate the t-distribution and the concept of degrees of freedom, consider the statistic

$$T = \frac{\hat{\mu} - \mu}{\hat{\sigma}/\sqrt{n}}$$

where $\hat{\mu} = \sum_{i=1}^{n} Y_i/n$ and $\hat{\sigma}$ is the corresponding estimate of σ. If $Y_1, \ldots, Y_n$ are normally distributed independent samples with mean μ and standard deviation σ, the T-statistic has a t-distribution with $n-1$ df. If the denominator contains σ instead of $\hat{\sigma}$, the distribution is standard normal.

Degrees of freedom are a function of the number of observations and the number of parameters that must be estimated from the data to construct a confidence interval, hypothesis test, or other model. A large number of degrees of freedom are desirable because this implies that the statistic of interest will be more certain. Conversely, the more parameters that need to be estimated from the data, the more degrees of freedom are lost. In the statistic above, one degree of freedom is lost because the standard deviation has to be estimated from the data.

The reason the t-distribution has a larger variance than the normal is the presence of $\widehat{\sigma}$ in the denominator of T. It varies from sample to sample. The two distributions (Figure 3.2) appear similar; however, there are noticeable differences in the tails. Consider $\Pr(T \geq t_{0.025,5}) = 0.025$, where 5 is the df and 0.025 is the probability in the right-hand tail. The quantile corresponding to this probability is $t_{0.025,5} = 2.571$. For $v = 30$ df, $t_{0.025,30} = 2.042$. A way to write this expression is $\Pr(T \geq t_{\alpha,}) = \alpha$, where α is the probability in the right-hand tail. It is shown as a subscript followed by the v, the df. Some algorithms require specification of the cumulative probability, $1 - \alpha$, while others require α when computing $t_{\alpha,v}$. A t-distribution with 30 df, denoted $t_{(30)}$, is for most applications indistinguishable from the standard normal. For example, $t_{0.025,30} = 2.042$, while $z_{0.025} = 1.960$.

3.4.2 χ^2 (Chi-Square) Distribution

The χ^2 *distribution* is used to make an interval estimate on a variance and in hypothesis testing. It results from summing the squares of independent standard normal variables. Specifically, if $Z_i \sim N(0,1)$, $i = 1, \ldots, v$ are independent, a random variable $Y = \sum_{i=1}^{v} Z_i^2 \sim \chi_v^2$, where $v > 0$ is the df. The notation $\chi^2(v)$ is also used. χ_v^2 is a right-skewed distribution, which tends to symmetry as v becomes larger. The χ^2 probability density functions for $v = 5$, 10, and 20 are shown in Figure 3.3. As with the t-distribution, a probability p can be specified and its inverse, y_p, determined.

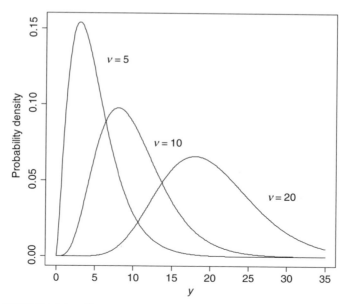

FIGURE 3.3 χ^2 probability density functions for $v = 5$, 10, and 20.

The probability density function for the χ^2 distribution is

$$f(y; v) = \frac{1}{2^{v/2}\,\Gamma(v/2)} y^{(v/2)-1} e^{-y/2} \qquad y > 0$$

where Γ is the gamma function. For example, let $y_1, y_2, \ldots, y_n$ be a random sample from a normal population $N(\mu, \sigma^2)$. Let $\text{SS}_Y = \sum_{i=1}^{n}(y_i - \bar{y})^2$. Then SS_Y/σ^2 is the sum of squares of n standard normal deviates and therefore is χ^2_{n-1}. Equivalently, $\hat{\sigma}^2(n-1)/\sigma^2$ is χ^2_{n-1}, where $\hat{\sigma}^2$ is the sample variance.

3.4.3 *f*-Distribution

The *f-distribution* results from a ratio of chi-square variates and is used to test the equality of variances. Let X and Y be independent chi-squared distributed random variables with v and ω df, respectively, written $X \sim \chi^2_v$ and $Y \sim \chi^2_\omega$. Then

$$F_{v,\omega} = \frac{\chi^2_v/v}{\chi^2_\omega/\omega} \qquad F \geq 0$$

has an *f*-distribution with v and ω df in the numerator (associated with X) and the denominator (associated with Y), respectively. The *f*-distribution is right-skewed but becomes less skewed as the degrees of freedom in the numerator and denominator approach equality (Figure 3.4). The *f*-distribution frequently arises as the null

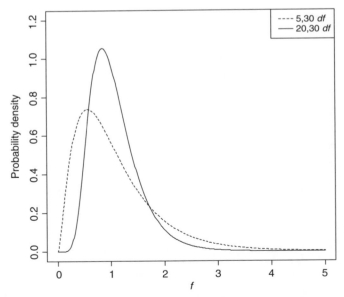

FIGURE 3.4 *f*-Probability density functions for (5,30) and (20,30) df.

hypothesis when testing model significance. If $v = 1$, then $\sqrt{F_{1,\omega}} \sim t_\omega$ or, in words, the square root of an f-distributed random variable with 1 and ω df is t-distributed with ω df.

As an example, consider independent samples of size n_1 and n_2 from a normal population with variance σ^2. Let $\widehat{\sigma}_1^2$ and $\widehat{\sigma}_2^2$ be the sample variance from samples n_1 and n_2, respectively. Then $\widehat{\sigma}_1^2/\sigma^2 \sim \chi_{n_1-1}^2$ and $\widehat{\sigma}_2^2/\sigma^2 \sim \chi_{n_2-1}^2$. Therefore, the statistic $F_{n_1-1,n_2-1} = \widehat{\sigma}_1^2/\widehat{\sigma}_2^2$ formed by their ratio has an f-distribution with $n_1 - 1$ and $n_2 - 2$ df.

3.5   CONFIDENCE INTERVAL ESTIMATE ON A MEAN

A *confidence interval* (CI) *estimate on a mean* μ is an interval ($k_1 \leq \mu \leq k_2$), where k_1 and k_2 depend on the sample mean, its variance, and a level of assurance called the *significance level* that the CI bounds μ. Initially, we illustrate the construction of an interval estimate using the classical frequentist approach. Subsequently, other approaches are introduced.

To compute a confidence interval on μ, the following assumptions are required:

- The observations are independent of each other and are drawn from a population with mean μ and variance σ^2.
- The sample mean is normally distributed.

The sample mean being normally distributed is not a strong assumption for moderate to large size random samples, due to the CLT. Independence of observations is more problematic. The independence assumption should always be questioned when sampling in space or over time because nearby observations are often correlated.

If k_1 and k_2 are finite, the interval is called a *two-sided confidence interval*. If either is infinite, it is a *one-sided confidence interval*. The choice is problem related and we give examples.

In interval estimation, the variance may be known or unknown. In practice, the variance is almost never known; however, the case of the known variance is presented because it is a building block for the more realistic case.

3.5.1   Variance Known: Frequentist Approach

The simplest case of interval estimation results from making the assumption that the variance σ^2 is known. For large samples or when adequate historical data are available, this assumption may be reasonable. In simulation the variance is known. When the variance is known, the two-sided interval estimate is

$$\widehat{\mu} - \frac{z_{\alpha/2}\sigma}{\sqrt{n}} \leq \mu \leq \widehat{\mu} + \frac{z_{\alpha/2}\sigma}{\sqrt{n}}$$

where $\widehat{\mu}$ is the sample mean, $z_{\alpha/2}$ is a function of the significance level, and $\widehat{\mu}$ is normally distributed. Finally, $\sigma/\sqrt{n}$ is the sampling error of $\widehat{\mu}$, sometimes called the standard error of the mean. It is a measure of variability of $\widehat{\mu}$ and is written $\sigma_{\widehat{\mu}} = \sigma/\sqrt{n}$. The subscript $\widehat{\mu}$ indicates that $\sigma_{\widehat{\mu}}$ is a measure of the variability of the sample mean, as opposed to the variability of a single random variable Y. The z_{α} represents a point on the normal distribution at a right-tail probability of $\alpha, 0 < \alpha < 1$, typically $0.001 < \alpha < 0.20$. If a random variable $Z \sim N(0,1)$, then $\alpha = P(Z \geq z_{\alpha})$; for example, $z_{0.05} = 1.645$.

Why choose a normal distribution to establish the significance level? It is because the statistic $Z = (\mu - \widehat{\mu})/\sigma_{\widehat{\mu}}$, when solved for μ, is $\mu = \widehat{\mu} + Z\sigma_{\widehat{\mu}}$, is centered at $\widehat{\mu}$, and is normally distributed.

A two-sided confidence interval is illustrated using rock type Yg from the water-well yield case study. The goal is to establish a 95% two-sided confidence interval on the Yg population mean. The 95% CI ($\alpha = 0.05$) is chosen arbitrarily. In a two-sided confidence interval, the interval estimate is bounded below and above. Typically, the probability in the tails is split evenly so that for $\alpha = 0.05$, $z_{0.05/2} = 1.960$. There is a 0.025 probability in each tail (Figure 3.5). Recall from Chapter 1 that $\widehat{\mu} = 9.64$ and $n = 81$. Assume that $\sigma = 11.40$. The standard error of the mean is $\sigma_{\widehat{\mu}} = \sigma/\sqrt{n} = 11.40/\sqrt{81}$. Thus, the 95% CI on μ is

$$\widehat{\mu} \pm z_{\alpha/2}\sigma_{\widehat{\mu}} = 9.64 \pm (1.960)(1.27) \quad \text{or} \quad (7.15, 12.13)$$

Of course, the interval $(7.15, 12.13)$ either brackets (includes) the true (population) mean or it does not.

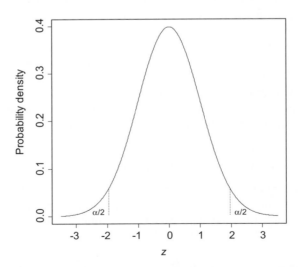

FIGURE 3.5   Normal distribution; the area to the right and left of the dashed vertical lines is $\alpha/2 = 0.025$; the dashed lines are located at $z_{0.025} = \pm 1.960$.

There are several ways to interpret a confidence interval. The traditional textbook frequentist interpretation of this interval is as follows. Imagine that a very large number of independent samples are taken, each of size 81, and that for each data set a 95% confidence interval is computed. If the assumptions are valid, 95% of these intervals will bracket the true mean. This, as Bayarri and Berger (2004) note, is the "imagining interpretation" and is misleading because such replication is not done except possibly in simulation. Neyman (1977) stated that the interpretation should be based on repeated use of a statistical model over a variety of real problems and differing data sets. The *frequentist principle* states that the long-run average actual accuracy should be at least as great as the long-run average reported accuracy. In the context of a 95% CI on the mean, there should be 95% coverage when used over different problems and data sets.

3.5.2   Variance Known: Bayesian Approach

The *Bayesian approach* provides an effective way to integrate prior beliefs and data, which is especially helpful when the data set is small. Bayes' theorem for discrete parameters is

$$h(\theta \mid y_1, \ldots, y_n) = \frac{f(y_1 \mid \theta) \cdots f(y_n \mid \theta) g(\theta)}{\sum_\theta f(y_1 \mid \theta) \cdots f(y_n \mid \theta) g(\theta)}$$

where $y_1, \ldots, y_n$ is the data set, $h(\theta \mid y_1, \ldots, y_n)$ is the conditional probability distribution of θ given the data, and the expression following the equal sign is the likelihood of the data given the parameter θ. The density function $g(\theta)$ represents the prior belief on the parameter θ. If the density f is continuous, the summation in the denominator above is replaced by an integral. An equivalent expression is

$$\text{posterior} \propto \text{likelihood} \times \text{prior}$$

Now revisit the example of a confidence interval on the mean of rock type Yg. Assume from prior information that $\mu \sim N(10, 4)$. Let x be an observation, and $y = \log_e(x)$ be normal with mean μ (unknown) and variance σ^2 is known. The known variance case is used for simplicity of presentation. The form a posterior distribution is $f(\mu \mid y) \propto f(y \mid \mu)\pi(\mu)$, where both the likelihood $f(y \mid \mu)$ and the prior distribution $\pi(\mu)$ are assumed to be normally distributed. Integrating with respect to μ yields the *Bayesian confidence interval*, often called a credible interval, defined as

$$\Pr(C \mid y) = \int_C f(\mu \mid y) d\mu \qquad \Pr(C \mid y) \geq 1 - \alpha$$

If the data are discrete, integration is replace by summation. In this example, the result is given in log units. The interpretation of a Bayesian confidence interval is as follows: The probability that μ lies in some interval C, given the data, is at least $1 - \alpha$. Some statisticians view this as a cleaner and more satisfying statement than is given by the frequentist approach, since it avoids the use of the repeated sampling argument employed by the latter group. A discussion of Bayesian data analysis and computation have been discussed in detail by Carlin and Louis (2000) and Gelman et al. (2004). Computations can be performed in WinBugs (WinBugs Project, 2005).

The Bayesian confidence interval or credible set is the same as the frequentist confidence interval if ignorance about the value of μ is assumed. Recall that in the Bayesian approach, μ is considered to be a random variable with a distribution. Ignorance of μ implies a prior distribution that is infinite and flat. An advantage of the Bayesian approach is that as new data become available, the interval estimate can easily be revised. For example, suppose that an additional 30 water-well yield data values for rock type Yg become available with sample mean $\widehat{\mu}_N = 10.3$ and variance $\widehat{\sigma}_N^2 = 156.25$, where the subscript N refers to the new data. When performing an update, Bayesians typically refer to precision, which is the inverse of the variance. In the original sample, the precision of $\widehat{\mu}$ is $81/11.40^2 = 0.623$. The precision of the new sample is $30/156.25 = 0.192$. The updated mean is

$$\widehat{\mu}_U = \frac{0.623}{0.623 + 0.192}\widehat{\mu} + \frac{0.192}{0.623 + 0.192}\widehat{\mu}_N$$

$$= (0.764)(9.64) + (0.236)(10.3)$$

$$= 9.79$$

where the precisions are the weights. Since the original observations have higher precision, the original mean, 9.64, is assigned the most weight. Since the prior (original) and new data means are normally distributed, the precision of the revised (posterior) distribution is the sum of the original plus new precision, $0.623 + 0.192 = 0.815$, and the new variance is its inverse, $1/0.815 = 1.22$. This additive relationship assumes known precisions but is approximately true for estimates. The updated CI is

$$\widehat{\mu}_U \pm z_{\alpha/2}\sigma_{\widehat{\mu}_U} = 9.79 \pm (1.960)(1.22) \quad \text{or} \quad (7.40, 12.18)$$

which is slightly shifted to the right and a bit narrower than the original CI.

In a Bayesian approach, the parameter μ is considered to be a random variable, and the user assigns a prior estimate of its distribution. The form of the prior depends on the Bayesian approach used. A classical Bayesian assigns a noninformation prior distribution, which is one that conveys the least amount of information. Typically, this is a distribution that is essentially flat with long tails. Reasons for choosing specific prior distributions have been given by Gelman et al. (2004). A subjective Bayesian chooses a prior distribution for μ that incorporates prior knowledge, which may result from previous studies or expert judgment. For example, there may be a reason to believe that the prior distribution on μ is $N(m = 9, s^2 = 10)$.

3.5.3 Variance Unknown

A frequentist or Bayesian confidence interval estimate is slightly more complicated when the variance is unknown. Initially, the frequentist approach is introduced. Instead of basing the CI on the normal distribution, it is based on Student's t-distribution because the statistic

$$T = \frac{\hat{\mu} - \mu}{\hat{\sigma}/\sqrt{n}}$$
$$= \frac{\hat{\mu} - \mu}{\hat{\sigma}_{\hat{\mu}}}$$

has a t-distribution. Note that $\hat{\sigma}_{\hat{\mu}}$ is an estimate of the standard error of the mean.

The rock type Yg example is used again. A 5% ($\alpha = 0.05$) level of significance is selected with a goal of estimating a two-sided CI. The significance probability is divided equally between the lower and upper tails. A probability of 0.025 is assigned to each tail. The 95% CI on μ (rock type Yg) is

$$\hat{\mu} \pm t_{\alpha/2,n-1}\hat{\sigma}_{\hat{\mu}}$$

where instead of $z_{\alpha/2}$, $t_{\alpha/2,n-1}$ is used. The value of $t_{0.05/2,81-1} = 1.990$. Thus, the 95% confidence interval on μ is

$$9.64 \pm (1.990)(1.27) \quad \text{or} \quad (7.11, 12.17)$$

which is slightly wider than when σ is assumed known. For a smaller sample size, a greater difference will be observed.

Construction of the Bayesian confidence interval is similar to the procedure outlined in the previous (variance known) section except that the t-distribution is substituted for the normal. Specifically, the updated CI is

$$\hat{\mu}_U \pm t_{0.05/2,80}\,\sigma_{\hat{\mu}_U} = 9.79 \pm (1.990)(1.22) \quad \text{or} \quad (7.36, 12.22)$$

which is slightly wider than the CI computed using the normal assumption.

3.5.4 Width of a Confidence Interval

Previously, a 95% confidence level has been computed on μ, the mean of the water-well yield for rock type Yg. The choice of 95% confidence is common but arbitrary. To achieve a greater level of confidence, for example, 99%, set $\alpha = 0.01$. Assume that $\sigma = 11.40$ is known; then $z_{0.001/2} = 3.29$. The 99% confidence interval is $9.64 \pm (1.27)(3.29)$, or 9.64 ± 4.18. Thus, there is more confidence (99% versus 95% in a frequentist sense) that the interval will bracket μ. The cost is a wider interval, which conveys less information. How can the level of confidence be increased without widening the interval? There are two ways: (1) increase the sample size, since the standard error of the mean is $\sigma/\sqrt{n}$, or (2) decrease σ via some nonstatistical procedure such as reducing measurement error. For example, to reduce error in temperature measurements, a digital thermometer can be used in place of the traditional mercury column device. Improved training of technicians may decrease error in the estimation of coal bed thickness.

3.5.5   One-Sided Confidence Intervals

When appropriate, a one-sided confidence interval is more informative than a two-sided interval. For example, a water company engaged in exploratory drilling may only be concerned with a yield that is at least as large as some minimally acceptable flow rate. Anything greater is a bonus; thus, a lower confidence interval should be computed. The example refers again to a confidence interval on the population mean of rock type Yg: namely,

$$\widehat{\mu} - t_{\alpha,n-1}\widehat{\sigma}_{\widehat{\mu}} \leq \mu \leq \infty$$
$$\widehat{\mu} - t_{0.05,80}1.27 \leq \mu \leq \infty$$
$$9.64 - (1.664)(1.27) \leq \mu \leq \infty$$
$$9.64 - 2.11 \leq \mu \leq \infty$$

where all of the probability (0.05 in this example) is assigned to the lower tail of the t-distribution. We leave it as an exercise for the reader to comment on why this interval is preferable to the two-sided confidence interval.

3.6   CONFIDENCE INTERVAL ON THE DIFFERENCE BETWEEN MEANS

Suppose that the problem is to obtain an interval estimate on the difference between two means, such as the difference between mean water-well yields from rock types Yg and Zc. To simplify the notation, the corresponding populations of water-well yield from these rock types are 1 and 2, respectively. Thus, the confidence interval on $\mu_1 - \mu_2$, which is estimated by $\widehat{\mu}_1 - \widehat{\mu}_2$, the difference in sample means, where observations $Y_{1,1}, Y_{1,2}, \ldots, Y_{1,n_1}$ and $Y_{2,1}, Y_{2,2}, \ldots, Y_{2,n_2}$ are judged to be independent within and between populations. Thus, the interval estimate is $k_1 \leq \mu_1 - \mu_2 \leq k_2$.

In the next two subsections, cases of variances known and unknown are examined from a frequentist perspective. Unknown variances can be equal or unequal.

3.6.1   Estimate on the Difference Between Means, Variances Known

For known variances, the variance of the difference between sample means (assuming independence) is

$$\sigma^2_{\hat{\mu}_1 - \hat{\mu}_2} = \text{Var}(\hat{\mu}_1 - \hat{\mu}_2)$$

$$= \frac{\sigma^2_1}{n_1} + \frac{\sigma^2_2}{n_2}$$

The two-sided $(1 - \alpha)$ % CI on $\mu_1 - \mu_2$ is

$$(\hat{\mu}_1 - \hat{\mu}_2) - z_{\alpha/2}\sigma^2_{\hat{\mu}_1 - \hat{\mu}_2} \leq \mu_1 - \mu_2 \leq (\hat{\mu}_1 - \hat{\mu}_2) + z_{\alpha/2}\sigma^2_{\hat{\mu}_1 - \hat{\mu}_2}$$

The statistics for the rock types Yg (subscript 1) and Zc (subscript 2) are shown in Table 3.1. Thus,

$$\hat{\mu}_1 - \hat{\mu}_2 = 9.64 - 11.81$$

$$\sigma^2_{\hat{\mu}_1 - \hat{\mu}_2} = \frac{\sigma^2_1}{n_1} + \frac{\sigma^2_2}{n_2}$$

$$= \frac{129.96}{81} + \frac{186.60}{171}$$

$$= 2.70$$

and the 95% CI on $\mu_1 - \mu_2$ is

$$-2.17 - (1.960)(2.70) \leq \mu_1 - \mu_2 \leq 2.17 + (1.960)(2.70)$$

$$3.12 \leq \mu_1 - \mu_2 \leq 7.46$$

3.6.2   Interval Estimate on the Difference Between Means, Variances Unknown and Assumed Equal

Variances are unknown and assumed to be equal. Subsequently, procedures to check for equal variances are examined. For this example, the values are those in Table 3.1 except that variances are assumed to be unknown. A two-sided CI is computed on $\mu_1 - \mu_2$, the difference between population means of rock types Yg and Zc,

TABLE 3.1   Summary Statistics of Water-Well Yield for Rock Types Yg and Zc

	Population	
	1	2
Rock type	Yg	Zc
Sample mean	9.64	11.81
n	81	171
Variance	129.96	186.60
Standard deviation	11.40	13.66
Std. error of the mean	1.27	1.04

respectively. Since equal variances are assumed, a pooled variance is computed as follow:

$$\widehat{\sigma}_p^2 = \frac{(n_1 - 1)\widehat{\sigma}_1^2 + (n_2 - 1)\widehat{\sigma}_2^2}{n_1 + n_2 - 2}$$

$$= \frac{(81 - 1)(129.96) + (171 - 1)(186.60)}{81 + 171 - 2}$$

$$= 168.48$$

where the subscript p means "pooled". Note that $(n_1 - 1)\widehat{\sigma}_1^2 = \sum_{i=1}^{n_1}(y_{1i} - \widehat{\mu})^2$ is the sum of squares (SS) of the observations from population 1, y_{1i}, about its sample mean $\widehat{\mu}_1$. A similar formula holds for population 2. The estimated standard error of the difference between means, $\widehat{\sigma}_{p(\widehat{\mu}_1 - \widehat{\mu}_2)} = \widehat{\sigma}_p/\sqrt{n_1 + n_2} = 0.821$. Since the variance is estimated, a t-distribution is required. The total number of observations is $n_1 + n_2$, but because the two variances are estimated from the data, 2 df are lost, leaving $n_1 + n_2 - 2$ for estimation of the CI. From an algorithm or table, $t_{0.025,250} = 1.969$. The two-sided $(1 - \alpha)\%$ CI on $\mu_1 - \mu_2$ is

$$(\widehat{\mu}_1 - \widehat{\mu}_2) - t_{\alpha/2,n_1 - n_2 - 2}\widehat{\sigma}_{p(\widehat{\mu}_1 - \widehat{\mu}_2)} \leq \mu_1 - \mu_2$$

$$\leq (\widehat{\mu}_1 - \widehat{\mu}_2) + t_{\alpha/2,n_1 - n_2 - 2}\widehat{\sigma}_{p(\widehat{\mu}_1 - \widehat{\mu}_2)}$$

For the difference between means of yields in rock type Yg and Zc, the 95% CI is

$$(9.64 - 11.81) \pm (1.969) \times (0.821) \quad \text{or} \quad (-3.79, -0.55)$$

This result suggests that $\mu_1 < \mu_2$ because the interval does not contain zero.

3.6.3   Confidence on the Difference Between Means, Variances Unknown and Unequal

If the population variances are assumed to differ or the investigator is unwilling to assume equal variances, *Welsh's t* may be used to obtain a variance estimate. The estimated variance of the difference between means using Welsh's t is

$$\widehat{\sigma}_W^2 = \frac{\widehat{\sigma}_1^2}{n_1} + \frac{\widehat{\sigma}_2^2}{n_2}$$

which is the same equation that is used when variances are assumed to be known, except that estimated variances replace known variances. The corresponding standard error is $\widehat{\sigma}_W = \sqrt{\widehat{\sigma}_W^2}$. The penalty for failure to assume equal variances is a loss of degrees of freedom. The df for Welsh's t computed using the Satterthwaite approximation is

$$\mathrm{df}_{SA} = \frac{(\widehat{\sigma}_w^2)^2}{(\widehat{\sigma}_{\widehat{\mu}_1}^2)^2/(n_1 - 1) + (\widehat{\sigma}_{\widehat{\mu}_2}^2)^2/(n_2 - 1)}$$

For the example shown in Table 3.1, assuming unknown variances, $\mathrm{df}_{SA} = 187.2$, where

$$\widehat{\sigma}_w^2 = \frac{\widehat{\sigma}_1^2}{n_1} + \frac{\widehat{\sigma}_2^2}{n_2} \qquad\qquad \widehat{\sigma}_{\widehat{\mu}_1}^2 = \frac{\widehat{\sigma}_1^2}{n_1} \qquad\qquad \widehat{\sigma}_{\widehat{\mu}_1}^2 = \frac{\widehat{\sigma}_1^2}{n_1}$$

$$= \frac{129.96}{81} + \frac{186.60}{171} \qquad\qquad = \frac{129.96}{81} \qquad\qquad = \frac{186.60}{171}$$

$$= 2.70 \qquad\qquad\qquad\qquad = 1.60 \qquad\qquad\qquad = 1.09$$

When variances are assumed to be equal, $\mathrm{df} = n_1 + n_2 - 2$, which for the example above is 250. The effect of reducing the df from 250 to 187.2 is minimal; however, a 25% reduction in df may be meaningful for smaller samples. Note that $t_{0.025,187.2} = 1.973$, whereas $t_{0.025,250} = 1.969$. Finally, the two-sided $(1 - \alpha)\%$ CI on $\mu_1 - \mu_2$ is

$$(\widehat{\mu}_1 - \widehat{\mu}_2) - t_{\alpha/2,n_1 - n_2 - 2}\widehat{\sigma}_w \leq \mu_1 - \mu_2 \leq (\widehat{\mu}_1 - \widehat{\mu}_2) + t_{\alpha/2,n_1 - n_2 - 2}\widehat{\sigma}_w$$

and for the difference between means of yields in rock type Yg and Zc, the 95% CI is

$$(9.64 - 11.81) \pm (1.973)\sqrt{2.70} \quad \text{or} \quad (-5.41, 1.07)$$

This result suggests that μ_1 may be equal to μ_2 because the CI contains zero. The CI for the previous variance unknown and assumed equal case does not include zero. The reason for this difference is due largely to the increased variance, and in a minor way to a loss of degrees of freedom in the unequal-variance case.

3.6.4 Investigating Equality of Variance

Equality of variance is important for many models where classical statistics are used. In regression (Chapter 4) a usual assumption is that variances of a response variable are equal across the domain of an explanatory variable. In time series and spatial statistics (Chapters 5 and 6, respectively), an assumption of equality of variance often is critical. Here, the assumption of equality of variance relates to hypothesis testing on means. If the population variances are equal, the resulting estimated variance, $\mathrm{Var}(\widehat{\mu}_1 - \widehat{\mu}_2)$, will have more degrees of freedom than if they are unequal, and more degrees of freedom yield a more powerful test. We need terms that describe equal and unequal variance: *Homoscedasticity* means that the population variances are equal; *heteroscedasticity* means that they are unequal. The classic test of homoscedasticity is that of Bartlett (Snedecor and Cochran, 1989). An alternative is Levene's test (Levene, 1960), which is less sensitive to the normality assumption.

Levene's Test   Suppose that there are $k \geq 2$ subgroups or populations. For a hypothesis test comparing two means, $k = 2$. Hypotheses involving more than two means are illustrated later, so a general form of *Levene's test* is presented. The null and alternative hypotheses for testing equality of variances are

$$H_0 : \sigma_1^2 = \cdots = \sigma_k^2$$
$$H_a : \sigma_i^2 \neq \sigma_j^2 \text{ for at least one pair } i \neq j$$

The hypothesis is that observations within and between the k subgroups are independent.

The test statistic is

$$L = \frac{(n - k) \sum_{i=1}^{k} n_i (\overline{Y}_{i.} - \overline{Y}_{..})^2}{(k - 1) \sum_{i=1}^{k} \sum_{j=1}^{n_i} (Y_{ij} - \overline{Y}_{i.})^2}$$

where n_i observations are from the ith subgroup, $n = n_1 + \cdots + n_k$, Y_{ij} is the jth observation from the ith subgroup, $\overline{Y}_{i.} = \sum_{j=1}^{n_i} Y_{ij}/n_i$ is the mean of observations in the ith subgroup, and $\overline{Y}_{..} = \sum_{i=1}^{k} \sum_{j=1}^{n_i} Y_{ij}/n$ is the overall or grand sample mean. The dot subscript notation is used in these and other computations to indicate "summed over." If the variances are equal, the test statistic L has an f-distribution. For significance level $1 - \alpha$, the test of equality of variance is rejected if $L > F_{(\alpha, k-1, n-k)}$, where $F_{(\alpha, k-1, n-k)}$ represents a value on the f-distribution corresponding to a significance level of α. The values $k - 1$ and $n - k$ are the degrees of freedom in the numerator and denominator of the F-statistic.

Note that:

- Levene's test can be defined for a trimmed mean and a median.
- Generally, equal sample sizes across groups are preferred when designing an experiment. An exception may be if there are perceived to be significant differences in variances.

When samples are small, graphic historical data and/or analogs may help to assess equality of variances. Alternatively, variances may be assumed to be unequal.

Graphical Procedures   Boxplots are an important graphical tool to help evaluate the homoscedasticity assumption. When a significant departure from the equal-variance assumption is found, a transformation to equalize variances may be required or a procedure that does not require a homoscedasticity assumption to be used.

The variable-width notched boxplot (Figure 3.6a) provides a visual comparison of the IQRs for rock types Yg and Zc. Clearly, both distributions are right (positively)-skewed; however, the distribution of water-well yield for rock type Zc is more skewed and its IQR (the height of the notched box) is greater. By trial and error, a fifth-root

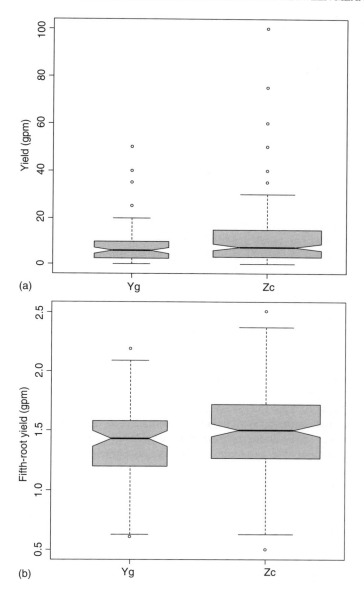

FIGURE 3.6   (a) Boxplots of water-well yield, rock types Yg and Zc. (b) Boxplots for the fifth-root transformation of water-well yield, rock types Yg and Zc.

transformation, $U = Y^{1/5}$, that yields approximate symmetry (Figure 3.6b) is found. A comparison of the IQRs between rock types Yg and Zc after transformation suggests that an equal-variance assumption is reasonable. An alternative is to use a Box–Cox transformation.

3.6.5   Bootstrap Confidence Intervals

Bootstrapping is a computer-intensive resampling procedure that generates uncertainty estimates of parameters where there is an unwillingness to assume a specific distributional form. A *bootstrap sample* is one obtained from an empirical distribution, usually with replacement, where the bootstrap sample size is often the same as the original data sample.

Previously, a confidence interval on μ for rock type Yg was estimated by assuming that the sample mean is normally distributed, which for $n = 81$ is reasonable. Suppose, instead, that $n = 10$ and little prior information exists on these data. For this smaller data set, an investigator may not be willing to make a normal assumptions. Instead of estimating $\overline{Y} \pm z_{\alpha/2}\mathrm{SE}(\overline{Y})$ by assuming that $\overline{Y}$ is standard normal, a new statistic, $z^* = (\overline{Y}^* - \mu)/\mathrm{SE}(\overline{Y}^*)$, is computed, where $\overline{Y}^*$ and $\mathrm{SE}(\overline{Y}^*)$ are the mean and standard error, respectively, generated by resampling. Replication of this process B times, where B may be at least 1000, creates a distribution of z^*'s from which a confidence interval can be obtained. Each bootstrap run requires a sample of size n taken with replacement. A simple approach to computing a 95% confidence interval is to construct a distribution of Yg from the entire bootstrap sample and find the 2.50 and 97.5 percentiles. A more accurate approach (Davison and Hinkley, 1997) is to normalize the data, compute a confidence interval, and then transform it into original units. This approach yields the bias corrected adjusted (BCA) confidence interval (see Davison and Hinkley, 1997). Table 3.2 shows the original Yg data and five bootstrap samples together with their corresponding means and standard errors. Below this are bootstrap results on the 10 Yg water-well yields. A 95% two-sided bootstrap confidence interval on the μ is 3.97 to 9.35. The confidence interval computed using a *t*-ratio on these 10 Yg water-well yields is 3.21 to 9.73. This result is slightly wider than the bootstrap estimate. Which interval estimate is correct? Clearly, the answer is unknown; however, the bootstrap procedure provides protection against outliers and skewed distributions. We recommend using the bootstrap estimate unless there is a reason to believe that these observations are normally distributed.

A complete discussion of bootstrap methods and applications is given in the book by Davison and Hinkley (1997). Many statistical packages have implemented bootstrapping and other resampling procedures. Bootstrapping is a powerful and valuable tool. It does not, however, replace deductive analysis and proper planning.

3.7   HYPOTHESIS TESTING ON MEANS

A *hypothesis* is a conjecture about the value of a parameter, in this section a population mean or means. One hypothesis is that the (population) mean water-well yield equals some chosen value, say 9 gpm, against the alternative that it is not equal to 9 gpm. Another hypothesis is that the (population) mean of yield from rock type Yg is equal to the mean yield from rock type Zc against the alternative that they are not equal. The computations to test hypotheses closely parallel those of confidence intervals. *Hypothesis testing* assists in making a decision under uncertainty: a decision to drill

TABLE 3.2 Resampling of Rock Type Yg

	Yg	Bootstrap Replication[a]				
		1	2	3	4	5
	15.00	10.00	7.00	10.00	10.00	6.00
	0.50	2.50	10.00	6.00	5.50	5.50
	2.50	7.00	6.00	5.50	5.50	2.50
	5.50	10.00	2.50	7.00	10.00	11.00
	6.00	15.00	7.00	6.00	11.00	11.00
	7.00	7.00	6.00	6.00	0.50	2.50
	6.00	11.00	1.20	10.00	15.00	1.20
	10.00	15.00	2.50	6.00	1.20	10.00
	1.20	5.50	5.50	5.50	6.00	0.50
	11.00	15.00	10.00	10.00	10.00	1.20
Mean	6.47	9.80	5.47	7.20	7.47	5.14
Std. error	1.44	1.37	1.01	0.62	1.44	1.33

Summary Statistics

Observed Mean	Bootstrap		
	Bias	Mean	SE
6.47	−0.04	6.43	1.44

BCA Percentiles

	2.50%	5%	95%	97.50%
Mean	3.97	4.21	8.90	9.35

[a]Bootstrap, $B = 1000$ replications.

a well, launch a satellite, or implement a toxic waste remediation program. A confidence interval provides information about uncertainty on a mean.

Perhaps no subject has created more controversy between members of the frequentist school and those of the Bayesian school than hypothesis testing. In addition, some statisticians believe that hypothesis testing is a meaningless activity designed only to give support to what the investigator already knows (see, e.g., Johnson, 1999). A detailed analysis of this controversy is beyond the scope of this book, but it is necessary to have some understanding of this important issue because different approaches to testing sometimes yield different results. It is clear that hypothesis testing is misused. However, when constructed properly, it can serve as a useful guide to decision making.

There are three basic schools of hypothesis testing: the frequentist approach established by Jerzy Neyman, the modified frequentist approach of R. A. Fisher, and the Bayesian approach pioneered by Harold Jeffreys. There is research to develop a unified approach (Bayarri and Berger, 2004) to hypothesis testing. Neyman's frequentist approach is presented first.

3.7.1 Hypothesis-Testing Procedure

Consider the following problem. A housing developer needs to demonstrate to a county planning board that the mean water-well yield in a proposed subdivision will exceed 9 gpm. Preliminary studies suggest that this may be true; however, a more scientifically based answer is needed. The following steps demonstrate how this conjecture is placed in a statistical hypothesis-testing context using the Neyman frequentist approach.

1. Construct a hypothesis. The hypothesis is

$$H_0: \mu = \mu_0$$
$$H_a: \mu > \mu_0$$

   where H_0 is called the *null hypothesis*, H_a is called the *alternative hypothesis*, and μ_0 is a constant, which for this example is 9 gpm ($\mu_0 = 9$). Implicit in H_0 is that if $\mu < 9$, H_0 will be accepted. An action, say development of the subdivision, will be taken only if $\mu > 9$. This is called a one-sided test because the alternative is in one direction. A test of $\mu = 9$ against the alternative that it is not is two-sided:

$$H_0: \mu = 9$$
$$H_a: \mu \neq 9$$

   H_0 typically represents the current state of nature or a historical value that the investigator hopes or expects to reject. The usual belief is that H_a is true. (This is one source of controversy discussed later.)
2. Decide on a sampling plan. A representative sample is needed for valid decision making (Chapter 1). A sampling plan involves collecting representative data. This is achieved using random sampling.
3. Select a test statistic. Since this is a test of a mean, a reasonable statistic is the sample mean.
4. Determine the distribution of the test statistic. The data are assumed to be normally distributed, or the sample size is large enough that the CLT holds. If σ is estimated from the data, the estimated standard error of the mean $\widehat{\sigma}_{\widehat{\mu}} = \sqrt{\widehat{\sigma}^2/n}$, as in the confidence interval computation. Therefore, the statistic $T = (\widehat{\mu} - \mu)/\widehat{\sigma}_{\widehat{\mu}}$ has a *t*-distribution with $n-1$ df. Sometimes an estimate of σ can be obtained from pilot studies, historical data, or expert judgment prior to testing a hypothesis. Prior information is valuable in planning an experiment.
5. Determine the sample size n. Sometimes historical data must be used. However, when possible it is desirable to determine the sample size needed for the test under consideration and collect appropriate data. In Section 3.7.4 we show how to determine the sample size. For now, assume that $n = 81$. The Yg samples are provided in Appendix I.

6. Select a rejection region. Typical choices of significance level α are 0.10, 0.05, 0.01, and 0.001, but careful consideration needs to be given to this decision. Since the alternative hypothesis is $\mu > 9$, the rejection region is in the right-hand tail of the t-distribution and is determined by α and the sample size. Some value of the sample mean larger than 9 will cause a rejection of H_0. For $\alpha = 0.05$ and $n = 81$, $t_{0.05,80} = 1.664$. H_0 is rejected if $T > 1.664$. If the alternative is $\mu \neq 9$, the rejection region will be in both tails, with a probability $\alpha/2$ in each tail.

7. Collect the data. The data in Appendix I are used to evaluate the hypothesis; however, assume that the data are new and have not been seen.

8. Compute the test statistic. $\hat{\mu} = 9.64$, $\hat{\sigma} = 11.40$; therefore, the test statistic is $T = 0.51$. On a $\hat{\mu}$ scale, the equivalent rejection region is $\hat{\mu} > k$ gpm, where $k = \mu + t_{\alpha,n-1}\hat{\sigma}_{\hat{\mu}}$. Since $t_{0.05,80} = 1.664$, $k = 11.11$. Thus, the acceptance region is $\hat{\mu} \leq 11.11$ and the rejection region is $\hat{\mu} > 11.11$.

9. Make a decision. Since $T = 0.51 \leq 1.664$ falls in the acceptance region or, equivalently, $\hat{\mu} = 9.64 \leq 11.11$, H_0 cannot be rejected. A key point is that "accepting the null" really means that there is no evidence to reject it.

10. Report the p-value. In this example it is 0.31. p-values are described in Section 3.7.2.

What is the practical consequence of this result? It is that the developer of the proposed subdivision cannot proceed because there is insufficient evidence to support an adequate water supply.

3.7.2 p-Values

Statistics cannot prove the truth or falsehood of a hypothesis, but a properly constructed and executed testing procedure provides evidence in support of or in opposition to the conjecture. In the water-well yield case study, the null hypothesis $\mu = 9$ is accepted. Does this mean that $\mu = 9$ is true? No, its acceptance means that there is not sufficient evidence to reject H_0 in favor of H_a: $\mu > 9$. Of course, H_0 may be true. Consider how likely it is that the sample can be from a distribution with $\mu = 9$ and have $\hat{\mu} = 9.64$. To determine this, compute

$$
\begin{aligned}
p\text{-value} &= \Pr(\hat{\mu} > 9.64 | \mu = 9) \\
&= \Pr\left(T > \frac{9.64 - 9}{1.27}\right) \\
&= \Pr(T > 0.504) \\
&= 0.31
\end{aligned}
$$

The p-value is the probability under the stated assumption that a more extreme result will be obtained. A result of 9.64 is more extreme than 9 in the direction of H_a.

A p-value of 0.31 implies that 31% of the time, a more extreme result will be obtained. For testing purposes, arbitrarily define as unusual an event that occurs once in 20 or fewer times. Since 0.31 is greater than 0.05, H_0 cannot be rejected. The acceptance or "failure to reject" and rejection regions for a t-distribution with 80 df and a p-value of 0.31 are shown graphically in Figure 3.7. The p-value is the area under the curve to the right of the vertical dotted line. Instead of $\widehat{\mu} = 9.64$, suppose that $\widehat{\mu} = 11.5$; then the p-value is 0.03 and H_0 is rejected at the 0.05 significance level. If the alternative is two-sided (i.e., $H_a \neq 9$), there will be two more extreme cases: $\Pr(\widehat{\mu} > 9 + 0.64 \mid \mu = 9)$ and $\Pr(\widehat{\mu} < 9 - 0.64 = 8.36 \mid \mu = 9)$. Since the t-distribution is symmetric, the

$$p\text{-value} = 2\Pr(\widehat{\mu} > 9.64 \mid \mu = 9)$$

$$= 2\Pr\left(T > \frac{9.64 - 9}{1.27}\right)$$

$$= 2\Pr(T > 0.504)$$

$$= 0.62$$

A critical assumption is that the sample mean of 9.64 is the result of chance alone; that is, it resulted from a random sample from the population with $\mu = 9$.

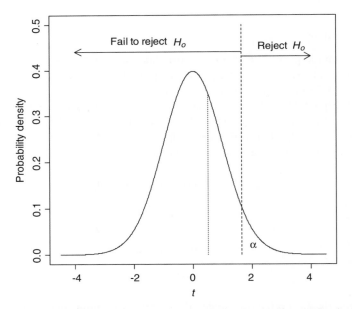

FIGURE 3.7 t-Distribution showing the rejection region at significance level α when H_0 is true.

3.7.3 Making a Decision

All of the components of a hypothesis test can be done correctly, yet a wrong decision can still be made. In fact, there are two possible incorrect decisions that can be made: type I and type II errors. A *type I error*, rejecting H_0 incorrectly, occurs with probability α, where α is selected prior to performing a hypothesis test. A *type II error*, accepting H_0 incorrectly, occurs with probability β, defined as

$$\beta = \Pr(\widehat{\mu} \text{ falls in acceptance region} \mid H_a \text{ true})$$

- *Consequences of making a type I error.* The null hypothesis is rejected incorrectly. For the water-well yield case study, the developer believes that there is sufficient water to develop a subdivision when in fact there is not, and after constructing houses he may be liable for failure to provide sufficient water.
- *Consequences of making a type II error.* The null hypothesis is accepted incorrectly. In the water-well yield case study, if the developer accepts the null hypothesis that the mean is less than or equal to 9.0 gpm when the yield is greater than 9.0 gpm, he may fail to develop a potentially lucrative subdivision.

The cost of making these types of errors are typically problem specific, and each case must be evaluated by estimating potential gain, loss, and risk. A decision table is shown in Table 3.3.

To compute the probability of a type II error β, a specific alternative must be defined. One consideration in specifying the alternative is the issue of statistical significance versus practical significance that was introduced in Chapter 1. In the water-well yield case study, for development to proceed there needs to be a yield greater than 9 gpm. Indeed, a yield of 9.1 gpm or greater will satisfy the legal requirements. However, a question of importance is: Is it practical or even possible to distinguish a yield with a population mean of 9.1 versus one of 9.0? Because of the inherent variability in yield, it may be extremely costly, if indeed it is possible, to take a large enough sample to distinguish 9.0 from 9.1 gpm. A more realistic alternative may be H_a: $\mu = 11$. Issues concerning the choice of a specific alternative are examined in later sections. If the alternative that $\mu = 11$ is accepted, there is reasonable assurance but not absolute certainty that $\mu > 9$. Now that an alternative is specified, the probability β that $\widehat{\mu}$ is in the acceptance region given that H_a is true can be determined. Recall that the acceptance region is $\widehat{\mu} \leq 11.11$ gpm at $\alpha = 0.05$.

TABLE 3.3 Hypothesis Decision Table

	Null Hypothesis	
Decision	True	False
Accept H_0	Correct decision	Type II error
Reject H_0	Type I error	Correct decision

Thus,

$$\begin{aligned}
\beta &= \Pr(\widehat{\mu} < \mu_0 + t_{\alpha,n-1}\widehat{\sigma}_{\widehat{\mu}} \mid \mu = \mu_a) \\
&= \Pr(\widehat{\mu} < 9 + \widehat{\sigma}_{\widehat{\mu}} t_{0.05,80} \mid \mu = 11) \\
&= \Pr(\widehat{\mu} < 9 + (1.664)(1.27) \mid \mu = 11) \\
&= \Pr(\widehat{\mu} < 11.11 \mid \mu = 11) \\
&= \Pr\left(T < \frac{11.11 - 11}{1.27}\right) \\
&= \Pr(T < 0.087) \\
&= 0.53
\end{aligned}$$

The probability β of making a type II error is 0.53 (the shaded region in Figure 3.8), against the alternative that $\mu = 11$. Now the question "How good is the test just performed?" can be answered. One way to do so is to compute the *power* of the test, denoted by π, which is defined as the probability of rejecting H_0: $\mu = 9$ versus H_a: $\mu = 11$ correctly, or expressed as a probability,

$$\begin{aligned}
\pi &= 1 - \beta \\
&= 0.47
\end{aligned}$$

This result implies that only 47 out of 100 times, on average, can this test differentiate between a 9-gpm and an 11-gpm average yield given $n = 81$ and $\sigma = 11.40$. Even

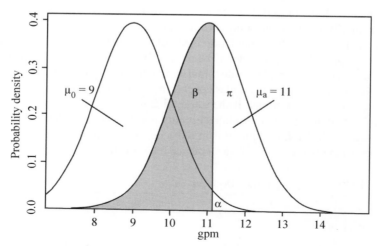

FIGURE 3.8 Acceptance, rejection, and power regions for water-well yield case study, $H_0 = 9$ versus $H_a = 11$.

when the true mean yield is 11 gpm, the null hypothesis $H_0: \mu = 9$ will be accepted 53% of the time. If $\mu = 11$, there will be sufficient water for the housing development. However, 53% of the time this test will fail to recognize it, causing a permit not to be issued. The distributions of means for $\mu_0 = 9$ and $\mu_a = 11$ are shown in Figure 3.8.

For exploratory work, a power of 0.50 may be adequate. For confirmatory work, many researchers suggest a power of at least 0.80. In this example, given the potential loss of development profit and relatively low cost of taking more samples, the developer should commission a larger sample size.

3.7.4 Determining the Sample Size

Instead of picking a sample size arbitrarily, suppose that a power of 0.80 is desired and the problem is to determine the sample size required. For a one-sided test the formula is

$$n = \left(\frac{z_\alpha + z_\beta}{d} \right)^2$$

where $d = (\mu_0 - \mu_a)/\sigma$ is the scaled difference between the null and alternative hypotheses. For the water-well yield case study, as in the previous example, $\alpha = 0.05$, $z_{0.05} = 1.645$, and $z_{0.20} = 0.842$. Since this computation is performed prior to collecting data, σ needs to be estimated by analogy or using historical data. Assume that $\sigma = 12$ gpm; then $d = (9 - 11)/12 = -0.167$, and therefore, $n = 222$. If $\alpha = 0.05$ and $\pi = 0.8$ and the desire is to detect a significant difference of 0.1 (i.e., $H_0: \mu = 9.0$ versus $H_a: \mu = 9.1$), then $d = (9.0 - 9.1)/12 = -0.0083$ and $n = 89,067$. This exceedingly large sample size is the price to be paid when a small difference relative to the variance needs to be detected. The investigator should decide in advance of conducting a study how small a difference needs to be detected. If it is not possible to obtain the n specified, the investigator may wish to consider a larger practical difference or not conduct the study.

For a two-sided alternative, α is replaced by $\alpha/2$. The approximate n is

$$n \doteq \left(\frac{z_{\alpha/2} + z_\beta}{d} \right)^2$$

which will yield a slightly larger value than the corresponding one-sided alternative.

3.7.5 Choosing α and β

An important practical question is: How are α and β chosen? There are losses (costs) associated with making type I and type II errors. A type I error, rejecting the null when it is true, may cause one to expend resources unnecessarily. Suppose a test is conducted on the null hypothesis that the mean level of contaminants is at or below a safe standard against the alternative that it exceeds it. Making a type I error may cause a manager to decide that a site which does not contain contaminants will need remediation. A type II error, incorrectly accepting the null hypotheses, may cause a manager not to clean up a polluted site, causing the company to incur significant penalties. It is important when formulating a hypothesis to consider the losses associated with each decision, plan the experiment accordingly, and not choose $\alpha = 0.05$ arbitrarily.

One of the more interesting real-life examples of how decision rules are formulated is not from the earth sciences but from the system of criminal justice in the United States. In criminal proceedings, the α level is set to an extremely small value, certainly far smaller than 0.001. This protects society against sending an innocent person to jail. H_0 is that the accused is not guilty against the alternative H_a that he is. Incorrectly rejecting H_0 puts an innocent person in jail, which society has decided should be an extremely rare event. In a criminal trial, the judge instructs the jury that to convict, they must be sure beyond a reasonable doubt. An even smaller α results if the judge states "beyond all doubt." A larger α will result, as it does in civil cases, when the judge uses the term "a preponderance of evidence." The penalty that a society pays for the choice of a small α in criminal cases is to allow a certain percentage of guilty persons to go free. This is a type II error, which occurs with probability β.

3.7.6   Neyman Versus Fisher Approaches

Consider the differences between the Neyman and Fisher approaches. In both, a test statistic T is chosen according to the procedures just described. However, in Neyman, an α value is chosen prior to conducting the test, the null hypothesis and a specific alternative are specified, and type I and type II error probabilities are computed. Fisher tests a null H_0: $\mu = \mu_0$ and reports a p-value, which he defines as a weight of evidence (Berger, 2003). The Bayesian approach is discussed later.

3.7.7   Comparing Two Means

Suppose that instead of testing a mean against a constant, the problem is to compare two means—for example, $\mu_1 = \mu_2$ against the alternative $\mu_1 \neq \mu_2$—where the first mean may be the mean yield from rock type Yg and the second from rock type Zc. As in the confidence interval example, subscript 1 refers to Yg and subscript 2 refers to Zc. A hypothesis is formulated as

$$H_0: \mu_1 - \mu_2 = d_0$$
$$H_a: \mu_1 - \mu_2 \neq d_0$$

where d_0 is a constant, often zero. Of course, H_a can be a one-sided alternative, either $H_a : \mu_1 - \mu_2 > d_0$ or $H_a : \mu_1 - \mu_2 < d_0$, depending on the problem. Let $d_0 = 0$ and $\alpha = 0.05$. Take the sample sizes as given previously, $n_1 = 81$ and $n_2 = 171$. Assume that these observations are independent. If they are paired, another test is more appropriate and it will be discussed subsequently. When an experiment can be designed, it is generally a good idea to take equal sample sizes, as this increases the power of the test. An exception is if the population variances differ significantly. In this case it may be desirable to take a larger sample from the population with the highest variance.

The same issues regarding inequality of variance as those discussed for confidence interval estimation apply to hypothesis testing. Assume that the population variances for the water-well yields from these two rock types are equal. The t-ratio test statistic is constructed as

$$T = \frac{\hat{\mu}_1 - \hat{\mu}_2}{\hat{\sigma}_{\hat{\mu}_1 - \hat{\mu}_2}}$$

$$= \frac{9.64 - 11.81}{1.75}$$

$$= -1.24$$

See Table 3.1 for the summary statistics. Since $t_{0.025,250} = 1.969$, the rejection region is $|T| > 1.969$. Since $T = -1.24$, H_0 cannot be rejected, and it is not possible to distinguish between μ_1 and μ_2.

3.8   BAYESIAN HYPOTHESIS TESTING

Consider the hypotheses

$$H_0: \mu = \mu_0$$
$$H_a: \mu \neq \mu_0$$

Let T be the sample mean from n observations, $Y_1, Y_2, \ldots, Y_n$. Of course, T is not restricted to being a mean but can be a median, variance, or other statistic; the distribution of the test statistic must be known or derived. Recall that in Bayesian statistics, the data are fixed and the parameters are unknown. Assume that H_0 and H_a are mutually excusive and exhaustive. By Bayes' theorem, the posterior probability of H_0 given the test statistic T, which is a function of the observed data, is

$$\Pr(H_0 \mid T) = \frac{\Pr(T \mid H_0)\Pr(H_0)}{\Pr(T)}$$

where $\Pr(H_0)$ denotes the probability of H_0 based on information obtained prior to conducting the experiment. The denominator,

$$\Pr(T) = \Pr(T \mid H_0)\Pr(H_0) + \Pr(T \mid H_a)\Pr(H_a)$$

is the probability of the data. If there is no preference for $\Pr(H_0)$, then $\Pr(H_0) = \Pr(H_a) = 0.5$. Of course, prior knowledge may suggest that one event is more likely to occur.

Since the two hypotheses (events) conditioned on T are mutually exclusive and exhaustive,

$$\Pr(H_a \mid T) = 1 - \Pr(H_0 \mid T)$$

The posterior odds ratio O_{ps} in favor of H_0 is the product of the prior odds ratio and the likelihood ratio:

$$O_{ps} = \frac{\Pr(H_0 \mid T)}{\Pr(H_a \mid T)} = \frac{\Pr(H_0)}{\Pr(H_a)} \frac{\Pr(T \mid H_0)}{\Pr(T \mid H_a)}$$

The ratio of posterior to prior odds is called the *Bayes factor K*, defined as

$$K = \frac{\Pr(T \mid H_0)}{\Pr(T \mid H_a)}$$

which depends only on the data. A simple decision rule is to accept H_0 if O_{ps} exceeds 1 and to reject it otherwise. This may be reasonable if the cost of accepting H_0 incorrectly is approximately the same as rejecting it incorrectly; but often this is not the case. Consider Table 3.4, which is a revised version of Table 3.3 except that instead of identifying types of errors, it identifies costs. $u(H_0 \mid H_a)$ is the cost of accepting H_0 when it is false and $u(H_a \mid H_0)$ is the cost of accepting H_a when it is false. Assume that the posterior probabilities of H_0 and H_a, which are called $\Pr(H_0 \mid T)$ and $\Pr(H_a \mid T)$, respectively, can be calculated. The expected values of the cost are

$$E(u, H_0) = (0)\Pr(H_0 \mid T) + u(H_0 \mid H_a)\Pr(H_a \mid T)$$
$$E(u, H_a) = u(H_a \mid H_0)\Pr(H_0 \mid T) + (0)\Pr(H_a \mid T)$$

If $E(u, H_0) \le E(u, H_a)$, H_0 is accepted, and if $E(u, H_0) > E(u, H_a)$, H_a is accepted. Some algebra shows that H_0 is accepted if

$$\frac{u(H_0 \mid H_a)}{u(H_a \mid H_0)} \le O_{ps}$$

TABLE 3.4   Cost of Making an Incorrect Decision

	Null Hypothesis	
Decision	True	False
Accept H_0	0	$u(H_0 \mid H_a)$
Reject H_0	$u(H_a \mid H_0)$	0

Returning to the example of the developer and water-well yield, consider again

$$H_0: \mu = 9$$
$$H_a: \mu = 11$$

and for simplicity assume that these are the only two possibilities. Further assume that $\Pr(H_0) = \Pr(H_a) = 0.5$ and the distribution of the y's (the water-well yield data) given μ is $N(\mu, 11.40^2)$ and that $\hat{\mu}|\mu \sim N(\mu, 11.40^2/n)$. The sample size $n = 81$, as before, and $\hat{\mu} = 9.64$. Recall that the probability density function of y is

$$f(y \mid \mu, \sigma^2) = \frac{1}{\sqrt{2\pi}\,\sigma} \exp\left[-\frac{(y-\mu)^2}{2\sigma^2}\right]$$

so the probability density function of $\bar{y} = \hat{\mu}$ is

$$f(\bar{y} \mid \mu, \sigma^2) = \left(\frac{n}{2\pi}\right)^{1/2} \frac{1}{\sigma} \exp\left[\frac{-n(\bar{y}-\mu)^2}{2\sigma^2}\right]$$

For this simple example,

$$\Pr(T \mid H_0) = f(y \mid 9, 11.40^2)$$
$$= \left(\frac{81}{2\pi}\right)^{1/2} \frac{1}{11.40} \exp\left[\frac{-(81)(9.64-9)^2}{(2)(11.40^2)}\right]$$
$$= 0.277$$

Similarly,

$$\Pr(T \mid H_a) = f(y \mid 11, 11.40^2)$$
$$= \left(\frac{81}{2\pi}\right)^{1/2} \frac{1}{11.40} \exp\left[\frac{-(81)(9.64-11)^2}{(2)(11.40^2)}\right]$$
$$= 0.177$$

Thus, the posterior odds ratio in favor of H_0 is

$$O_{\text{ps}} = \frac{\Pr(H_0 \mid T)}{\Pr(H_a \mid T)} = \left(\frac{1/2}{1/2}\right)\left(\frac{0.277}{0.177}\right)$$
$$= 1.56$$

Assuming that $u(H_0 \mid H_a)$ is approximately equal to $u(H_a \mid H_0)$, the decision is to accept H_0. There are two loss functions: one to accept H_0 given that H_a is true, the second to accept H_a given that H_0 is true. If the former is twice the latter, the decision will be to accept H_a since $2 > 1.56$, which is the value of O_{ps}.

3.9 NONPARAMETRIC HYPOTHESIS TESTING

When data sets are small and little supplementary information is available, it is difficult to know if the assumptions required of a t-test—the normal distribution of means and independence—are valid. An alternative procedure is a *nonparametric test* called the *Wilcoxon test* (also known as the *Mann–Whitney rank sum test*). A Wilcoxon one-sample test and a Wilcoxon two-sample test are nonparametric versions of the one- and two-sample t-tests. Nonparametric tests typically use ranked data instead of original raw data. The *Wilcoxon one-sample test* addresses the question: Is the distribution of some random variable symmetric about a conjectured mean μ? The more interesting application is the two-sample test. For illustrative purposes, suppose that there are only 10 water-well yield observations from rock types Yg and Zc in the sample. Again, for simplicity of notation, refer to these as groups 1 and 2, respectively, and their associated sample sizes as n_1 and n_2. The data and associated rank ordering are given in Table 3.5. The null and alternative hypotheses are

$$H_0 : \text{Gp1} = \text{Gp2}$$
$$H_a : \text{Gp1} \neq \text{Gp2}$$

Gp1 and Gp2 are used to indicate identical distributions, possibly differing only in a shift of location (medians). The *Wilcoxon two-sample test* attempts to detect a shift in location if it exists. For the inequality alternative specified, H_0 will be rejected if Gp1 > Gp2 or Gp2 > Gp1. Of course, a one-sided alternative can be specified if the situation warrants it. Note that the test inference may be incorrect if the assumption of identical distributions is significantly violated.

TABLE 3.5   Subset of Water-Well Yield Data and Ranks for Rock Types Yg and Zc

Rock Yg Yield (gpm)	Yg Yield Rank	Rock Zc Yield (gpm)	Zc Yield Rank
15.0	17.0	3.0	6.0
0.5	1.0	6.0	10.5
2.5	5.0	1.5	3.0
5.5	9.0	5.0	7.5
6.0	10.5	40.0	20.0
7.0	14.0	5.0	7.5
6.2	12.0	2.0	4.0
10.0	15.0	6.5	13.0
1.2	2.0	16.0	18.0
11.0	16.0	20.0	19.0
r_1	101.5		

The computational procedure is as follows:

1. Rank-order the combined data set of 20 observations with the smallest observation assigned rank 1. In case of ties, use an average of the ranks. For example, the seventh and eighth observations are 6 gpm, so each is assigned a rank of 7.5.
2. Let $r_1 = 101.5$ be the sum of the ranks of the smaller sample size. When the sample sizes are equal, the choice is arbitrary. The ranks in Gp1 are summed.
3. If $\min(n_1, n_2) < 10$, exact tables or an algorithm are required. Otherwise, the normal approximation is usually satisfactory.
4. The mean and standard deviation for Gp1 are $\mu_1 = [n_1(n_1 + n_2 + 1)]/2 = 105$ and $\sigma_1 = \sqrt{[n_1 n_2(n_1 + n_2 + 1)]/12} = 13.23$, respectively.
5. For a one-sided alternative, if H_0 is true, the most extreme result, r_1, being in the lower tail, will be $\Pr(R_1 \leq 101.5) \doteq \Pr(Z \leq (101.5 - 105)/13.23)$. Therefore, a one-sided p-value $\doteq 0.40$ and the two-sided p-value $= (2)(0.40)$, or 0.80. If the significance level is $\alpha = 0.05$, H_0 will not be rejected.

Rank sum tests are useful when the normal assumption cannot be verified. They provide protection against outliers. When used with normal data, nonparametric tests typically have at least 80% of the power of a T- or Z-statistic and are superior to these statistics when the normal assumption is violated. A significant violation causes bias and/or loss of power in T- or Z-tests. Another advantage of nonparametric tests is that they negate the need to transform data to achieve normality. A note of caution: When many values are equal, the power of a rank test declines. Rank tests react to differences in shape between the two distributions and perform best when the two distributions have approximately the same shape.

3.10   BOOTSTRAP HYPOTHESIS TESTING ON MEANS

As noted previously, bootstrapping is a resampling method that is especially useful for analyzing small samples when calculations are difficult and/or when assumptions required for other types of testing may be questionable. We illustrate both a nonparametric and a parametric bootstrap procedure.

The *nonparametric bootstrap procedure* is outlined below. The problem is to compare yield means between rock types Yg and Zc using a nonparametric bootstrap test. Assume that the shapes of the underlying distributions are identical. To perform a nonparametric bootstrap hypothesis test:

1. Formulate the null and alternative hypotheses, which for this example are H_0: $\mu_1 = \mu_2$ versus the alternative H_a: $\mu_1 \neq \mu_2$.
2. Select a test statistic. For this example it is $T = \widehat{\mu}_1 - \widehat{\mu}_2$, where $\widehat{\mu}_1 = 6.47$ and $\widehat{\mu}_2 = 10.50$. Thus, $T = -4.03$.
3. Find a distribution $\hat{F}_0$ under the null hypothesis. Since the shape of the two distributions is assumed to be identical, a reasonable approximation to the

empirical distribution function (EDF) is the pooled EDF of the samples from rock types Yg and Zc.

4. Resample from $\widehat{F}_0$ R times. For this example $R = 999$, $n_1 = 10$, and $n_2 = 10$. Each sample is of size $n_1 + n_2 = 20$ with replacement.

5. Compute the means μ_{1r}^* and μ_{2r}^* using the first n_1 observations for μ_{1r}^* and the remaining n_2 observation for μ_{2r}^*. Compute and store the difference between sample means $t_r^* = \mu_{1r}^* - \mu_{2r}^*$, $r = 1, \ldots, R$.

6. For the two-sided alternative the p_{boot}-value $= 2\,\text{Pr}^*(T^* \geq |T| \,|\widehat{F}_0)$ (a bootstrap p-value) or its approximation $\widetilde{p}_{\text{boot}}$-value $= (1 + \sum_{r=1}^{R} I_{|t_r^*| \geq |T|})/(R + 1)$, where t_r^*, $r = 1, \ldots, R$ are estimates of T from the R bootstrap samples and $I_{|t_r^*| \geq |T|}$ is an indicator function, equal to 1 if $|t_r^*| \geq |T|$ and 0 otherwise. If the alternative is H_a: $\mu_1 > \mu_2$, the $\widetilde{p}_{\text{boot}}$-value $= (1 + \sum_{r=1}^{R} I_{t_r^* \geq T})/(R + 1)$.

The distribution of the differences in bootstrap means of water-well yields is shown in Figure 3.9. The down arrow points to the test statistic T. The approximate bootstrap p-value is.

$$\widetilde{p}_{\text{boot}}\text{-value} = \frac{1 + 27}{999 + 1}$$

$$= 0.028$$

If a significance level $\alpha = 0.05$ is selected, H_0: $\mu_1 = \mu_2$ will be rejected.

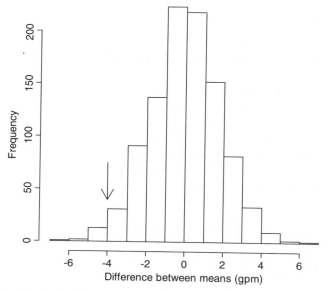

FIGURE 3.9 Distribution of bootstrap differences between means of Yg and Zc yields assuming that H_0 is true.

The *parametric bootstrap procedure* is similar to the nonparametric procedure except that a distributional form is assumed and fit to the data. For example, in the preceding example it may be reasonable to assume that under H_0: $\mu_1 = \mu_2$ the distribution is normal. Unless the variance is known, it will need to be estimated from the data (the combined Yg–Zc data set in the preceding example). The R independent samples of size n are taken from the fitted null model, $f_0(y|N(\widehat{\mu}_i, \widehat{\sigma}_p^2))$, $i = 1, 2$. As in step 5 of the nonparametric bootstrap test, $t_r^* = \mu_{1r}^* - \mu_{2r}^*$ is computed. The remainder of the simulation proceeds in the same manner as above. If there is evidence, usually historical, to suggest that a specific parametric distribution may be appropriate, the parametric approach is preferred; otherwise, we recommend using the nonparametric method.

3.11   TESTING MULTIPLE MEANS VIA ANALYSIS OF VARIANCE

Suppose that the problem is to compare the means of four principal rock types: Yg, Ygt, Ymb, and Zc. In *analysis of variance* (ANOVA) parlance, the rock types are called treatments. A typical null hypothesis is H_0: $\mu_1 = \mu_2 = \mu_3 = \mu_4$ and a usual alternative hypothesis is H_a: $\mu_i \neq \mu_j$, $i \neq j$, where i and j are the integers 1 to 4 in this example. H_a says that if at least two means differ from each other, H_0 is rejected. The model is

$$Y_{ij} = \mu + \alpha_i + \varepsilon_{ij}$$

where Y_{ij} is the jth observation ($j = 1, \ldots, n_i$) in the ith treatment group; μ is the grand mean (a parameter common to all treatments); $\alpha_i = \mu_i - \mu$, the effect due to the ith treatment (in the water-well yield case study, an effect due to the ith rock type); and ε_{ij} is the random deviation of the jth observation in the ith treatment group. Samples from the treatment groups are assumed to be independent and $\varepsilon_{ij} \sim N(0, \sigma^2)$. The latter assumption implies homogeneity: namely, that $\sigma_1^2 = \sigma_2^2 = \sigma_3^2 = \sigma_4^2$. An equivalent way to represent the hypothesis of equality of means against the alternative that two or more means differ is

$$H_0: \alpha_1 = \alpha_2 = \alpha_3 = \alpha_4 = 0$$
$$H_a: \alpha_i \neq 0$$

ANOVA partitions the total variability (see Table 3.6) into variability accounted for by the four means (the model) and that which is left over (the error or residual). If a

TABLE 3.6   Analysis of Variance Table

Source	Sum of Squares	df	Mean Square	F-ratio
Model	SSM	$k-1$	MSM = SSM/$(k-1)$	MSM/MSE
Error	SSE	$n-k$	MSE = SSE/$(n-k)$	
Total	SST	$n-1$		

specific alternative is of interest, such as $\mu_1 = (\mu_2 + \mu_3 + \mu_4)/3$, it should be incorporated into the model prior to conducting the test. More than one alternative can be part of the model. Nonparametric procedures serve as an alternative when normality assumptions are questionable.

3.11.1   Partitioning Variability

Before proceeding with the test of hypothesis described above, the assumptions—namely that the population variances are equal, the samples are independent, and the sample means are normally distributed—need to be examined. Unless the data set is large, these assumptions are examined using graphical procedures, theory, and analogs. If the assumptions are met, an ANOVA procedure may be used to test H_0 versus H_a. The basic components of an ANOVA table (Table 3.6) are the *sum of squares total* (SST), the *sum of squares error* (SSE), and the *sum of squares due to the model* (SSM). SSM and SSE are the components of SST partitioned by the model. The mean of the ith treatment group is defined as $\overline{Y}_{i.} = \sum_{j=1}^{n_i} Y_{ij}/n_i$, where n_i is the number of observations in the ith treatment group. Note that a dot in the subscript implies "summed over." For example, given the set $\{Y_{ij}, i = 1, \ldots k, j = 1, \ldots, n_i\}$, $Y_{i.} = \sum_{j=1}^{n_i} Y_{ij}$. The sums of squares are defined as follows:

$$SST = \sum_{i=1}^{k}\sum_{j=1}^{n_i}(Y_{ij} - \overline{Y}_{..})^2$$

$$SSM = \sum_{i=1}^{k}\sum_{j=1}^{n_i}(\overline{Y}_{i.} - \overline{Y}_{..})^2$$

$$SSE = \sum_{i=1}^{k}\sum_{j=1}^{n_i}(Y_{ij} - \overline{Y}_{i.})^2$$

where n_i, $i = 1, \ldots, k$ are the number of samples in the ith treatment group. Some algebra will show that $SST = SSM + SSE$. This is a type of conservation of energy law applied to statistical models. If the sample means, the $\overline{Y}_{i.}$'s, are all equal, SSM is zero and the total variability in the Y's is SSE, which represents the deviations of observations from their corresponding group mean.

3.11.2   Degrees of Freedom and Mean Squares

The model (rock type) has 3 df since there are four groups (rock types) and 1 df is lost because the grand mean μ is estimated from the data. It is $\overline{y}_{..} = 12.05$. For k groups, the model has $k - 1$ df. The df for SST are $n - 1$, where $n = \sum_{i=1}^{k} n_i$. An easy way to find the df of SSE is by subtraction: $n - 1 - (k - 1) = n - k$. The *mean square due to the model* (MSM) is a standardized SSM, $MSM = SSM/(k - 1)$. The *mean square error* (MSE) is a standardized SSE, $MSE = SSE/(n - k)$.

TABLE 3.7   ANOVA Table for Water-Well Yield by Rock Type Transformed

Source	Sum of Squares		df	Mean Square		F-Value	p-Value
Model	SSM	5.4	3	MSM	1.80	12.86	3.87×10^{-8}
Error	SSE	79.5	567	MSE	0.14		
Total	SST	84.9	570				

			Table of Means			

Rock Type	Yg	Ygt	Ymb	Zc	Grand
Mean y	9.64	18.35	9.66	11.81	12.05
Mean $y^{1/5}$	1.41	1.63	1.17	1.50	1.47
n	81	115	204	171	571

3.11.3   Model Significance

The *F-statistic* is a ratio, MSM/MSE. It may be viewed as a signal-to-noise ratio, with the model (MSM) being the signal and the error (MSE), being the noise. If the signal is strong relative to the noise, it will be heard or, equivalently, the model will account for a statistically significant percentage of the variability. The larger the F-statistic, the more likely it is that two or more of the means differ from each other. A small F-statistic indicates that the variability accounted for by the model is small in comparison to the residual (the unexplained variability or noise). If H_0 is true, the F-statistic will have an f-distribution with $k - 1$ and $n - k$ df. H_0 is rejected if the F-statistic exceeds a given value, as determined from the f-distribution. (The F-statistic is also referred to as the *F-ratio*.)

As noted previously, the water-well yield data are highly right-skewed. Instead of using the raw data, a fifth root $y^{1/5}$ yields symmetry and is used as the response variable. The four rock types—Yg, Ygt, Ymb, and Zc—are the treatments or levels of the single factor, rock type. Select $\alpha = 0.05$, which determines the rejection region. The ANOVA table (Table 3.7) shows the results. A p-value of 3.87×10^{-8} implies a highly significant result, and it is concluded that at least two means are different. At this point, which pairs of means differ is unknown.

Model assumptions include homogeneity of variance and normally distributed errors. A boxplot of residuals by rock type (Figure 3.10) suggests that these assumptions are reasonable.

3.12   MULTIPLE COMPARISONS OF MEANS

After determining that there is a statistically significant difference between population means, the investigator needs to determine where the differences occur. The concept of a two-sample *t*-test was introduced earlier. At this point it may be reasonable to ask: Why not use it? The problem is that the probability of rejecting the null hypothesis simply by chance (where real differences between population means fail to exist) increases as the number of pairwise tests increase. In the example

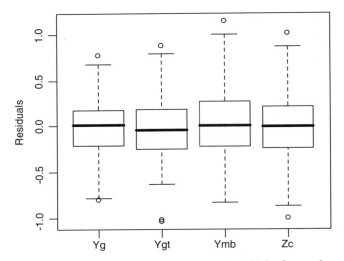

FIGURE 3.10 Boxplots of residuals; water-well yield for four rock types.

above, there are four population means that result in $(4-1)! = 6$ comparisons: $\mu_1 = \mu_2, \mu_1 = \mu_3, \ldots, \mu_3 = \mu_4$. If one assumes that pairs are independent (a conservative assumption) and the test on each pair (called a *pairwise test*) is made at the 0.05 significance level, the probability that at least one pair will be declared significant by chance is the complement of the probability that all of the nulls will be accepted. With six comparisons, the probability that all will be accepted is $(1 - 0.05)^6$, and therefore the probability of at least one type I error is $1 - (1 - 0.05)^6 = 0.26$. Note that $(1 - 0.05)$ is the probability of accepting the null for one test. In general, if the probability of a type I error on any single comparison is α, the probability of making at least one type I error on the entire set of tests (called the experiment) is no greater than $1 - (1 - \alpha)^k$. This is the *multiple comparison of means* problem. The multiple comparison problem is analogous to the situation where one examines a data set in several ways. Many times, a pattern will appear and the question "Is the pattern real or is it a function of sampling variability that is discovered through an exhaustive examination of the data?" should be asked. There are numerous ways to deal with this problem (Hsu,1996). Three methods are presented here.

3.12.1 Bonferroni's Method

Bonferroni's method is a simple procedure that can be applied to equal and unequal sample sizes. If a decision is made to make m pairwise comparisons, selected in advance, the Bonferroni method requires that the significance level on each test be α/m. This ensures that the overall (experiment wide) probability of making a type I error is less than or equal to α. In the example above, the significance level on each test is $0.05/6 = 0.0083$. Comparisons can be made on means (such as $\mu_2 = \mu_3$) and specified linear combinations of means (contrasts), that is, $(\mu_1 + \mu_2)/2 = (\mu_3 + \mu_4)/2$. Bonferroni's method is sometimes called the Dunn method. A variant on the

Bonferroni approach is the *Sidak method*, which yields slightly tighter confidence bounds. If the treatments are to be compared against a control group, Dunnett's test should be used. Clearly, the penalty that is paid for using Bonferroni's method is the increased difficulty of rejecting the null on a single comparison. The advantage is protection against a type I error when making multiple comparisons.

3.12.2 Tukey's Method

Tukey's method provides $(1 - \alpha)$ 100% simultaneous confidence intervals for all pairwise comparisons. Tukey's method is exact when sample sizes are equal and is conservative when they are not. As in the Bonferroni method, the Tukey method makes it more difficult to reject H_0 on a single comparison, thereby preserving the α level chosen for the entire experiment.

Table 3.8 compares pairs of means in the water-well yield case by rock type. The difference between means, the 95% confidence intervals (lower and upper endpoints), and the adjusted *p*-values are given. The *p*-values presented are adjusted for multiple comparison of means. Four pair of means are significantly different at the 0.05 level. Perhaps the most important information for a well driller is that the Ygt yield is greater than Zc, so he will drill in Ygt unless it is more costly to do so. Specifically, Zc $-$ Ygt $= -0.129$ with the lower and upper endpoints of -0.245 and -0.013, respectively. Since this interval does not include zero and the associated *p*-value is 0.023, it may be concluded that the means differ, with Ygt being greater.

3.12.3 Scheffé's Method

Scheffé's method applies to all possible contrasts of the form

$$C = \sum_{i=1}^{k} c_i \mu_i$$

where $\sum_{i=1}^{k} c_i = 0$ and k is the number of treatment groups. Thus, in theory an infinite number of contrasts can be defined.

TABLE 3.8   Tukey Multiple Comparison of Means, Water-Well Yields from Four Rock Types[a]

Comparison	Difference	Lower Endpoint	Upper Endpoint	*p*-Value Adj.
Ygt–Yg	0.219	0.079	0.359	0.000
Ymb–Yg	-0.041	-0.167	0.086	0.842
Zc–Yg	0.090	-0.040	0.220	0.283
Ymb–Ygt	-0.260	-0.372	-0.147	0.000
Zc–Ygt	-0.129	-0.245	-0.013	0.023
Zc–Ymb	0.131	0.031	0.231	0.005

[a]Transformed data; response variable is $y^{1/5}$.

3.12.4   Discussion of Multiple Comparison Procedures

If H_0 in an ANOVA is rejected, it is proper to go to a multiple comparison procedure to determine specific differences. Occasionally, statistically significant differences may be found by a multiple comparison procedure when H_0 is not rejected because the multiple comparison procedure may be more powerful, that is, have greater ability to detect smaller differences than the F-statistic in the ANOVA model. Since this can occur, some statisticians recommend that the user proceed directly to a multiple comparison procedure and ignore the ANOVA hypothesis test. We recommend that ANOVA results be examined but that the use of a multiple comparison procedure be considered even when H_0 is not rejected.

An obvious question is, "Which procedure should I pick?" The basic idea is to have the narrowest confidence bounds for the entire experiment (set of comparisons), consistent with the contrasts of interest. Equivalently, in hypothesis-testing parlance, it is desirable to make it as easy as possible to reject H_0 on a single comparison while preserving a predetermined significance level for the experiment. If all pairwise comparisons are of interest, the Tukey method is recommended; if only a subset is of interest, Bonferroni's method is a better choice. If all possible contrasts are desired, Scheffé's method should be used. However, because all possible contrasts must be considered in Scheffé's method, rejecting the null hypothesis is extremely difficult. As with most statistical procedures, no single method works best in all situations.

3.13   NONPARAMETRIC ANOVA

When uncertainty about the assumption of normality exists, a nonparametric alternative to ANOVA called the *Kruskal–Wallis* (K-W) *test* is available. Instead of using observed values, the K-W procedure uses ranks and then compares the ranks among the treatment groups. The null and alternative hypotheses

$$H_0: \mu_1 = \mu_2 = \mu_3 = \mu_4$$
$$H_a: \mu_i \neq \mu_j \text{ at least one } i \neq j$$

if there are $k=4$ populations. Next obtain samples n_i, $i=1,\ldots,k$. Rank the combined samples as in the Wilcoxon two-sample example (Table 3.5). The K-W test statistic is

$$H = \frac{12}{n(n+1)} \sum_{i=1}^{k} \frac{R_i^2}{n_i} - 3(n+1)$$

where $n = \sum_{i=1}^{k} n_i$ and R_i is the sum of the ranks in the ith treatment group. If H_0 is true, H is approximately χ^2(chi-square) with $k-1$ df. H_0 is rejected if $H > \chi^2_{\alpha,k-1}$.

For the example with $k = 4$ treatments, $H = \chi^2_{0.05,4-1} = 7.81$. If $H > 7.81$, the null hypothesis of equality of the four population means will be rejected.

3.14   PAIRED DATA

In previous discussions on estimating confidence intervals or testing hypotheses on differences of means, the observations between populations are assumed to be independent. In the water-well yield case study, a given observation in rock type Yg is not associated with an observation in rock type Zc. However, consider the situation where the well driller wants to gain greater confidence in his ability to measure yield correctly. Instead of taking just one measurement, the driller takes two measurements, Yg-1 and Yg-2 from the same well (Table 3.9). There are now 20 measurements, but is the sample size 20? The answer is no, because the first observation taken on a given well is not independent of the second observation taken on the same well. Taking repeated samples is an important way to determine measurement error. It does not increase the overall sample size. The sample size is 10. The driller has chosen to take two observations on each sample unit. How can such data be analyzed? A confidence interval and/or test of a hypothesis is performed on pairwise differences (Table 3.9).

The procedure is similar to computing the confidence interval on a mean except that it is computed on a mean difference μ_d. The values of the sample mean $\overline{d}$ and standard error $\mathrm{SE}(\overline{d})$ are given in Table 3.9. Note that $\mathrm{SE}(\overline{d})$ is not the difference between standard errors of Yg-1 and Yg-2 but, rather, is computed from the difference between means (Table 3.9). If $\overline{d}$ is normally distributed, then

$$T = \frac{\overline{d} - 0}{\mathrm{SE}(\overline{d})}$$

TABLE 3.9   Pairs of Yg Data

Well	Yg-1	Yg-2	Difference: Yg-1 − Yg-2
1	15.0	13.0	2.0
2	0.5	1.0	−0.5
3	2.5	2.5	0.0
4	5.5	6.0	−0.5
5	6.0	7.0	−1.0
6	7.0	6.0	1.0
7	6.0	6.0	0.0
8	10.0	11.0	−1.0
9	1.2	1.0	0.2
10	11.0	12.0	−1.0
Mean	6.47	6.55	−0.08
Std. error	1.44	1.38	0.31

and T is t-distributed with $n-1$ df. The 95% confidence interval on μ_d is

$$\overline{d} \pm t_{\alpha/2,n-1}SE(\overline{d})$$

$$-0.08 \pm (t_{0.025,9})(0.31)$$

$$-0.08 \pm (2.262)(0.31)$$

$$(-0.78, 0.62)$$

To help determine independence between observations, consider Table 3.9. If the Yg-1 data can be permuted without altering the problem, the observations will be independent. In this example, it is not possible to do so because $(Yg-1)_i$ is associated with $(Yg-2)_i$, $i=1,\ldots,10$. The second sample is a repeated sample on the same experimental unit and thus is dependent.

3.15   KOLMOGOROV–SMIRNOV GOODNESS-OF-FIT TEST

Goodness-of-fit tests are used to judge the adequacy of a given statistical model. They take various forms, and several will be introduced in later chapters. One important application is where the statistical model is a distribution. In hypothesis testing, a critical assumption often is the form of the underlying distribution. Numerous tests exist to see if a sample comes from a population with a specific distributional form. A test that has several attractive features is the *Kolmogorov–Smirnov* (K-S) *goodness-of-fit test*. One type of K-S test is based on a comparison between an empirical cumulative distribution function and a continuous distribution of known form. It is not appropriate for a discrete distribution, but other goodness-of-fit tests exist for this purpose. The K-S test is constructed as follows:

1. Let $\{y_{[1]}, \ldots, y_{[n]}\}$ be a sample data set of size n in ascending order.
2. Let the empirical cumulative distribution function be defined as

$$F_n(y) = \begin{cases} 0 & y < y_{[1]} \\ i/n & y_{[i]} \leq y < y_{[i+1]}, \quad i = 1, \ldots, n-1 \\ 1 & y_{[n]} \leq y \end{cases}$$

3. Let $G(y)$ be a theoretical cumulative distribution with known parameters.
4. The K-S test is the maximum difference D between the empirical and theoretical distributions at $y_{[i]}$ over all i. Specifically,

$$D = \max_{1 \leq i \leq n}[G(y_{[i]}) - (i-1)/n, \ i/n - G(y_{[i]})]$$

5. The null hypothesis H_0 is that the sample follows the specified distribution G against the alternative H_a that it does not.

6. A large value of D is evidence against the null hypothesis. The null hypothesis that the empirical distribution is from a specified form is rejected at the α level if $\sqrt{n}\,D > K_\alpha$ where n is the sample size and K_α is a point, typically in the right-hand tail of the K-S probability density function.

7. Critical regions and p-values are computed in most algorithms.

A major advantage of the K-S test is that a test can be performed against any known distribution. It is also an exact test and does not depend on the choice of cell size. A major disadvantage is that the parameters of the distribution must be known. If they are estimated from the data, the usual case, critical values are usually conservative. More exact critical values must be estimated by simulation. One-sided alternatives are possible. The K-S test can also be used to compare two empirical distributions.

Consider the water-well yield data for rock type Yg (Appendix I). From the histogram (Figure 3.1a) it is clear that the distribution of these 81 water yields is right-skewed. A fifth-root transform yields a distribution that is approximate normal. The step function in Figure 3.11 is the empirical cumulative distribution; the smooth curve is the normal distribution with mean and standard deviation estimated from the fifth root of Yg water-well yields. The test statistic $D = 0.073$ and the p-value $= 0.78$ for the two-sided alternative that the Yg data are not from a normal distribution with the same mean and standard deviation as the sample. The $y_{D=0.078}^{1/5} \doteq 1.59$ (see the arrow in Figure 3.11). Because parameters are estimated from the data, the p-value serves only as a guide. However, since the test is conservative, it is reasonable to conclude that the

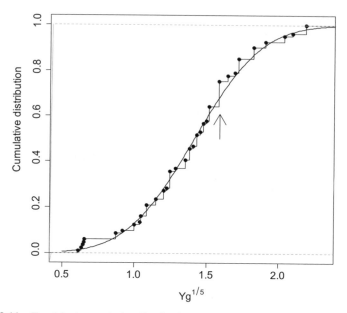

FIGURE 3.11 Empirical cumulative distribution (step function) versus normal cumulative distribution (curved line), water-well yield data, rock type Yg; the arrow indicates maximum D.

transformed sample is likely to be drawn from a normal distribution with $\mu = 1.414$ and $\sigma = 0.366$. The K-S test is not exact when there are ties, as with the Yg sample, because water yields are reported to only two significant digits. Simulation procedures involve sampling from the theoretical distribution of interest for a range of parameter values, estimating the parameter values and then computing the D values. Critical values are obtained from the K-S distribution.

A second type of K-S test is used to compare two empirical distributions. The test statistic $D = \left| G_1(y_{[i]}) - G_2(y_{[i]}) \right|$, where G_1 and G_2 are empirical distribution functions. Note that if the sample sizes differ, interpolation is required so that a comparison can be made. Usually, this feature is incorporated into the algorithm. The null hypothesis of equality of the distributions is rejected at the α level if $\sqrt{n_1 n_2 / (n_1 + n_2)} D > K_\alpha$, where n_1 and n_2 are the sample sizes and K_α is a point, typically in the right-hand tail of the K-S probability density function.

3.16   COMMENTS ON HYPOTHESIS TESTING

A major criticism of hypothesis testing is that frequently it is used to "prove" what the investigator already knows to be true (Johnson, 1999). In our experience this occurs most often when testing a mean against an alternative, such as H_0: $\mu = \mu_0$ versus H_a: $\mu = \mu_a$, where scientific evidence or common sense clearly supports the alternative. Sometimes this occurs when a person views the experimental results and then picks a significance level to give the conclusion desired. A corollary is the investigator who shops around to find the test that will yield the desired result and then reports only the results of that test. Even more insidious are those that select that subset of the data that is consistent with the desired results. Despite abuses, there are many situations where the outcome of an experiment, given uncertainty, is not obvious and hypothesis testing can be of valuable assistance to a decision maker.

The results of a test of hypothesis test are often misinterpreted. This is especially true of the p-value, which is presented as evidence against the null hypothesis. A major criticism by Berger (2003) is that the p-value overstates the evidence against the null. If a p-value is 0.05, the frequentist interpretation is that there is a 5% chance that the null hypothesis will be rejected incorrectly. So in a long series of tests, where in each case the p-value is near 0.05, the expectation is that 5% of the nulls are true. However, Berger (2003) demonstrates that this is not the case. He finds that among the subset of tests that have p-values near 0.05, at least 22% of the nulls are true. In addition, Sterne and Smith (2001) note that:

- The p-value is not the probability that the null is true.
- 1 – the p-value is not the probability that the alternative is true.
- A p-value is based on a chance event that has not occurred (observations in the tails). It cannot be used simultaneously to judge the likelihood of that event.

It is important to remember that the p-value, as presented by Fisher, is a measure of strength of evidence against the null hypothesis and not a probability.

As noted, there are significant differences in using the frequentist approach versus the Bayesian approach to hypothesis testing. There is a general recognition that at some level everyone is a Bayesian because science is not done in a vacuum and every investigator has some prior information on the problem at hand. A major difference between the approaches is that the frequentist computes the probability of some function of the data given a parameter (or set of parameters), whereas the Bayesian is concerned with the probability of various plausible hypotheses given the data. An advantage of the Bayesian approach is that the investigator can incorporate prior knowledge into the procedure in a formal way. The Bayesian approach also allows incorporation of additional data and expert judgment without a complete revaluation of the model. In addition, $\Pr(H_0 \mid \text{data})$ reflects real expected error rates; p-values do not. A formal procedure exists to evaluate $\Pr(H_0 \mid \text{data})$. Computational methods exist to find an approximate solution when an exact analytic solution is not possible. $\Pr(H_0 \mid \text{data})$ is the same as the type I error probability, conditional on observing data with the same strength of evidence as the actual data.

There are also nonparametric equivalents of models used by frequentists and Bayesians. The Wilcoxon test is a nonparametric equivalent of the one-way ANOVA used to compare means. Computer-intensive methods such as the bootstrap allow relaxation of some traditional classical assumptions.

A key question is: Which approach and/or method should be used? Unfortunately there is no single easy answer. When experiments are designed and appropriate assumptions satisfied, the frequentist approach is usually satisfactory. The Bayesian approach has advantages in interpretation and when experiments are not designed. Nonparametric tests work well for small samples and/or when assumptions cannot be verified. Often, the problem with a presentation of results in the scientific literature rests not with the choice of a test but rather with failure to state assumptions and with faulty interpretation.

3.17   SUMMARY

The focus of this chapter is on interval estimation and hypothesis testing of means. A sample mean, by itself, contains little useful information because its variability is unknown. Confidence intervals provide that information. However, when using classical statistics, computation of a confidence interval requires that the distribution of the statistic, in this case the mean, is known. Fortunately, the central limit theorem states that under relatively mild conditions, means are normally distributed for a large sample, regardless of the distributional form of the data. Means of normally distributed data are normally distributed for any sample size. An important assumption is that samples be representative of the population of interest, which implies that observations be independent and drawn at random. An exception is when paired samples are considered.

Three important distributions used in interval estimation and hypothesis testing are the t, chi-squared, and f. All can be constructed from the normal distribution and have degrees of freedom. In any interval estimation or testing situation, there are initially as many degrees of freedom as data values. Subsequently, degrees of freedom are lost due to parameters that must be estimated from the data. This loss is a penalty paid for using the data multiple times.

When constructing a confidence interval on a mean with unknown variance (the usual situation), a t-distribution is used. The interval width is a function of the estimated variance, the sample size, and the level of significance α (selected in advance). The width of the interval can be decreased by increasing α (thus having less confidence), increasing the sample size, or decreasing the variance by a nonstatistical procedure. An interpretation of this frequentist confidence interval estimate states that in repeated sampling, $(1 - \alpha)\%$ of the interval estimates bracket the true mean. Two alternatives to this approach are the Bayesian credible interval and the bootstrap interval estimate. The Bayesian approach treats the parameter of interest as a random variable. It combines an estimate of the distribution of the parameter with data to form a posterior distribution, from which the credible set is computed. The bootstrap procedure is based on resampling and allows the user to relax the normality assumption. Both of these procedures are especially useful when data sets are small. In addition to computing a confidence interval on a mean, Bayesian and bootstrap estimates can be computed on a difference of means or other linear combinations of means. Although most confidence interval estimates assign an equal amount of probability to both tails, this is not a requirement. They can also be one-sided.

Statistical hypothesis testing is a formal procedure used to make a decision under uncertainty. Hypothesis tests can be performed on individual means, pairs of means, multiple means, and linear combinations of means. An important consideration in hypothesis testing is that the significance level and testing procedures need to be decided on in advance of collecting data; otherwise, the test will not be valid. In hypothesis testing, a null hypothesis is often selected to represent the status quo, and an alternative hypothesis to reflect what one thinks or would like to see happen. Decisions made under uncertainty can be wrong, even when all procedures have been performed correctly. The null hypothesis can be rejected when it is correct, a type I error; or it can be accepted when false, a type II error. The complement of the probability of a type II error is the power of the test. In determining power, a practical significant difference needs to be established. The computational procedures for testing a mean against a constant, or for comparing two means, are almost identical to those used in interval estimation. When more than two means are to be compared, analysis of variance (ANOVA) is required. The model is defined by the null hypothesis, which may state, for example, that $\mu_1 = \mu_2 = \mu_3$ against an alternative that one or more pairs differ. A measure of model significance from an F-statistic is the variability accounted for by the model compared to that left over (the residual). Rejecting the null hypothesis implies that one or more pairs of means are unequal; detecting which pair or pairs is a multiple comparison problem. This problem may cause an increase in the probability of making a type I error. Three procedures that alleviate this problem are the Bonferonni, the Tukey, and the Scheffé. In addition to ANOVA, which requires a normality assumption on the distribution of means, there is a nonparametric procedure called the Kruskal–Wallis test. All of the interval estimates and tests involving two or more means assume homogeneity of variance.

Statistical tests and boxplots should be used to investigate this assumption. There is controversy between frequentist and Bayesian approaches to hypothesis testing. The Bayesian approach models parameters as variables conditioned on the data, while the frequentist approach considers parameters as fixed. A key result in frequentist approach is the p-value, which is a measure of the weight of evidence but is often misinterpreted as an error probability. The K-S test provides a way to compare a sample with a known distribution or to compare two empirical distributions. Finally, we admonish the reader not to "shop around" for a statistical test that works nor to omit part of a data set that does not conform to a preconceived conjecture.

EXERCISES

3.1 Suppose we determine that the 95% confidence interval on the mean of uranium is 2.95 to 3.22 ppm based on 1657 stream sediment observations. Can we determine if the mean uranium content is 3.0 versus the alternative that it is not (at the 5% significance level) without additional computation? Explain. What is the duality between hypothesis testing and confidence interval construction?

3.2 Refer to Exercise 3.1. Can the true mean be 3.3? Explain.

3.3 What is the trade-off between constructing a 90% confidence interval versus a 99% confidence interval?

3.4 Why is it desirable, when appropriate, to formulate a one-sided alternative in a test of hypothesis as opposed to a two-sided alternative?

3.5 Suppose that a test of hypothesis yields a p-value of 0.40. Should H_0 be accepted or rejected? Explain.

3.6 What is wrong with making a comparison of means using the t-ratio *and* after computations are completed, deciding on the level of significance?

3.7 Compare the advantages and disadvantages of a rank sum test over a test using a t-ratio.

3.8 A biologist is studying algae concentration in five randomly selected lakes. She takes two samples from each lake. Is it possible for her to see if differences in algae content exist among lakes? Explain.

3.9 A model partitions variability. Explain why this is important.

3.10 Suppose that we wish to compare means from five populations. Using ANOVA we determines that $SST = 160$ and $SSE = 50$. Can we conclude that some means differ based on the fact that $SST/SSE = 3.2$?

3.11 When a test statistic is created, why is it necessary to know its distribution?

3.12 Why is it not appropriate to establish a decision rule after viewing the data?

3.13   Is it better to compute a confidence interval or test a hypothesis? Explain.

3.14   Because of the danger of methyl mercury (MeHg) poisoning due to fish consumption from lakes in northern Georgia, a researcher is asked to conduct a survey to determine the amount of daily fish consumption. EPA guidelines suggest that anything in excess of 17.5 g of fish per day is excessive. What type of test needs to be formulated? What factors should be considered when deciding on a sample size?

3.15   Suppose that we have a random sample of 40 arsenic soil samples and the task is to compute a confidence interval on the mean. Why does the central limit theorem help with this task? What should we do if it is not appropriate? Explain.

3.16   In performing a comparison between two means, why is it helpful to be able to assume equal variance?

3.17   Is the expression "you can prove anything with statistics" true or false? Comment.

3.18   A preliminary analysis of historical data indicates that to test an arsenic level of 5.0 ppm against a specific alternative of 5.2 ppm with a power of 80% and a significance level of 5%, a sample of size 120 is needed. This is more than the project can afford. What are the alternatives and their implications?

3.19   Why is it important to report the standard error or, equivalently, the standard deviation and sample size whenever a mean is reported.

3.20   Construct an example to show that the Wilcoxon two-sample test is sensitive to differences in shapes between the two distributions.

3.21   Explain why in the case of the small-sample water-well yield problem (Table 3.5), the results of the Wilcoxon nonparametric test and the bootstrap differ.

4 Regression

4.1 INTRODUCTION

Regression, as used in statistics, is a model that relates a variable to one or more additional variables. The term comes from Sir Francis Galton, who while studying seeds, noticed that the median diameter of the offspring of larger seeds was smaller than that of their parents, whereas the median diameter of the offspring of the smaller seeds was larger than that of their parents. He later observed the same phenomenon when studying the heights of fathers and sons. Galton called this tendency to revert to the mean or median size regression toward the mean or simply regression. In the following coal quality case study, we explain the variability in energy content to other variables, including ash, moisture, sulfur, and mercury.

The primary variable of interest is called the *response variable*. The variables that are used to attempt to explain variability in the response variable are called *explanatory variables*. Some books refer to the response variable as the *dependent variable* and the explanatory variables as *independent* or *predictor variables*.

We begin with simple linear regression (SLR), which has one explanatory variable, then develop multiple regression for more than one explanatory variable, and conclude with a section on nonlinear regression modeling. Important components of the process are formulating a model, fitting the model to the data, and model evaluation. Numerous textbooks provide additional insight into regression analysis, notably those of Draper and Smith (1998), Montgomery et al. (2001), Ryan (2008), and Seber and Lee (2003).

Linear regression is a special case of generalized linear models (McCullagh and Nelder, 1989). In Chapter 8, special cases where the response variable is discrete are examined.

4.2 PITTSBURGH COAL QUALITY CASE STUDY

Worldwide increases in demand for energy, declines in the discovery of conventional oil and gas resources, and concern about air and water quality, coupled with enormous coal reserves have heightened interest in coal quality. Coal-fired power plants produce

Statistics for Earth and Environmental Scientists, By John H. Schuenemeyer and Lawrence J. Drew

TABLE 4.1 Pittsburgh Coal Quality Data Set

Obs.	Heat (Btu/lb)	Moisture (%)	Ash (%)	Sulfur (%)	Mercury (ppm)	Obs.	Heat (Btu/lb)	Moisture (%)	Ash (%)	Sulfur (%)	Mercury (ppm)
1	13701	2.60	5.80	3.60	0.09	32	13633	1.00	11.31	1.60	0.14
2	14003	2.70	6.00	0.40	0.19	33	13980	1.90	7.30	1.00	0.03
3	13543	2.30	6.30	2.10	0.29	34	12783	2.96	10.20	3.96	0.15
4	13711	2.70	6.50	2.50	0.40	35	13278	3.00	9.40	4.80	0.20
5	12700	6.10	6.60	1.50	0.07	36	14300	1.90	4.40	0.60	0.05
6	13559	2.70	6.70	0.90	0.12	37	13554	3.60	6.40	1.70	0.19
7	13412	2.44	6.82	1.29	0.18	38	12114	2.80	13.50	3.60	0.16
8	13731	2.60	7.00	2.90	0.08	39	14115	1.00	7.50	1.80	0.22
9	13762	1.60	7.30	2.70	0.12	40	13454	2.90	6.10	3.10	0.50
10	13965	1.50	7.40	1.60	0.21	41	13600	2.11	8.36	1.42	0.02
11	13276	2.70	7.50	2.00	0.41	42	11867	3.97	15.28	1.02	0.06
12	13090	2.98	7.62	1.46	0.14	43	13170	1.63	12.10	1.20	0.06
13	13564	1.10	7.90	2.70	0.23	44	13010	5.78	6.59	0.79	0.00
14	13115	2.40	8.00	3.20	0.16	45	13180	2.90	7.60	2.00	0.14
15	12922	3.77	8.04	1.17	0.17	46	13168	1.50	10.00	5.20	0.22
16	13514	1.10	8.40	2.80	0.07	47	13607	1.80	7.20	2.70	0.10
17	13175	2.75	8.40	1.99	0.26	48	13488	2.20	6.30	1.00	0.07
18	13671	0.70	8.90	3.30	0.07	49	13189	3.30	6.90	3.60	0.10
19	12926	2.90	9.30	2.60	0.14	50	13360	4.40	5.80	1.50	0.55
20	12964	2.50	9.50	3.70	0.25	51	13141	3.70	7.10	1.90	0.28
21	12712	3.50	9.85	2.98	0.17	52	12677	3.07	10.57	2.22	0.31
22	13240	2.50	10.10	2.70	0.22	53	13761	2.40	6.70	1.40	0.09
23	12600	4.10	10.10	2.90	0.30	54	12618	3.02	10.90	2.90	0.36
24	13243	1.70	10.20	2.60	0.57	55	13240	3.00	7.30	1.90	0.22
25	13180	1.40	10.30	3.80	0.21	56	12933	3.10	9.00	3.00	0.12
26	12645	2.70	10.60	3.20	0.65	57	12630	4.80	8.20	2.30	0.15
27	13101	3.50	10.70	4.00	0.74	58	13545	1.18	9.85	1.92	0.16
28	12508	3.14	10.98	3.18	0.30	59	13147	2.90	6.90	1.50	0.10
29	13498	1.49	11.72	0.89	0.05	60	11563	2.05	20.36	4.40	0.51
30	12025	2.68	14.26	3.60	0.20	61	13512	2.70	8.50	3.60	0.30
31	12061	3.50	16.10	4.00	0.25	62	11803	7.10	10.90	1.00	1.00

over half of the electricity consumed in the United States. The presence of sulfur, mercury, and arsenic in coal are a major problem. In addition, high moisture content and ash can seriously degrade the heat content of coal.

The coal quality data used in examples in this chapter are taken from the Pittsburgh coal bed (Bragg et al., 1998). Data for this coal bed (Table 4.1) consist of 62 observations and up to 105 variables in each observation. Five key variables are chosen: heat of combustion, ash, moisture, sulfur, and mercury.

4.3 CORRELATION AND COVARIANCE

The concepts of correlation and covariance describe association between two random variables and, strictly speaking, belong in Chapter 7. However, the concept of correlation is used in the study of regression and thus is introduced here.

Covariance is a measure of linear association between two random variables Y_1 and Y_2 in units of $Y_1 \times Y_2$. For a bivariate sample (Y_{1i}, Y_{2i}), $i = 1, \ldots, n$, the covariance is estimated as

$$\widehat{\text{Cov}}(Y_1, Y_2) = \sum_{i=1}^{n} \frac{(Y_{1i} - \overline{Y}_1)(Y_{2i} - \overline{Y}_2)}{n - 1}$$

where $\overline{Y}_1$ and $\overline{Y}_2$ are the sample means. If the units of Y_1 and Y_2 are ppm, the units of $\widehat{\text{Cov}}(Y_1, Y_2)$ are ppm^2. The estimated covariance is also represented as $\hat{\sigma}_{Y_1, Y_2}$ or $\hat{\sigma}_{1,2}$. If the variables are in different units (e.g., one in parts per million and the other in meters), it rarely makes sense to compute the covariance. However, there is a standardized measure of covariance called the *correlation* which is dimensionless. The *population correlation coefficient*, a standardize covariance, is defined as

$$\rho_{Y_1, Y_2} = \frac{\sigma_{Y_1, Y_1}}{\sqrt{\sigma_{Y_1}^2 \sigma_{Y_2}^2}} \qquad -1 \leq \rho_{Y_1, Y_2} \leq 1$$

and the *sample correlation coefficient* is defined as

$$\hat{\rho}_{Y_1, Y_2} = \frac{\hat{\sigma}_{Y_1, Y_1}}{\sqrt{\hat{\sigma}_{Y_1}^2 \hat{\sigma}_{Y_2}^2}} \qquad -1 \leq \hat{\rho}_{Y_1, Y_2} \leq 1$$

$$= \frac{\sum_{i=1}^{n}(Y_{1i} - \overline{Y}_1)(Y_{2i} - \overline{Y}_2)}{\sqrt{\sum_{i=1}^{n}(Y_{1i} - \overline{Y}_1)^2 \sum_{i=1}^{n}(Y_{2i} - \overline{Y}_2)^2}}$$

Often, $\hat{\rho}_{Y_1, Y_2}$ is written as r_{Y_1, Y_2} or simply $\hat{\rho}$ or r if there is no ambiguity about the variables of interest. This measure of correlation is sometimes called *Pearson's correlation coefficient*, to distinguish it from a nonparametric measure called *Spearman's correlation coefficient*, which will be introduced shortly.

Pearson's correlation coefficient is illustrated using constructed data. Several correlation patterns are illustrated in Figure 4.1. Four important facts concerning correlation are:

- *Correlation does not imply cause and effect.* There are many spurious correlations, and sometimes the correlation between two variables is related to a third variable (a concept called partial correlation, defined later in the section).
- The Pearson correlation coefficient is sensitive to outliers (this is investigated further later).
- The Pearson correlation coefficient is influenced by highly skewed distributions.
- Correlation is a measure of linear association.

Spearman's correlation coefficient can be computed using the same formula as Pearson's except that in place of the random variables $\{Y_{1i}, Y_{2i}, i = 1, \ldots, n\}$, their

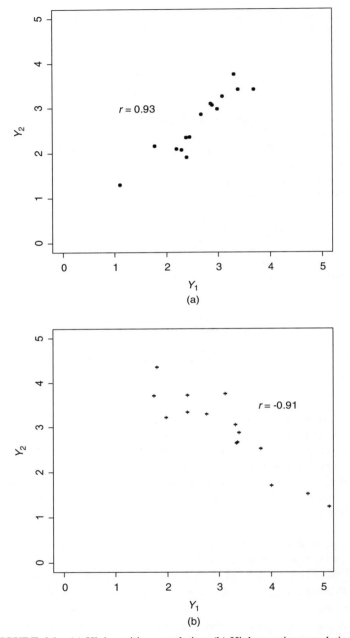

FIGURE 4.1   (a) High positive correlation. (b) High negative correlation.

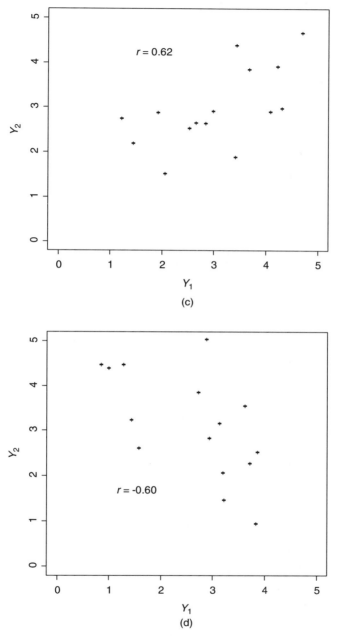

FIGURE 4.1 (*Continued*) (c) Moderate positive correlation. (d) Moderate negative correlation.

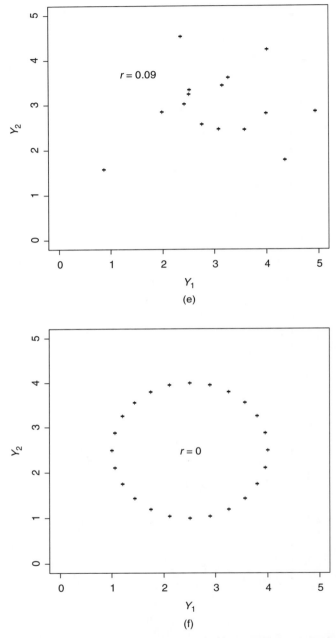

FIGURE 4.1 (*Continued*)   (e) Minimal correlation. (f) No correlation, nonlinear pattern.

TABLE 4.2   Rank of Subset of Pittsburgh Coal Quality Data

Ash (%)	Moisture (%)	Rank Ash	Rank Moisture
5.80	2.60	1	2
6.00	2.70	2	3.5
6.30	2.30	3	1
6.50	2.70	4	3.5
6.60	6.10	5	5

ranks are used (Table 4.2). If ties exist, such as the two moisture values of 2.70, each rank is the average of their two ranks. For this example, Pearson's correlation coefficient is 0.58 and Spearman's is 0.62. In the section on outliers, the performance of the two types of correlation coefficients in the presence of outliers is examined.

*Partial correlati*on is a measure of the relationship between two variables, say X and Y, when some dependence exists between X and/or Y and a third variable, Z. Failure to examine this conditional relationship may over- or understate the linear relationship between X and Y. The population partial correlation between X and Y given Z is defined as

$$\rho_{XY|Z} = \frac{\rho_{XY} - \rho_{XZ}\,\rho_{YZ}}{\sqrt{1 - \rho_{XZ}^2}\sqrt{1 - \rho_{YZ}^2}}$$

4.4   SIMPLE LINEAR REGRESSION

It may be postulated, based on theory and/or empirical observations that there is an inverse linear relationship between Btu and ash. One way to address this question is to formulate a model in the form of an equation. This model needs to be fit to data and then evaluated. The three major components of regression are model specification, fitting or estimating model parameters, and evaluating the model.

4.4.1   Model Specification and Assumptions

The model for simple linear regression (SLR) is $Y = \beta_0 + \beta_1 X + \varepsilon$, where Y is the response variable and X is the explanatory variable. In SLR there is only one explanatory variable, X. The model parameters β_0 and β_1 (the intercept and slope, respectively) must be estimated. In a "good" model, ε is random error or deviation from the true regression line $\mu_{Y|X} = \beta_0 + \beta_1 X$, where $\mu_{Y|X}$ is the mean of Y given X (recall that "|" is read as "given"). Four key assumptions are made when this form of model is adopted.

1. The true mean values $\mu_{Y|X}$ for X in the domain of interest lie on a straight line. This is the line shown in Figure 4.2.

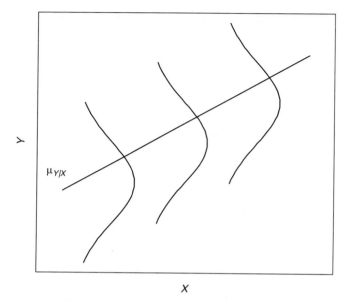

$\mu_{Y|X}$

Y

X

FIGURE 4.2   Theoretical simple linear model showing a true regression line and distributions of the Y's about this line.

2. $\mathrm{Var}(Y \mid X) = \sigma^2$, the variance of $Y \mid X$, is constant across the X's. This is the homoscedasticity or equal-variance assumption. Note that the distributions shown in Figure 4.2 all have the same spread.
3. The ε's are normally distributed (the distributions shown in Figure 4.2 are normally distributed and centered about the true regression line).
4. The ε's are independently distributed.

Investigation of the validity of these assumptions is discussed later.

4.4.2   Properties of Estimators

Let $\widehat{\beta}_0$ and $\widehat{\beta}_1$ be estimates of β_0 and β_1, respectively, and $\widehat{\sigma}$ be an estimate of the error σ. There are numerous ways to make estimates, and estimators have many properties. An important question is: How good is an estimator? A goal of any study is to produce good estimates. What is meant by good? Some of the important properties include:

- *Lack of bias.* A statistic is unbiased if it is correct on average. Most commonly used statistics are unbiased or nearly unbiased, but this is not true of all statistics.
- *Minimum variance.* A statistic has minimum variance if it has the smallest variance of all possible estimators. Different estimators will have different variances.

- *Consistency.* As the sample size increases, the statistic approaches the population parameter. If this is not a property, increasing the sample size is not useful. The statistics used in this book are consistent.

Other important properties include efficiency and sufficiency, which are beyond our scope in this book. It is not always possible simultaneously to achieve the optimum result for each property of a statistic. Sometimes compromises must be made, as in the case of unbiasedness and minimum variance.

Other considerations for the choice of estimator include the type and amount of data available. Sometimes estimates need to be made using hard data (measurements), expert judgment, or a combination of both.

Correlation Versus Regression   We digress slightly here to show that the regression model can arise in two different contexts, regression and correlation. To keep the notation simple, we consider the SLR model; however, this discussion applies equally to models with more than one explanatory variable.

In regression, the explanatory variable X is considered to be fixed: that is, known without error. Typically, the X's will be obtained from a designed experiment in which the X's are chosen first and then Y's are obtained for the given X's. For example, an ecologist may wish to investigate possible changes in dissolved oxygen content in a stream at 50-m intervals from a point source of pollution. The X's, the distances downstream, are established first. At each of the X's, one or more measures of dissolved oxygen will be taken. There is only one random variable, the dissolved oxygen. Rarely is X measured without error; however, in many applications the error is small enough so as not to influence the results of the analysis. In the dissolved oxygen problem, it is unlikely that an error of 1 m will make any difference in the results.

For the Pittsburgh coal quality case study, it is impossible to design an experiment to obtain predetermined values of ash: for example, at 5%, 10%, and 15%. Rather, a coal sample is obtained and each element analyzed for heat content (Btu), ash, and other attributes. This yields a set of multivariate observations. However, the relationship between Btu and ash is postulated by the model $Y = \beta_0 + \beta_1 X + \varepsilon$, and the expected value of Y is often written $E[Y] = \beta_0 + \beta_1 X$. How is this possible, given that X and Y both appear to be random variables? Clearly, values of X are not selected by design. The interpretation is that the expected value of Y is conditioned on X as indicated by

$$E[Y|X = x] = \beta_0 + \beta_1 X$$

Thus, this model is conditioned on the specific values of the given X's. It may not be valid for different X's. If the X's are to be considered random variables as opposed to the results conditioned on a given x, a measurement error model (Fuller, 1987) needs to be used. When both X and Y are random variables, their linear association can be estimated via the covariance or correlation coefficient, as described previously.

4.4.3   Fitting the Model to Data: Least-Squares Procedure

Consider how ash content may affect the Btu. In a regression context, Btu is defined as the response variable and ash as the explanatory variable. A scatter plot of Btu versus ash (Figure 4.3) suggests that a linear model may provide a reasonable fit. The reader will see how to quantify the term reasonable using graphical and statistical tools. The observations identified on this plot correspond to those listed in Table 4.1 and are discussed in the context of regression diagnostics.

In Chapter 1, a graphical approach to fitting is used. The obvious problem with this approach is its nonuniqueness and arbitrariness. Most often, a better way is to fit a model to data using a formal statistical procedure or algorithm of, which there are many. A classical procedure in common use is called *least squares* (LS). The theory was probably developed independently by Carl Friedrich Gauss and Adrien-Marie Legendre (Stigler, 1986) in the late eighteenth century; however, sixteenth century astronomers, including Tycho Brahe, probably knew of and used the procedure. The least-squares method of fitting has well-understood statistical properties and yields satisfactory results when model assumptions are satisfied. Subsequently, other approaches to fitting a model are examined.

Let $\widehat{Y}_i = \widehat{\beta}_0 + \widehat{\beta}_1 X_i$, $i = 1, \ldots, n$, represent the fitted model, which is the equation of a straight line. Define the ith residual $e_i = Y_i - \widehat{Y}_i$ to be an estimate of ε_i, the true error at the ith observation Y_i; e_i is often referred to as a *raw residual* to distinguish it from other transformed residuals, which are introduced in subsequent sections. The

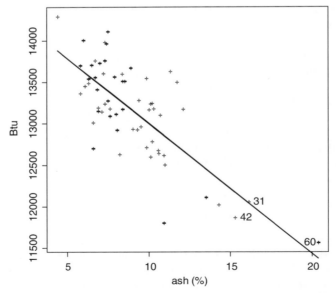

FIGURE 4.3   Scatter plot of Btu versus ash plot, Pittsburgh coal quality case study. The line is a least-squares fit.

estimated model parameters $\widehat{\beta}_0$ and $\widehat{\beta}_1$ obtained by LS are those that minimize $\text{SSE} = \sum_{i=1}^{n} e_i^2$, the residual sum of squares. The parameter estimates are

$$\widehat{\beta}_1 = \frac{S_{XY}}{S_{XX}}$$
$$\widehat{\beta}_0 = \overline{Y} - \widehat{\beta}_1 \overline{X}$$

where

$$S_{XX} = \sum_{i=1}^{n} (X_i - \overline{X})^2 \qquad S_{XY} = \sum_{i=1}^{n} (X_i - \overline{X})(Y_i - \overline{Y}) \qquad S_{YY} = \sum_{i=1}^{n} (Y_i - \overline{Y})^2$$

Estimates of the variability of $\widehat{\beta}_0$ and $\widehat{\beta}_1$ are

$$\text{Var}(\widehat{\beta}_0) = \widehat{\sigma}^2 \left(\frac{1}{n} + \frac{\overline{X}^2}{S_{XX}} \right)$$
$$\text{Var}(\widehat{\beta}_1) = \frac{\widehat{\sigma}^2}{S_{XX}}$$

Their square roots are often referred to as the *standard error* (SE) of $\widehat{\beta}_0$ and $\widehat{\beta}_1$, respectively.

In the Pittsburgh coal quality case study, a goal is to investigate the relationship between Btu (y) and ash (x), where ash is the explanatory variable. The fitted model is $\widehat{y}_i = 14{,}572 - 157x_i$ (Figure 4.3). The $x = 0$ is not shown; therefore, the intercept $\widehat{\beta}_0 = 14{,}572$ is not visible. Now that an estimate of model parameters and error has been obtained, the question that must be asked is: Is this a satisfactory model?

4.4.4   Model Evaluation

Model evaluation is critical. It is necessary to know if the model is satisfactory for the job asked of it. Every statistical model and estimation technique carries with it a set of assumptions. In least-squares regression, these assumptions include normally distributed independent errors with constant variance. Other models may have different, and in some cases, less restrictive assumptions. When assumptions fail to hold, statistical tests of significance may be invalid and model estimates can be seriously biased. A challenge in model evaluation is that all assumptions are not of equal importance. With experience one learns that some assumptions may be relaxed without doing significant harm to model results, whereas others can be critical.

Methods to evaluate a model include an examination of statistics produced by the model-fitting procedure (least squares in the Pittsburgh coal quality case study). This investigation involves formal hypothesis testing and examination of descriptive statistics and graphs. Graphs are essential. One needs to view a variety of model evaluation tools because no single statistic or graph gives complete information on the quality of a model. Model evaluation/diagnostic tools are discussed below.

Analysis of Variance (ANOVA)   When means are compared in Chapter 3, an ANOVA table is presented which shows the partitioning of total variability between means (the model) and within means (the error). An SLR model partitions the total variability between that which is accounted for by the regression model, called the *sum of squares due to regression* (SSR), and the sum of squares due to error (SSE). Begin by writing the total sum of squares (SST; also previously defined as S_{YY}) as

$$\text{SST} = \sum_{i=1}^{n} \left(Y_i - \overline{Y} \right)^2 = \sum_{i=1}^{n} \left[\left(Y_i - \widehat{Y}_i \right) + \left(\widehat{Y}_i - \overline{Y} \right) \right]^2$$

Some simple algebra shows that SST = SSR + SSE, where

$$\text{SSR} = \sum_{i=1}^{n} \left(\overline{Y} - \widehat{Y}_i \right)^2 = \sum_{i=1}^{n} \left(\overline{Y} - \widehat{\beta}_0 - \widehat{\beta}_1 X_i \right)^2$$

$$\text{SSE} = \sum_{i=1}^{n} \left(Y_i - \widehat{Y}_i \right)^2 = \sum_{i=1}^{n} \left(Y_i - \widehat{\beta}_0 - \widehat{\beta}_1 X_i \right)^2$$

SST is a function of the Y's and thus is independent of the model. SSR expresses the variability of the fitted values about the grand mean $\overline{Y}$. Finally, SSE expresses the variability of the observed values (the Y_i's) about their corresponding fitted values (the $\widehat{Y}_i$'s). If all of the observations are on the line $\widehat{Y}_i = \widehat{\beta}_0 + \widehat{\beta}_1 X_i$, SSE $= 0$.

Mean-Square Error and Degrees of Freedom   Each sum of squares (ANOVA table, Table 4.3) is divided by its degrees of freedom to obtain mean squares. There are 62 observations in this example, and the response variable is Btu. A degree of freedom is lost because the grand mean $\overline{Y}$ is estimated from the data, leaving 61 df to estimate SST. The need first to estimate $\overline{Y}$ is a constraint on the model. There is 1 df assigned to the model because it contains one parameter, β_1. This parameter is estimated from the data. What about the intercept, β_0? $\widehat{\beta}_0 = \overline{Y} - \widehat{\beta}_1 \overline{X}$, so this loss of a degree of freedom is accounted for because $\widehat{\beta}_0$ is related linearly to $\overline{Y}$. The concept of degrees of freedom is expanded in the discussion of multiple regression.

A problem with interpretation of the sum of squares is that it is a function of a number of parameters and observations. Standardized versions of these quantities are called mean squares. Estimates of mean squares for the SLR model are

$$\text{MSR} = \frac{\text{SSR}}{1}$$

$$\text{MSE} = \frac{\text{SSE}}{n-2}$$

TABLE 4.3   Regression Output, Pittsburgh Coal Quality Case Study with the Fitted Regression Line Shown in Figure 4.3

Residuals

Minimum	Q_1	Median	Q_3	Maximum
-1056.77	-279.35	-58.58	257.45	837.65

Coefficients

Terms	Value	Std. Error	p-Value
(Intercept)	14,572.49	163.64	0.000
Ash	-157.13	17.51	0.000

Residual standard error: 381.9 on 60 df
R^2 0.573
R^2_{adj} 0.566

ANOVA Table ($n = 62$)

Terms		Sum of Squares	df		Mean Square	F-Statistic	p-Value
Ash	SSR	11,746,912	1	MSR	11,746,912	80.54	0.000
Residuals	SSE	8,751,433	60	MSE	145,857		
Total	SST	20,498,345	61	MST	336,038		

where MSR is the *mean square due to regression* and MSE (also denoted $\hat{\sigma}^2$), the mean-square error, is an estimate of the variability of the data about the regression line. Its square root is called the *residual standard error* (Table 4.3) and is 381.9.

Model Significance   The hypotheses to test model significance are:

H_0: The model *does not* account for statistically significant variability in the response variable Y.

H_a: *It does* account for statistically significant variability in Y.

The test statistic is F-statistic $= $ MSR/MSE. When H_0 is true, there is no regression; the explanatory variable fails to explain a statistically significant proportion of the variability in Y; MSR and MSE each estimate σ^2_Y, the variance of Y; and the F-statistic has an f-distribution. In the Btu–ash example there are 1 and 60 df for MSR and MSE, respectively.

If H_0 is rejected and the investigator concludes that the model accounts for a statistically significant percent of the variability in the Y's, MSR $>$ MSE. For the example (Table 4.3), the F-statistic $= 80.54$. The p-value $= \Pr(f > F$-statistic $\mid H_0$ true$) \sim 0$. Thus, it is highly unlikely that an F-statistic of 80.54 will arise from an f-distribution with 1 and 60 df. Therefore, it can be concluded that this model is highly significant and ash, the X variate, accounts for a statistically significant proportion of the variability in Y, the Btu. Note that $F_{(0.05, 1, 60)} = 4.00$, where 0.05 is the significance level and 1 and 60 are the df.

Think of the F-statistic as a signal-to-noise ratio. The higher the signal (the variability accounted for by the model) is to the noise (the random error), the more likely it is that the model is significant. When driving across a sparsely populated region at night, it is often possible to pick up a weak radio station clearly. This station is never heard if it is located in a major industrial area with unshielded wires and other background noise. Thus, it is not the absolute strength of the signal (MSR) that is important but the strength in relation to the noise (MSE).

A related goodness-of-fit statistic is the *coefficient of determination*, which is the fraction of total variability accounted for by the model. It is defined as

$$R^2 = \frac{\text{SSR}}{\text{SST}} = 1 - \frac{\text{SSE}}{\text{SST}}$$

where $0 \leq R^2 \leq 1$. $R^2 = 0.573$ for the regression of Btu on ash (Table 4.3). For SLR, R^2 is numerically equal to the square of $r_{X,Y}$, the sample correlation between X and Y. Note that where there is no ambiguity; the symbol R^2 will be used instead of the term coefficient of determination. When an additional explanatory variable is added to the model, as in multiple regression, R^2 may increase even when the new variable is not statistically significant. To compensate for this bias, the *adjusted R^2*,

$$R^2_{\text{adj}} = \frac{\text{MST} - \text{MSE}}{\text{MST}}$$

is computed, where MST is the mean-square total and MSE is the mean-square error. Note that $R^2_{\text{adj}} \leq R^2$ and R^2_{adj} can be negative; an example of R^2_{adj} is presented in the discussion of multiple regression. For the Btu–ash example $R^2_{\text{adj}} = 0.566$.

Is the F-statistic or R^2 sufficient to imply a good model? No. The model looks promising but further investigation is needed, as outlined in the following steps.

Fitted Value Plot   A useful evaluation tool is a scatter plot of $(\widehat{Y}_i, Y_i)$, $i = 1, \ldots, n$, the fitted value versus the observed value. A "good fit" implies that the observations cluster around a $45°$ line, as they do for the Btu–ash regression example (the $45°$ line in Figure 4.4). More scatter about this line implies a lower-quality fit (a higher R^2). Clearly, a perfect fit implies that all observations are on the $45°$ line. Data point scatter not about the $45°$ line may suggest the presence of outliers or an incorrect choice of a model.

Residuals   A way to investigate the assumptions that are made about the error is to plot residuals versus the fitted values e_i versus $\widehat{Y}_i$, $i = 1, \ldots, n$, where $e_i = Y_i - \widehat{Y}_i$. Residual plots are extremely informative because if the model is formulated and fitted properly, the residuals should be random. The scatter should be approximately constant across the domain of fitted values and no systematic variability in the pattern of residuals should be seen. When plotted against the fitted values, the

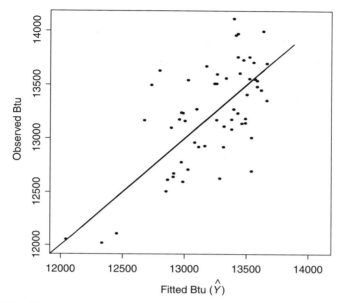

FIGURE 4.4   Observed Btu versus fitted Btu from SLR model, Pittsburgh coal quality case study.

residuals should be randomly distributed $N(0, \sigma^2)$. Several types of residuals are introduced following a discussion of leverage points.

Leverage Points   The ith leverage point measures the distance between X_i and $\overline{X}$, $i = 1, \ldots, n$, where X is the explanatory variable (ash in the Pittsburgh coal quality case study). The *leverage point* is a measure of how far away the observation X_i is from the main body of the data (Belsley et al., 1980). A high leverage point has the potential to exert undue influence on regression results. The leverage value h_i is defined as

$$h_i = \frac{1}{n} + \frac{(X_i - \overline{X})^2}{\text{SSX}}$$

where $\text{SSX} = \sum_{i=1}^{n} (X_i - \overline{X})^2$. In the subsequent discussion of multiple regression, additional explanatory variables are added to the regression model. The h_i will represent the distance between the ith observation and the means of all of the explanatory variables. A leverage point h_i is considered high if it is greater than $4(p + 1)/n$, where p is the number of explanatory variables. For SLR, $p = 1$. In this example, only observation 60 exceeds the $(4)(2)/62 = 0.13$ leverage point guideline (Figure 4.3) with a leverage value of $h_{60} = 0.29$. The next two highest leverage points are observations 31 and 42, which have leverage values of 0.12

and 0.10, respectively. These three observations are the most distant from the main body of the X's.

Modified Residuals   The *modified residual* is $r_i = (Y_i - \widehat{Y}_i)/\sqrt{1 - h_i}$, where h_i is the ith leverage point. Modified residuals are preferred over raw residuals because they are pairwise independent, whereas raw residuals are not. Since the errors (the ε's) are assumed to be independent, their estimates should be independent. It can be shown that the $\mathrm{Var}(e_i|X) = \sigma^2(1 - h_i)$ and therefore the modified residual has constant variance. It is also in the same units as the raw residual. Unless there are high leverage points, there is little difference between modified and raw residuals.

Standardized Residuals   An alternative to the raw residual e_i is the *standardized residual*. Unfortunately, there is not a universally accepted definition. It may be defined as $e_{\mathrm{sta}, i} = e_i/\widehat{\sigma}$; the raw residual divided by the residual standard error, which of course has variance 1. However, it is also defined as $e_{\mathrm{sti}, i} = e_i/(\widehat{\sigma}\sqrt{1 - h_i})$, which is also called an *internally studentized residual*. A bit of algebra shows that $\widehat{\mathrm{Var}}(e_i) = \widehat{\sigma}^2(1 - h_i)$.

Studentized Residuals   The ith *externally studentized residual* is defined as

$$e_{\mathrm{ste}, i} = \frac{e_i}{\widehat{\sigma}_{[i]}\sqrt{1 - h_i}}$$

where $\widehat{\sigma}_{[i]}$ is estimated by omitting the ith residual. When attempting to determine if the ith observation is an outlier, it should be omitted from computation, because potential outliers influence the estimate of the residual standard error. The use of externally studentized residuals is recommended, even though in most instances all forms of residuals yield similar results. Outliers may be easier to identify in a plot of studentized or standardized residuals than in a plot of raw residuals, because the number of standard deviations away from zero is readily viewed.

A plot of studentized residuals versus fitted Btu values is shown in Figure 4.5. Visually, the residual data appear to be random and normally distributed; however, fitted values are concentrated above 12,600 Btu. Observations 5, 29, 32, and 62 may be outliers. As noted, observation 60 is a high leverage point and observations 31 and 42 are marginally high leverage points.

Cook's Distance   Among the tools used to detect influence points is *Cook's D_i* (Cook's distance for the ith observation). Computing the model parameters using the n observations and then computing model parameters, omitting the ith observation, yields Cook's D_i. It is a measure of the changes in the estimated model coefficients due to the ith observation (Belsley et al., 1980). A $D_i > 1$ is considered large. All of the observations used in the Btu–ash regression have $D_i < 1$; however, the usual suspects, observations 5, 29, 32, 60, and 62, stand out (Figure 4.6).

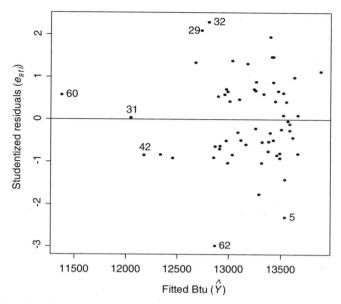

FIGURE 4.5   Fitted Btu versus studentized residuals, Pittsburgh coal quality case study data.

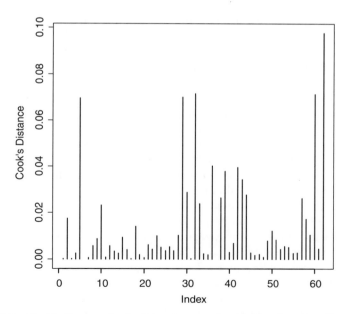

FIGURE 4.6   Cook's distance plot for Btu–ash regression, Pittsburgh coal quality case study.

DFFIT $\text{DFFIT}_i = e_{\text{ste},i}\sqrt{h_i/(1-h_i)}$ is similar to Cook's distance. *DFFIT* measures the difference between a response predicted from the model with and without inclusion of the ith observation. It is considered large if $|\text{DFFIT}_i| > 2\sqrt{(p+1)/n}$, where p is the number of explanatory variables. For the Btu–ash example, $p=1$ and $n=62$ and $2\sqrt{(p+1)/n} = 0.36$. Observations 5, 32, 60 and 62 have $|\text{DFFIT}| > 0.36$, indicating that further investigation may be warrented.

DFBETA $\text{DFBETA}_{j,i} = \left(\widehat{\beta}_j - \widehat{\beta}_{j[i]}\right)/\left(\text{Var}(\widehat{\beta}_{j[i]})\right)^{1/2}$, where $\widehat{\beta}_j$ is an estimate of the jth regression coefficient using all of the data and $\widehat{\beta}_{j[i]}$ is the same, estimate except the ith observation is removed. A value $|\text{DFBETA}_{j,i}| > 2$ is considered large. *DFBETA* is similar to DFFIT except that it is computed on each regression parameter. The largest $|\text{DFBETA}| = 0.37$. The authors caution that definitions of DFFIT and DFBETA and guidelines for further investigation vary.

Comments on the Detection of Unusual Points   We strongly encourage the use and judicious interpretation of the procedures discussed previously. Two main concerns in data analysis and model fitting are influence points and outliers. First, consider influence points. In the ideal data set, each observation should exert approximately the same influence on the model fit. However, observation 60 (Figure 4.3) is of interest because its ash value is far from the rest of the data. Such observations often influence model fit significantly because a small change in their Btu can affect a greater change in the model parameters or predicted response than can a corresponding small change in Btu at ash values near the main body of the data. A concern is that if Btu measured at observation 60 is in error or subject to large variability, the estimated model parameters will change more drastically than if other observations are in error. The DFFIT procedure identified observation 62 as having a significant effect on the response predicted. So what is to be done? It appears that a significant percentage of the variability in the model is due to observation 60. We leave it as an exercise to the reader to see how much. Recall in the SLR model that ash accounts for only 57% of the variability in Btu. Clearly, if observation 60 is suspect on other grounds, such as being from another population (say, a different coal seam), an argument can be made that there is no statistically significant relationship between Btu and ash as described by the model chosen. However, even if it is believed that observation 60 is reasonable, results should be interpreted cautiously. Leverage, DFFIT, DFBETA, and Cook's distance can be computed using the function influence.measures in the R-project.

As stated previously, an outlier is an observation that is located away from the main body of the data. Why should one care about outliers? There are three general reasons: (1) outliers may result from a mistake, such as a transcription or recording error, (2) they may indicate a mixed population that needs to be separated prior to further analysis, and (3) they may unduly influence model estimates. Typically, observations corresponding to residuals that are at least two to three standard deviations from zero may be considered outliers (Rousseeuw and Leroy, 1987). However, in a sample of size 100, five residuals, on average, are expected to be at least two standard deviations from zero. When dealing with univariate data such as

Btu, outliers may be able to be spotted using a boxplot. In regression, outliers are not the only potentially troublesome observations. For example, the fitted regression line (Figure 4.3) passes close to observation 60, so in a regression context these do not show up as outliers (Figure 4.5). Conversely, observation 62 (Figure 4.3) will not show up as an outlier when considering only the distribution of Btu; however, it will be an outlier in a regression context.

Key points regarding outliers are:

- They may or may not be influence points.
- The larger the data set, the less the influence of individual outliers.
- What appears to be an outlier in a small data set may not be an outlier in a large data set. For example, in a data set of sample size 20, an observation located three standard deviations from the mean will be somewhat unusual; the same observation will not be so unusual in a data set of size 100.
- Outliers often result from measurement error.
- Outliers can result from a mixture of populations.
- What appears as an outlier in a small sample may be an observation drawn from a highly skewed distribution.

What should be done with suspected outliers and influence points? The following are recommended:

- Examine the raw data if possible. See if there is evidence that a mistake has been made, such as a recording error or equipment failure; if so, it is proper to delete and/or correct the data value.
- Determine if the outlier is from another population; if so, attempt to separate the populations or delete the outlier.
- Determine if the data are from a skewed distribution using analog, theory, simulation, or other techniques; if so, consider a normalization transformation such as a log.
- Consider a robust procedure, which will down-weight outliers. Robust procedures are discussed later in the chapter.
- When it is necessary and proper to delete an outlier or influence point, the investigator needs to document it and provide a justification. Observations must not be removed to satisfy an investigator's conjecture or to provide an outcome desired by a sponsor.

Figure 4.7 shows data sets constructed to illustrate the effect of outliers on least-squares procedures. In Figure 4.7a, a single outlier far removed from the main body of the data can induce a high correlation. The Pearson and Spearman correlations for the nine observations in the lower left cluster are 0.11 and -0.05, respectively. However, with the addition of a tenth observation at (8,8) they become 0.96 and 0.24, respectively. An outlier can also reduce the correlation (Figure 4.7b). The Pearson and Spearman

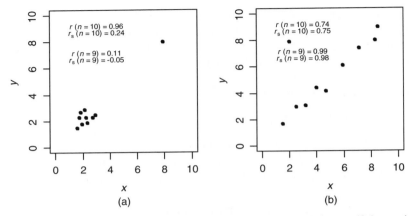

FIGURE 4.7   Effect of outliers on correlation; r is the Pearson correlation coefficient; r_s is the Spearman correlation coefficient.

correlations for the nine observations along a northeast sloping line are 0.99 and 0.98, respectively. However, a tenth observation in the northwest corner of the plot causes the correlations to be reduced to 0.749 and 0.75, respectively. We leave it as an exercise for the reader to see why there is agreement between both types of correlation in Figure 4.7b.

Normal Probability Plot   There are a number of ways to aid in ascertaining the distribution of the error structure, including the histogram, the boxplot, and numerous tests of normality. However, the graph the authors prefer is a *normal probability plot*, which is a special case of a q-q (quantile–quantile) plot, with the quantile on the vertical axis being the data and the quantile on the horizontal axis being the normal scores. A normal score transformation works as follows. Let $e_1, \ldots, e_n$ be the data set. In this example the e_i's are Studentized residuals, however, they can be any variable. A *normal score* $q_{(i)}$ is a value on the standard normal axis such that $\Pr(Q \le q_{(i)} | Q \sim \text{normal}) = (i - 0.25)/(n + 0.25)$. The 0.25 used in this formula is a correction factor. In the ash–Btu regression example, $n = 62$. The normal score $q_{(1)}$ is found by solving $\Pr(Q \le q_{(1)} | Q \sim \text{normal}) = (1 - 0.25)/(62 + 0.25) = 0.012$. Thus, $q_{(1)} = -2.257$ since $\Pr(Q \le -2.257 | Q \sim \text{normal}) = 0.012$. The remaining $q_{(2)}, \ldots, q_{(31)}$ are found in a similar fashion. Now sort the e_i's in ascending order, denoted by the subscripted parentheses, and plot $(q_{(i)}, e_{(i)})$, $i = 1, \ldots, 62$. If the observations plot close to a straight line, one can reasonably assume that the data, in this case the e_i's, are normally distributed. Statistically significant deviations from normality usually occur in the tails of the distribution. A correlation coefficient between $e_{(i)}$ and $q_{(i)}$ can be computed and tested for significance (Johnson and Wichern, 2002).

The normal probability plot of studentized residuals for the Btu–ash SLR model is shown in Figure 4.8a. For comparison, a normal probability plot of 62 observations from a normal distribution is shown in Figure 4.8b, and a normal probability plot of 62 observations from a lognormal distribution is shown in Figure 4.8c. Recall that the lognormal is a right-skewed distribution. The residual plot (Figure 4.8a) and the plot

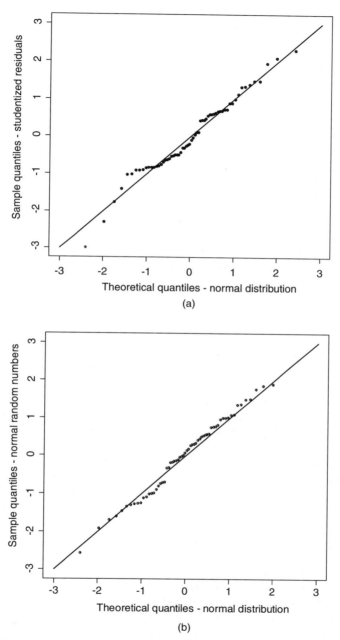

FIGURE 4.8   (a) Normal probability plot of studentized residuals from the Btu–ash SLR model. (b) Normal probability plot of a sample of size 62 from a normal population.

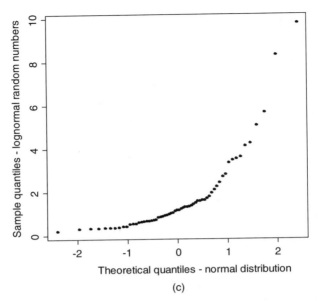

FIGURE 4.8 (*Continued*) (c) Normal probability plot of a sample of size 62 from a lognormal population.

using a sample from a normal distribution (Figure 4.8b) are similar. Normally distributed data will lie on the solid line to within sampling variability. The normal probability plot made using data from a lognormal distribution (Figure 4.8c) shows significant departure from a straight line. Only data near the center of the distribution lie close to the line. A plot such as that observed in Figure 4.8c usually indicates a sample from a skewed distribution; occasionally, it may reflect a mixture of populations. If lognormal data are suspected, a log transformation is justified, followed by a normal probability plot.

The main reason for the use of the normal probability plot in regression is to see if the normality assumption is valid, because this assumption is required for testing model significance, computing confidence estimates on the model parameters, and other tests. Normal probability plots can also help to identify outliers and may serve as a guide to suggest a transformation when departures from normality are observed.

Violations of Assumptions Illustrated by Residual Plots Residual plots help verify model assumptions. One critical assumption is homoscedasticity of errors. A violation is shown in the artificial data constructed for Figure 4.9. The fan-shaped pattern indicates that the variance appears to increase as $\widehat{Y}$, the fitted Y, gets larger. This can occur when the measuring device error is calibrated as a percentage of the quantity being measured. These errors can also occur in time series, due to a change in a measurement device or station location.

The expectation is that a model will account for the systematic variability in the data. Patterns that are observed in a residual plot may indicate that the model has not

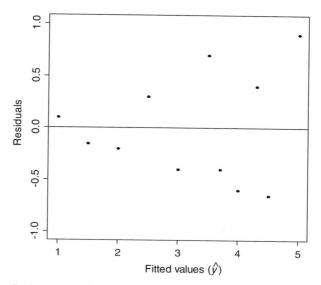

FIGURE 4.9   Residual plot showing a violation of the assumption of homoscedastic errors.

accounted for all of the important systematic variability in the data. An example of a systematic pattern of residuals is shown using the artificially constructed data in Figure 4.10. This pattern of residuals suggests that a quadratic term is needed in the model. Therefore, a better candidate model may be $Y = \beta_0 + \beta_1 X + \beta_2 X^2 + \varepsilon$.

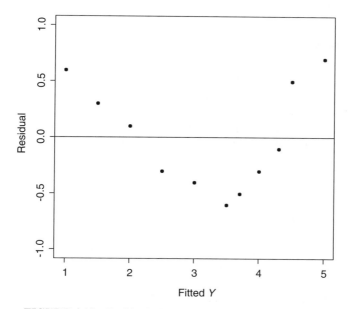

FIGURE 4.10   Residual plot showing systematic variability.

Time Series   Time series analysis is discussed in Chapter 5. However, in some regression applications, time (or distance) is an explanatory variable. Such models include precipitation over time, nitrogen content of soil along a transect, or geophysical measurements. When time or distance is an explanatory variable, nearby responses (precipitation or nitrogen content) are often correlated. Spatial correlation is a violation of the independence of errors assumption. When correlated errors exist, a different model of the error structure needs to be formulated. If correlated errors are suspected, the Durbin–Watson statistic (Durbin and Watson, 1950) on the residuals should be used. Problems of correlated errors are covered in more detail in Chapters 5 and 6. Even when time is not thought to be important enough to be incorporated into the model as an explanatory variable, the response variable should be plotted against time.

4.4.5   Transformation of the Response Variable

Until now errors have been assumed to be normally distributed. What is to be done if the Y's are skewed? As noted earlier, the distribution of many earth science variables are right-skewed, so it is sometimes necessary to make a log or other transform in the response variable prior to analysis. We recommend that transformations be used judiciously because interpretation is difficult in the transformed space and back-transforming estimates to the original space can result in bias. Occasionally, the X's and Y's need to be transformed if both are highly right-skewed. When the data are believed to be lognormal, the appropriate transformation is $W = \log(Y)$. However, for other right-skewed data, a Box–Cox or root transformation should be considered, especially when one or more of the Y's are zero.

4.4.6   Extrapolating

Extrapolating is predicting outside the domain of the data. In the Btu–ash example, the domain of ash is from 4.40 to 20.36%. Can the Btu for an ash content outside this domain reasonably be predicted? The answer is, perhaps. Such a prediction relies heavily on the assumption that the same linear relationship that is observed holds outside the domain of ash. This is a risky assumption unless it can be supported by theoretical considerations or analogy. The relationship between Btu and ash may be nonlinear outside the observed domain.

4.4.7   Confidence Intervals

The regression model $\widehat{Y}_i = \widehat{\beta}_0 + \widehat{\beta}_1 X_i$ can be used to estimate the mean of Y at X_0, denoted by $\widehat{\mu}\{Y \mid X_0\} = \widehat{\beta}_0 + \widehat{\beta}_1 X_0$. It can also be used to estimate Y for a single occurrence of X_0 denoted by $\widehat{Y}_0 = \widehat{\beta}_0 + \widehat{\beta}_1 X_0$. Point estimates will be the same in both situations. Confidence intervals are illustrated for each situation.

Confidence Interval on an Estimate of the Mean   Suppose that it is desirable to estimate the mean Btu and corresponding CI at 12% ash. Think of this as a mean that will be obtained from repeated sampling of Y at $X_0 = 12$. The point estimate of the mean result obtained at $X_0 = 12$ into the fitted model is $\hat{\mu}\{Y \mid X_0\} = 12,687$ Btu. A confidence interval on $\mu\{Y \mid X_0\}$ is constructed as follows. First, the standard error

$$SE(\hat{\mu}\{Y \mid X_0\}) = \hat{\sigma}\sqrt{\frac{1}{n} + \frac{(X_0 - \overline{X})^2}{SSX}}$$

where $SSX = \sum_{i=1}^{n}(X_i - \overline{X})^2$ and $\hat{\sigma} = \sqrt{MSE}$. $\hat{\sigma}$ is an estimate of the error standard deviation, sometimes referred to as the *residual standard error*. The $(1 - \alpha)100\%$ confidence lines are therefore

$$\hat{\mu}\{Y \mid X_0\} \pm t_{\alpha/2,\, df}SE(\hat{\mu}\{Y \mid X_0\})$$

The df used in constructing confidence intervals and hypothesis testing in regression are almost always the df error, which in this example is 60. The standard error of the estimated mean for the Btu–ash example is

$$SE(\hat{\mu}\{Y \mid X_0 = 12\}) = 381.9\sqrt{\frac{1}{62} + \frac{(12 - 8.926)^2}{475.8}} = 72.45$$

The 95% confidence interval on $\{Y \mid X_0\}$ for $X_0 = 12$ is

$$12,687 \pm (t_{0.025,60})(72.45)$$

$$12,687 \pm (2.000)(72.45)$$

$$(12,542,\ 12,832)$$

A frequentist interpretation is that there is a 95% chance that the interval (12,542, 12,832) brackets the true mean at $X_0 = 12$. In Figure 4.11 the CI is narrowest at an ash value of 8.926%, its mean, and increases for smaller and larger values of ash.

Prediction Interval The *prediction interval* is a confidence interval on an estimate of an observation. Suppose that one wants to determine the confidence interval on a single Btu response corresponding to an $X_0 = 12\%$ ash. The point estimate is the same as in the previous example, 12,687 Btu, but the confidence interval on the latter estimate is wider than the former because of the added uncertainty associated with the estimate of a single point. The standard error is

$$SE(\hat{Y}_0) = \hat{\sigma}\sqrt{1 + \frac{1}{n} + \frac{(X_0 - \overline{X})^2}{SSX}}$$

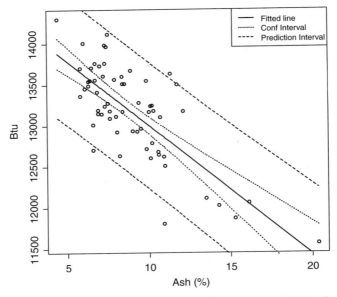

FIGURE 4.11   Fitted regression line, 95% confidence interval, and prediction interval on the mean, Btu–ash regression example.

Consider the standard error rewritten as

$$SE(\widehat{Y}_0) = \sqrt{\widehat{\sigma}^2 + \frac{\widehat{\sigma}^2}{n} + \frac{(X_0 - \overline{X})^2 \widehat{\sigma}^2}{SSX}}$$

The error components reflect the variability associated with a single occurrence of a random variable (Y_0), its mean $(\overline{Y}_0)$, and the location of X_0 relative to $\overline{X}$. The variability associated with an estimate of Y at $X_0 = 12\%$ ash is

$$SE(\widehat{Y}_0) = 381.9 \sqrt{1 + \frac{1}{62} + \frac{(12 - 8.926)^2}{475.8}} = 388.7$$

which is over five times the standard error that results from an estimate of the mean. Thus, the 95% confidence interval on Y_0 for $X_0 = 12$ is

$$12,687 \pm (t_{60,0.025})(388.7)$$
$$(11,907, 13,464)$$

a larger interval than for the estimate of the mean (Figure 4.11).

The prediction interval is appropriate when the interest is in the resultant value, in the example above it is Btu for a given value of ash. Consider an example of estimating how high to build a levy by examining historical flood data. The mean high-water mark at a given point in the river may not be of much value. The interest is in a confidence interval (one-sided in this case) for the height of floodwater at a given location as opposed to a confidence level on a mean.

4.5   MULTIPLE REGRESSION

Frequently, there is more than one explanatory variable. To address this problem, a *multiple regression* model $Y = \beta_0 + \beta_1 X_1 + \cdots + \beta_k X_k + \varepsilon$ is formulated, where k is the number of explanatory variables. Many of the model procedures for formulation, fitting, and evaluation are the same as those in SLR.

4.5.1   Model Formulation

In the Btu–ash example, suppose that moisture content, sulfur, and mercury are also included as explanatory variables. The new model is formulated as

$$Y = \beta_0 + \beta_1 X_1 + \beta_2 X_2 + \beta_3 X_3 + \beta_4 X_4 + \varepsilon$$

where X_1 represents ash (%), X_2 moisture (%), X_3 sulfur (%), and X_4 mercury (ppm). A set of scatter plots called a *matrix plot* or *splom plot* is shown in Figure 4.12. In addition to the ash–Btu correlation, there appears to be a weak correlation between Btu and moisture.

4.5.2   Estimation

Least squares is used to estimate model parameters and error. The output from multiple regression is similar to that of simple linear regression. The estimates for β_0, the intercept, and β_1, β_2, β_3, and β_4 are shown in Table 4.4, where, as before, the response variable is Btu. Corresponding to these estimates are the standard error and p-value resulting from H_0: $\beta_i = 0$ versus H_a: $\beta_i \neq 0, i = 0, 1, \ldots, 4$. Recall that a small p-value indicates statistical significance. The intercept, ash, and moisture are significant at less than 0.0005. Sulfur and mercury are significant at the 0.015 and 0.605 levels, respectively. Mercury does not explain any statistically significant component of Btu. Because the intercept β_0 is statistically significantly different from zero, the regression line almost certainly does not pass through the origin.

4.5.3   Model Evaluation

Model evaluation is a critical component of the modeling process. It is accomplished in part by statistics and graphics that have been used to formulate the model. Also, it is highly desirable to use new data to examine model performance.

ANOVA Table   The ANOVA table (Table 4.4) provides basic information on the goodness of fit of the model. The partitioning of variability is the same as in SLR: that $SST = SSR + SSE$. However, SSR is the sum of squares accounted for by all four of

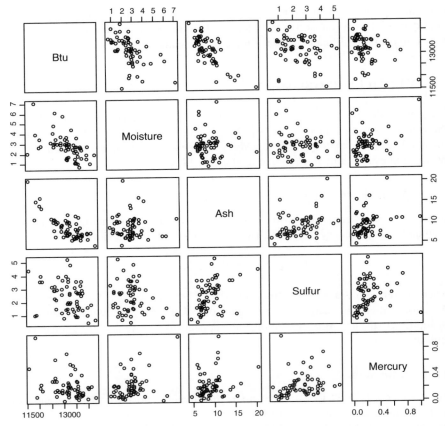

FIGURE 4.12   Matrix plot of multiple regression variables, Pittsburgh coal quality case study (Table 4.1).

the explanatory variables. A measure of model significance is again given by the F-statistic $= \text{MSR/MSE} = 114.68$, where $\text{MSR} = 4{,}558{,}211$ and $\text{MSE} = 39{,}746$. Since the p-value is less than 0.0005, the model is highly significant. The model sum of squares ($\text{SSR} = 18{,}232{,}843$) is partitioned into components associated with each explanatory variable [i.e., SSR (ash) $= 11{,}746{,}912$]. The fractions of model variability accounted for by the explanatory variables are: ash $11{,}746{,}912/18{,}232{,}843 = 0.644$, moisture 0.342, sulfur 0.013, and mercury 0.001. The variability accounted for by each variable is dependent upon the order that the variable enters the model since these explanatory variables are not independent of each other.

Graphs and Regression Diagnostics   Graphs are an essential component of model evaluation. Some graphs are the same as those used in SLR, including a plot of Y_i versus $\widehat{Y}_i$ (Figure 4.13a) and $\widehat{Y}_i$ versus e_i (Figure 4.13b). The residuals plotted against the fitted values (Figure 4.13b) are examined for patterns, outliers, influence points,

TABLE 4.4   Multiple Regression Output Using the Pittsburgh Coal Quality Case Study Variables (Table 4.1)

Residuals

Minimum	Q_1	Median	Q_3	Maximum
-333.20	-156.54	-42.18	117.15	473.43

Coefficients

	Value	Std. Error	p-Value
(Intercept)	15,423.32	111.25	0.000
Ash	-150.98	10.16	0.000
Moisture	-280.04	23.17	0.000
Sulfur	-64.81	25.94	0.015
Mercury	79.93	153.52	0.605
Residual standard error		199.40	
R^2		0.89	
R^2_{adj}		0.88	

ANOVA Table ($n = 62$)

Terms	Sum of Squares	df	Mean Square	F-Value	p-Value
Model	18,232,843	4	4,558,211	114.68	0.000
Ash	11,746,912	1	11,746,912	295.55	0.000
Moisture	6,237,884	1	6,237,884	156.95	0.000
Sulfur	237,272	1	237,272	5.97	0.018
Mercury	10,775	1	10,775	0.27	0.605
Residuals	2,265,502	57	39,746		
Total	20,498,345	61			

and correlations that may indicate violations of model assumptions. The curved line is from a lowess regression, a local smoother to be explained shortly, which is designed to help identify possible trends in residuals as a function of fitted values. The mild curvature of the lowess regression (Figure 4.13b) is not a cause for concern. The scatter appears random. Observations are considered to be high leverage if the leverage value $h_i > 4(p + 1)/n = 0.32$, where p is the number of explanatory variables. Only observation 62 satisfies this criterion. Observations 42 and 60 have marginally high values (Figure 4.13b). The highest values of Cook's distance (Figure 4.13c), observation 27, is only 0.25 and thus is not large enough to affect model estimates significantly. DFBETAs are computed for the four explanatory variables and the regression intercept. There are none for which |DFBETA| is greater than 1. A normal probability plot on residuals (Figure 4.13d) helps verify the normality assumption of the errors, namely that the residuals lie on the 45° line to within sampling error. As noted, a significant departure may indicate that distribution of the residuals (estimated errors) is skewed or outliers are present. There

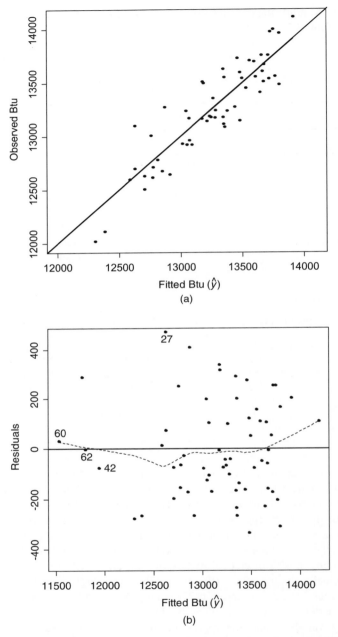

FIGURE 4.13   (a) Observed Btu versus fitted Btu from multiple regression model, Pittsburgh coal quality case study. (b) Residual e_i's versus fitted Btu for the multiple regression model, Pittsburgh coal quality case study; a lowess smoother is indicated by the dashed line.

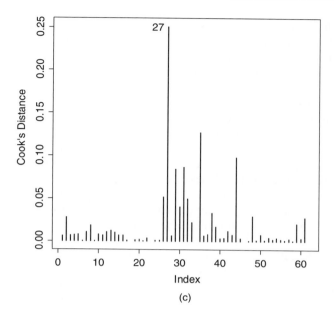

(c)

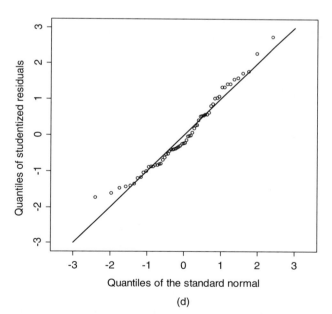

(d)

FIGURE 4.13 (*Continued*)   (c) Cook's distance for the multiple regression model, Pittsburgh coal quality case study. The horizontal axis is the observation number. (d) Normal probability plot of residuals for the multiple regression model, Pittsburgh coal quality case study.

is a slight departure from the line corresponding to the largest negative residuals, indicating that perhaps the left tail of the distribution is not as long as that of a normal; however, this deviation is judged to be too small to be significant. None of the observations noted seems troublesome enough to invalidate the model results, but additional investigation may be warranted. No values of DFFIT (not shown) exceed the cutoff of $|\text{DFFIT}_i| > 2\sqrt{(p+1)/n} = 0.58$.

In SLR, the relationship between X and Y can be seen by constructing a scatter plot. In multiple regression, plotting Y versus the $X_j, j = 1, \ldots, p$ explanatory variables is also recommended. A way to show multiple pairwise scatter plots on the same graph is by using a matrix plot (Figure 4.12). However, scatter plots in multiple regression can be misleading because a plot of Y_i versus $X_{ij}, j = 1, \ldots, p$ for the jth variable, excludes the effects of the other $p - 1$ variables included in the model. A way to show the relationship between X_j and Y, conditioned on the other $p - 1$ variables in the model, is by use of a partial residual plot (also called a *partial regression* or *component plus residual plot*). It is a plot of $e_i + \widehat{\beta}_j X_{ij}$ versus $X_{ij}, i = 1, \ldots, n$, where e_i is the ith residual from the full model and $\widehat{\beta}_j$ is the estimate from the full model. The simple linear regression of X_j (the explanatory variable) on $e + \widehat{\beta}_j X_j$ (the response variable) has the same slope as that of the full model. A scatter plot of sulfur versus Btu is shown in Figure 4.12. The corresponding partial residual plot is shown in Figure 4.14. The

> A *partial residual plot* attempts to show the relationship between the response variable and an explanatory variable in the context of a model.

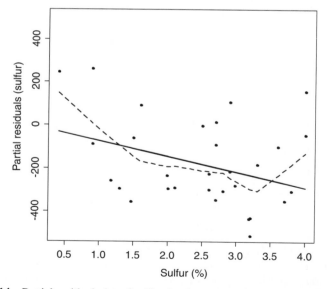

FIGURE 4.14   Partial residual plot of sulfur for the multiple regression model, Pittsburgh coal quality case study; the solid line is the linear regression of partial residuals on sulfur; the dashed line is a lowess smoother.

curved dashed line in this display is a lowess smooth and "follows the data." (Lowess smoothing will be explained soon.) The parabolic nature of this curve suggests that the relationship between sulfur and Btu given the other variables may be nonlinear, and an additional term that is a function of sulfur may be needed. However, as this is a small data set, the model as formulated appears to be reasonable. The straight line is a linear fit and shows Btu declining as sulfur increases in the context of the full model.

Overfitting the Data   Consider a small subset of the Pittsburgh coal quality case study data (the circles shown in Figure 4.15). One could naively choose to fit the data with a polynomial model

$$Y = \beta_0 + \beta_1 X + \beta_2 X^2 + \beta_3 X^3 + \beta_4 X^4 + \varepsilon$$

where in this example Y is Btu and X is ash. Not surprisingly, since there are five observations, a fourth-order polynomial accounts for 100% of the variability in Y and therefore the residuals are all zero. The fitted model is

$$\widehat{Y} = 457,650 - 258,910X + 54,390X^2 - 4890X^3 + 159X^4$$

and the curve it traces is shown in Figure 4.15. While the model fits perfectly at the data points observed, it is clearly inappropriate elsewhere, especially from 9

> *Overfitting* occurs when random variation in the data is modeled.

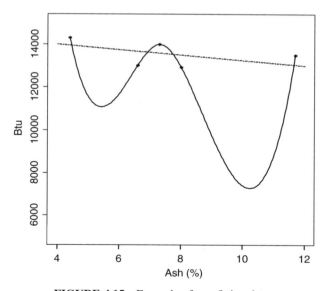

FIGURE 4.15   Example of overfitting data.

to 11% ash, unless the data are without error, and this model can be supported by other considerations. This small data set appears to support at most a SLR model (dashed-dotted line, Figure 4.15). Although the reader may not make this mistake, it is easy to add terms to a model, perhaps X^2, $\log(X)$ or $\cos(X)$, until it has been overparameterized. There are several ways to protect against this problem.

Cross-Validation   An important goal of most modeling efforts is prediction. One way to evaluate prediction is to plot the fitted response versus the observed response, as in Figure 4.13a. The points should plot along a 45° line to within sampling error. A way to summarize these data is to compute the sum of squared due to prediction (SSP$_r$), defined as

$$\text{SSP}_r = \sum_{i=1}^{n_j} (Y_i - \widehat{Y}_i)^2$$

and its normalized equivalent, the mean-square error due to prediction (MSP$_r$), where $\text{MSP}_r = \text{SSP}_r/n$. The $\{Y_i, i = 1, \ldots, n\}$ are the same data as those used to estimate the model. The subscript r refers to resubstitution of these data into the model. It is also referred to as the *apparent error* (Davison and Hinkley, 1997). An obvious problem with MSP$_r$ is that it is biased downward. Two procedures, which reduce the chance of overfitting the data, are holdout and cross-validation. Both involve hiding a portion of the data set, which is not used in model fitting. Splitting data sets is a first step toward the world of data-intensive computation and robust analysis, which include jackknifing and bootstrapping.

In the *holdout* or *split-sample procedure*, the data set is split randomly into two parts. One subset is used as the training set to initialize model parameters and the other subset is used to compute errors of prediction, defined as $e_{Pi} = Y_{Pi} - \widehat{Y}_{Pi}$, where Y_{Pi} and $\widehat{Y}_{Pi}$ are the observed and predicted response variables in the prediction set. The problem with this procedure is that results can be influenced by the way in which the data are split.

A better procedure is *cross-validation*. The procedure is outlined and then illustrated using the Pittsburgh coal quality data:

1. Divide the data into K subsets of approximately equal size. This is usually done randomly. If there is concern about ensuring that each subset is similar according to some criterion, a statistical design can be used to partition the data. The number of model parameters is a factor in selecting K because the desire is to maintain as many degrees as possible for estimating the error. A split that results in fewer than 30 observations in the training set is not recommended.

2. Each subset is removed from the sample data and a prediction made using the remaining data. Thus, there are K models. Corresponding to each model is a data set not used in that estimation.

3. Estimate the response variable in the corresponding omitted data set for each model.

4. The prediction error is an average of results made from each of the k "omitted subsets."

The advantage of this procedure, called *K-fold cross-validation*, is that all the data are ultimately used in estimation. The *sum of squares due to prediction* (SSP_j) is predicted on the jth omitted data set,

$$SSP_j = \sum_{i=1}^{n_j} (Y_{ij} - \widehat{Y}_{i[j]})$$

where Y_{ij} is the ith observation from data set j, $\widehat{Y}_{i[j]}$ is the fitted response variable using the explanatory variables from the jth (the omitted) subset (denoted by brackets), and n_j is the number of observations in the jth data set. Therefore,

$$SSP_K = \sum_{j=1}^{K} SSP_j$$

and $MSP_K = SSP_K/n$.

For the Pittsburgh coal quality case study data (Table 4.1), the randomly chosen split is shown in Table 4.5. As before, the response variable is Btu, and the explanatory variables are ash, moisture, sulfur, and mercury. The multiple regression model is fit separately to each data set (group). The $SSP_5 = 2,921,914$ and $MSP_5 = 47,128$. Contrast this with the apparent error $SSP_r = 2,265,502$ and $MSP_r = 36,540$, both slightly lower, as expected, than the fivefold cross-validation results. Of course, results may still be a function of the observations selected in the K groups. One way to address this problem is to repeat the K-fold cross-validation and average results. A special case of this procedure is the *leave-one-out cross-validation* (LOOCV), which uses n training sets of size $n - 1$. So, for example, if Y_j is left out, the model will be estimated using the other $n - 1$ observations and the predicted value $\widehat{Y}_{[j]}$ compared with Y_j. The latter procedure may be more computationally intensive than the K-fold procedure, is less biased, but may have greater variance.

TABLE 4.5   Observations by Group: Fivefold Cross-Validation

Group	Observations													
1	5	45	4	17	54	18	58	35	28	24	34	62		
2	49	25	15	27	44	3	57	36	48	2	8	59		
3	32	30	50	11	56	47	16	7	12	51	33	23		
4	6	39	20	53	1	42	55	61	37	10	22	38		
5	29	52	40	43	19	41	46	14	13	9	21	26	31	60

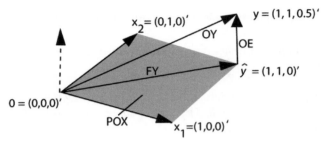

FIGURE 4.16   Projection of **y** into the plane of the **x**'s.

4.5.4   Geometric Interpretation of Multiple Regression

Regressing Y on X is a projection of Y into the space of X. To visualize this projection, consider the following simple example with $n = 3$ observations. The sample model is

$$y_i = \beta_1 x_{i1} + \beta_2 x_{i2} + \varepsilon_i \qquad i = 1, 2, 3$$

For simplicity the intercept term is omitted. Previous initial exploratory scatter plots have been of (Y, X_1), (Y, X_2), and (X_1, X_2). To illustrate the geometry of multiple regression a plot is constructed in the three-dimensional sample space with the origin $0 = (0,0,0)'$, $\mathbf{y} = (1,1,0.5)'$, $\mathbf{x}_1 = (1,0,0)'$, and $\mathbf{x}_2 = (0,1,0)'$ (Figure 4.16). The prime following the parentheses indicates the transpose. In general, this plot is in $n > 3$ dimensions. The points $\mathbf{x}_1$, $\mathbf{x}_2$, and 0 form a plane, called POX. Recall that any three noncollinear points form a plane. The vector made by the 0 to **y** is $\overrightarrow{OY}$. A perpendicular (vector $\overrightarrow{OE}$) extends from **y** to the POX plane. The point in POX is $\hat{\mathbf{y}} = (1,1,0)'$: the fitted values due to regression. In vector notation,

$$\overrightarrow{OY} = \overrightarrow{FY} + \overrightarrow{OE}$$

which corresponds to SST = SSR + SSE. Note that $\overrightarrow{FY}$ and $\overrightarrow{OE}$ are at right angles: that is, orthogonal to each other. In this example, $R^2 = 0.89$, and e(residual) = $(\mathbf{y} - \hat{\mathbf{y}}) = (0,0,0.5)'$.

4.5.5   Dependency Between Explanatory Variables

In the Pittsburgh coal quality multiple regression example, variables are entered into the regression equation in the following order: ash, moisture, sulfur, and mercury. Suppose that the explanatory variables are inserted sequentially into the equation in a different order: for example; moisture, sulfur, mercury, and ash. Does one expect the percentage of SSR accounted for by each of these variables to be the same? Consider ash. The sum of squares accounted for by ash with the new ordering is 8,778,129 (Table 4.6). Previously, it was 11,746,912. Why is there such a difference? In the first model, 11,746,912 is the unconditional contribution that ash makes in reducing the variability of Btu because it is the first variable entered into the model. In the second

TABLE 4.6   Effect of Different Order of Entry of Explanatory Variables, Pittsburgh Coal Quality Case study

First Model		Second Model	
Variable Order	Sequential Sum of Squares	Variable Order	Sequential Sum of Squares
Ash	11,746,912	Moisture	5,775,620
Moisture	6,237,884	Sulfur	3,545,904
Sulfur	237,272	Mercury	133,190
Mercury	10,775	Ash	8,778,129
Total	18,232,843	Total	18,232,843

model, the 8,778,129 component of variability accounted for by ash is conditioned upon the amount of variability accounted for previously by moisture, sulfur, and mercury.

Another way to see this is to view the correlation among explanatory variables (Table 4.7). An overlap between the explanatory power of ash, moisture, sulfur, and mercury is indicated because all correlations among these variables are different from zero. Of course, in a sample some deviation from zero may be sampling variability; however, in a sample of size 62, four correlations are statistically significant, at least at the 10% level. Conversely, suppose that all the off-diagonal elements are zero, which implies an identity matrix. Does the order of entry of the variables matter? An example of what happens when two highly correlated explanatory variables are chosen is presented in the exercises.

An *identity matrix* has ones on the diagonal

In certain instances, the dependency among pairs of explanatory variables becomes so large (approaching a correlation of plus or minus one) that numerical instability is created in the model-fitting algorithms, and the standard errors of some parameters become unreasonably large. *Multicollinearity* occurs when two or more explanatory variables are collinear or nearly collinear. Variables that are highly collinear make almost identical contributions to reducing the variability in the response variable, so that the presence in a model of one, given the other, contributes

TABLE 4.7   Pearson Correlations Among Explanatory Variables (Lower Triangular Matrix) and p-Values (Upper Triangular Matrix), Pittsburgh Coal Quality Case Study

	Ash	Moisture	Sulfur	Mercury
Ash	1	0.834	0.001	0.068
Moisture	−0.027	1	0.200	0.013
Sulfur	0.411	−0.165	1	0.092
Mercury	0.233	0.315	0.216	1

almost nothing. Multicollinearity may not be a problem if the goal is to predict; however, it may be a problem if the goal is to understand the contribution of explanatory variables.

When regression is used in an exploratory manner to study observational data and numerous explanatory variables are included in the model, multicollinearity almost always results. The original U.S. Geological Survey Coal Quality Data Set for the Pittsburgh Coal Bed (Bragg et al., 1998) contains over 100 variables on each observation, some of which are highly collinear. For example, the correlation between sulfur and pyretic sulfur is 0.94. Dependencies can be introduced by including a term and its power—for example, X and X^2—in polynomial regression. Dependency can also be introduced by including a cross-product term such as $X_1 X_2$. Such terms may be appropriate, however, one should be aware of the introduction of dependencies.

An important diagnostic tool for assessing multicollinearity in regression is the *variance inflation factor* (VIF). Consider the model

$$Y = \beta_0 + \beta_1 X_1 + \beta_2 X_2 + \beta_3 X_3 + \beta_4 X_4 + \varepsilon$$

The VIF for the jth coefficient on centered data (see the guidelines below) is defined as

$$\text{VIF}(\widehat{\beta}_j) = \frac{1}{1 - \tilde{R}_j^2}$$

where $\tilde{R}_j^2$ is the centered coefficient of determination when variable X_j is regressed on the remaining independent variables. Thus, to determine, say, $\text{VIF}(\widehat{\beta}_2)$ in the model above, X_2 is regressed on (X_1, X_3, X_4), written briefly as $\text{VIF}_j = \text{VIF}(\widehat{\beta}_j)$. As a guide, VIF < 5 implies no collinearity problem, VIF between 5 and 10 indicates moderate collinearity, and VIF 10 or greater indicates a serious problem. For additional details on variance inflation factors, see the work of Belsley (1991) and Gros (2003). The VIFs for ash, moisture, sulfur, and mercury are, respectively, 1.2, 1.2, 1.3, and 1.2, implying that no collinearity problems exist. The reciprocal of VIF is called the *tolerance*.

The following options serve as a guide for reducing multicollinearity:

- Avoid the problem by designing the experiment. This is discussed in Chapter 9.
- Use theory to reduce the number of explanatory variables.
- Center the data. *Centered data* are obtained by subtracting a measure of location, usually the mean, from each observation.
- Remove highly correlated variables. For example, one may choose to include either sulfur or pyretic sulfur, as they are highly correlated.
- Use an automated system, such as stepwise regression, to select a smaller subset of explanatory variables. Stepwise regression is discussed later in the chapter.
- Use principal components analysis, which is a multivariate procedure described in Chapter 7, to reduce the number of explanatory variables by constructing a smaller number of linear composites.

4.5.6   Categorical Explanatory Variables

In the Pittsburgh coal quality case study, all of the variables are continuous. However, it is possible in multiple regression to include categorical variables. First, consider a simple example where a "type of coal" variable is present, the single explanatory variable is ash, and the response variable is Btu. The type of coal is lignite or not lignite. The model can then be represented as

$$Y = \beta_0 + \beta_1 X_1 + \delta_1 D_1 + \varepsilon$$

where X_1 represents the continuous variable, ash, and D_1 represents the categorical variable, type of coal. *Categorical explanatory variables* are sometimes referred to as *dummy variables* (Hardy, 1993). The type of coal can be coded arbitrarily as 1 if the sample is in lignite and 0 if it is not. Assuming that lignite is significant, δ_1 is expected to be negative because lignite has less heat content and the true models may look as shown in Figure 4.17. In this example the dummy variable is binary. Note that the only difference between the models with and without lignite present is the shift δ_1. The slope β_1 remains the same. Second, consider a more complex example where the type of coal has levels: lignite, subbituminous, bituminous, and anthracite being the explanatory variables. Since these categories reflect increasing heat content, a type of coal variable can be included in the model with $1 = $ lignite, $2 = $ subbituminous, $3 = $ bituminous, and $4 = $ anthracite. This may yield satisfactory results, but it assumes that the differences in Btu are the same from levels 1 to 2 as from 2 to 3 and 3 to 4. The advantage of this coding scheme is that only one degree of freedom is used. An

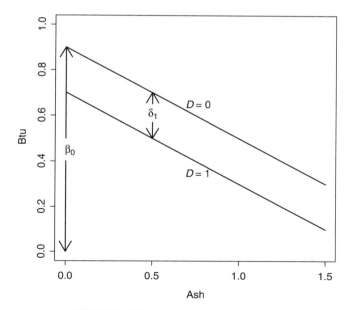

FIGURE 4.17   Dummy variable model.

alternative is to code these coal types as present and absent, using, say, a 0 for absent
and a 1 for present. So in this example, $D_1 = 1$ with lignite present and 0 when absent,
$D_2 = 1$ when subbituminous is present and 0 when absent, and $D_3 = 1$ when
bituminous is present and 0 when absent. When anthracite is present, $D_1 = D_2 = D_3$
$= 0$, so that the presence or absence of anthracite should not be coded. The major
disadvantage of this scheme is that 3 df are lost (one for each dummy variable), which
may be a problem for a small data set. However, for nominal data this is the
appropriate coding scheme. If coal type is the only explanatory variable, a model
with this coding scheme is of the form

$$Y = \beta_0 + \delta_1 D_1 + \delta_2 D_2 + \delta_3 D_3 + \varepsilon$$

If, for example, lignite is present, its effect will be added or subtracted from Y,
depending on whether δ_1 is positive or negative. In general, if the explanatory variable
has m levels, it is coded in $m - 1$ variables, each with two levels. Of course, as in the
previous example, continuous variables can be included in the model.

An alternative way to code dichotomous variables uses -1 or $+1$. For example, if
lignite is absent and δ_1 is positive, its affect will be subtracted from Y. The situation
where the response variable is categorical or ordinal is discussed in Chapter 8.

4.5.7 Including or Removing Variables from the Model Based on Significance Tests

When a multiple regression model is constructed, variables based on theory,
results of exploratory data analysis, and intuition are often included. It is common
that not all of these variables will be statistically significant. Before a model is put in
service, should these nonsignificant variables be eliminated? Here are some guidelines:

- If the sample size is large, omit variables that are not statistically significant.
- If the sample size is small, err on the side of retaining variables.
- If theory suggests that a variable should be present, err on the side of retaining it.

There is a potential loss associated with either action. The coefficient of mercury is
estimated to be 79.93 with an associated p-value of 0.605 (Table 4.4), clearly not
significant. The sign of the coefficient is positive, indicating that larger values of
mercury are associated with larger values of Btu. If, based on theoretical considerations, mercury is left in the equation and the relationship between Btu and mercury is
positive, we may have acted correctly. However, if it is left in and the true relationship
is negative (remember that the 95% confidence interval on the mercury coefficient
includes zero), the model has been weakened because a term has been included that
has the wrong sign. As a check, we recommend that the reader use judgment to
estimate the signs of the model coefficients prior to fitting the model. If a decision is
made to remove some variables, what is the next step? Options include reporting the
model using the statistically significant coefficients from the full model or reestimating the model after insignificant terms have been removed. There is controversy on

this subject. We recommend reestimating the parameters and redoing the model evaluation. However, one must always be aware that multiple passes through the same data sets change significance levels.

4.6 OTHER REGRESSION PROCEDURES

When a large number of candidate explanatory variables are present and little theory is available to guide their selection, an automated process to include and remove variables from the regression equation may be required. In addition, LS regression results may be suspect when assumptions are not satisfied. A complete discussion of these procedures is beyond the scope of this book, but some are mentioned and the reader is encouraged to compare LS with these procedures.

4.6.1 Variable Selection Procedures

A host of procedures and variants fall under the category of variable selection procedures. Backward selection, forward selection, and stepwise regression are presented.

Backward Selection   *Backward selection* begins with the full model: All candidate explanatory variables are included; then the least significant one is removed. Think about this in the context of the Pittsburgh coal quality case study, where the first model is

$$Y = \beta_0 + \beta_1 X_1 + \beta_2 X_2 + \beta_3 X_3 + \beta_4 X_4 + \varepsilon$$

Next, a comparison is made between the SSR value for the full model and that with X_i $i = 1, \ldots, 4$, removed. If the X_i variable that accounts for the smallest increase in SSR is not statistically significant, it is removed. For example, if X_4 is removed, the algorithm proceeds to compute three regressions by omitting X_1, X_2, and X_3 in turn. Stoppage occurs when all of the variables in the model are statistically significant.

Forward Selection   *Forward selection* is the inverse of backward selection. The algorithm evaluates four SLR models (in the example above). The explanatory variable that correlates most highly with the response variable is incorporated in the model. Next, the algorithm computes regressions containing two explanatory variables: the one just mentioned and all other variables. If the second variable added accounts for some statistically significant reduction in variability of the response variable, it remains in the model and then the algorithm proceeds to examine three variable regressions. When all the variables have been added or the variable added does not increase SSR by a statistically significant amount, the algorithm stops.

Stepwise Regression   *Stepwise regression* is a combination of the forward and backward procedures. The procedure typically begins as in forward selection; however, at the second and subsequent steps, the algorithm examines the variables that have been included to see if they are still statistically significant. If not, the algorithm removes them. The specific statistical tests used to enter and/or remove variables differ with implementation of the algorithm, but the user is usually afforded some control over the level of entry and removal.

Comments on Variable Selection   All variable selection procedures can yield useful results; all can give misleading results; and none is optimal. In most statistical packages, the variable selection procedures fall under the category of stepwise regression. Changing, entering, and removing a criterion can achieve backward and forward selection. We strongly suggest that when using a variable selection procedure, the final model be examined for consistency with theory. There are instances when, for theoretical or other considerations, variables retained by the model should be removed and those omitted from the model should be included.

4.6.2   Bootstrap Estimation

Bootstrapping is a resampling procedure used to obtain estimates of parameters (Davison and Hinkley, 1997). Its use is illustrated with the model $Y = \beta_0 + \beta_1 X + \varepsilon$. In contrast to a parametric model, where assumptions such as normality of errors and homogeneity of variance are required, bootstrapping does not require these assumptions. In small samples, these assumptions can be difficult to verify. Two formulations of bootstrap regression procedures are available, depending on the data and model. They are an assumption that the model is from a bivariate distribution and the model is designed with the X's assumed to be fixed in advance.

Model from a Bivariate Distribution   Suppose that the simple linear regression model originates from a bivariate sample of Btu and sulfur. Let B be the number of bootstrap estimates (replications) and n the sample size. A larger value of B yields more precise estimates. B should be at least 999. If this is considered a simulation experiment, B is akin to the number of times that a simulation will be run. Let $i_1, \ldots, i_n$ be index numbers for the $(x_1, y_1), \ldots, (x_n, y_n)$ pair, so, for example, i_3 refers to (x_3, y_3). Let i_k^* denote the kth bootstrap sample. The asterisk is used to represent bootstrap values. An algorithm to obtain bootstrap estimates of the parameters and error by resampling the (x, y) pairs is:

1. For $b = 1, \ldots, B$:
   a. Randomly sample $i_1^*, \ldots, i_n^*$ with replacement from $\{1, \ldots, n\}$. Since sampling is with replacement, the same (x, y) pair may be selected multiple times by chance.
   b. Let $x_j^* = x_{i_j^*}$ and $y_j^* = y_{i_j^*}$ for $j = 1, \ldots, n$; this generates a new set of $(x_j^*, y_j^*), j = 1, \ldots, n$.
   c. Fit least-squares regression to $(x_1^*, y_1^*), \ldots, (x_n^*, y_n^*)$, yielding estimates $\widehat{\beta}_{0,b}^*, \widehat{\beta}_{1,b}^*$, and $\widehat{\sigma}_b^{*2}$.
2. End the bth bootstrap loop. Save the estimates.
3. Compute statistics from the B bootstrap estimates.

The mean bootstrap estimates for the Btu example are $\widehat{\beta}_{0,b}^* = 14,625$ and $\widehat{\beta}_{0,b}^* = -161$ for $B = 1000$. Because this is a well-behaved data set, where the normality assumptions are reasonably well satisfied, the bootstrap estimates are the same as the least-squares estimates. This procedure makes no assumptions about the distribution or homogeneity of errors or about the conditional distribution of Y given $X = x$.

Designed Model   *Designed modeling* is sometimes called *model-based resampling.* Suppose that the X's are fixed in advance of obtaining the Y's and have negligible measurement error, implying that the variability is only in the Y's. The form of the model is $Y \mid X = \beta_0 + \beta_1 X + \varepsilon$, where the distribution of ε is assumed to be independent of X. Instead of resampling from the (x,y) pairs, as in the preceding example, resampling is from the distribution of modified residuals. Recall the modified residual $r_i = (y_i - \widehat{y}_i)/\sqrt{1 - h_i}$, where h_i is the ith leverage point.

The algorithm is as follows:

1. Fit the model to the data.
2. Compute the raw residuals, $e_i = y_i - \widehat{y}_i$, $i = 1, \ldots, n$ and corresponding h_i.
3. Compute the modified residuals r_ℓ.
4. For $b = 1, \ldots, B$:
   a. For $i = 1, \ldots, n$:
      i. Set $x_i^* = x_i$ to represent the bootstrap sample.
      ii. Randomly sample e_i^* from $r_j - \bar{r}$, $j = 1, \ldots, n$, where $\bar{r} = \sum_{j=1}^n r_i/n$.
      iii. Set $y_i^* = \widehat{\beta}_0 + \widehat{\beta}_1 x_i^* + e_j^*$.
   b. Fit a least-squares regression to $(x_1^*, y_1^*), \ldots, (x_n^*, y_n^*)$, yielding $\widehat{\beta}_{0,b}^*$, $\widehat{\beta}_{1,b}^*$, and $\widehat{\sigma}_b^{*2}$. Save the estimates.
5. Compute statistics for the B bootstrap estimates.

4.6.3   Robust Regression

Robust regression refers to a class of methods that are less sensitive to outliers and other assumptions than least squares regression. One concern with least-squares regression is the possibility of significant bias in estimated parameters due to the presence of outliers and departures from the normality assumption. In LS estimation, parameter estimates are those that minimize $\sum_{i=1}^n (Y_i - \widehat{Y}_i)^2$, where $\widehat{Y}_i = \beta_0 + \sum_{j=1}^k \beta_j X_{ji}$. This is a squared error or L_2 loss function (Figure 4.18, dotted line). In Figure 4.19, $L_2 = z^2/2$. As noted previously, a major criticism is that it assigns greater weight to high prediction errors. An alternative is the L_1 loss function, defined as $\sum_{i=1}^n |Y_i - \widehat{Y}_i|$. This, like the L_2 loss, is a convex function (Figure 4.18, solid line). In Figure 4.18, $L_1 = |z|$. An advantage over L_2 is that differences between relative errors are treated equally. A third alternative is the

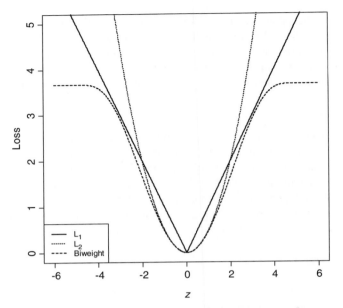

FIGURE 4.18 L_1, L_2, and Tukey biweight loss functions.

Tukey biweight function (Mosteller and Tukey, 1977), defined as

$$
\rho(z) = \begin{cases} \dfrac{k^2}{6}\left\{1 - \left[1 - \left(\dfrac{z}{k}\right)^2\right]^3\right\} & \left(\dfrac{z}{k}\right)^2 < 1 \\[2ex] \dfrac{k^2}{6} & \text{otherwise} \end{cases}
$$

where ρ represents the loss function and k is a tuning parameter. A k value of 4.685 produces 95% efficiency when the data are normal and can be used to adjust the upper bound on the loss consistent with the nature of the problem. A bounded loss function, the Tukey biweight, is shown by the dashed line in Figure 4.18. The upper bound on the loss function is 3.666. Contrast this with the upper bound on the L_1 and L_2 loss functions for $z = 4$. The loss function can be stated more generally as

$$
\min_{\beta_j} \sum_{i=1}^{n} \rho\left(\frac{Y_i - \widehat{Y}_i}{s_0}\right)
$$

where s_0 is a scale factor. This loss function finds parameter estimates, the $\widehat{\beta}_j$'s, that minimize the sum of square deviations over the parameter space, the β_j's. The choice of a loss function is subjective and problem-specific (Henning and Kutlukaya, 2007).

Robust procedures are insensitive to small departures from idealized assumptions. A procedure is called *resistant* if it is minimally influenced by outliers. Sometimes the terms *robust* and resistant are used interchangeably.

A variety of robust regression procedures have been developed (Andersen, 2008), each with somewhat different properties. As noted, an advantage a robust regression has over LS is the former's ability to retain but down-weight outliers. Typically, the user can control how much down weighting to allow, which of course introduces some subjectivity into the modeling process.

4.6.4   Local Regression Smoothing (Loess and Lowess)

An application of loess regression is the curved dashed line in Figure 4.14. *Loess regression* is a nonparametric smoother that "follows" the data. It fits a model $Y = f(X_1, X_2, \ldots, X_k) + \varepsilon$ without assuming a specific functional form such as a linear model. Each estimate of Y is a function of the X_i's in the neighborhood of the observation to be estimated, with the closest observations receiving more weight than the more distant observations. A loess smoother requires specification of a *span*, which is the percent or total number of observations to be included in the estimated regression at observation i. The weight at this observation is

$$w_i = \left(1 - \left| \frac{x - x_i}{d(x)} \right|^3 \right)^3$$

where x is the smoothed response, x_i is the nearest neighborhood along the span, and $d(x)$ is a distance along the span. The difference between a loess smoother and a *lowess smoother* is in the regression. The former uses a second-degree polynomial, whereas the latter uses a first-degree polynomial. Because these nonparametric models follow the data, they are useful as both an exploratory and evaluation tool to see if a model has been selected that is consistent with the data. Results of these smoothers should be interpreted cautiously when sample sizes are small because smoothers may follow influence points too closely, as opposed to following significant trends in the data.

4.7   NONLINEAR MODELS

With the exception of the loess (and lowess) regression, the models that have been discussed are linear models. That is, they are linear in the parameters and may be represented as

$$Y = \beta_0 + \sum_{k=1}^{p} \beta_k X_k + \varepsilon$$

Although these are still extremely useful, many relationships in the earth sciences are nonlinear.

The negative exponential equation

$$Y = \beta_0 e^{-\beta_1 X} \varepsilon$$

is a nonlinear model because the parameters β_0 and β_1 and the error ε enter into the model in a nonlinear manner. However, this model is *transformably linear*, because it is linear in the log

$$\log(Y) = \log(\beta_0) - \beta_1 X + \log(\varepsilon)$$

If the error term is assumed to be additive, as opposed to multiplicative, as in $Y = \beta_0 e^{-\beta_1 X} + \varepsilon$, the model cannot be linearized with a simple transformation but must rely on nonlinear estimation procedures. Logistic regression, where the response is binary, is another example of a nonlinear model to be discussed in Chapter 8.

How does nonlinear estimation work? Think of hiking up a high mountain with the goal of reaching the summit. The summit in a numerical problem will be achieved when the parameter estimates that minimize some loss function are found. For example, one starts up the slope, goes some number of paces, gets his or her bearings, and perhaps changes direction. This process is repeated many times until the summit is finally reached. Unlike multiple regression, a numerical solution is solved by a series of steps (iterations). The algorithm begins at some starting point; finds a direction to move, usually by computing a local slope; moves some distance; checks for a solution; and either stops or finds a new direction to move. The direction of a move is usually determined by a differential or difference equation. The specific steps in *nonlinear modeling* are outlined as follows:

1. Formulate the model.
2. Estimate an initial value for each model parameter. Often, this can be achieved using graphs, a search through the solution (parameter) space on a course grid, analogs, and/or expert judgment. Sometimes, reducing the model to a simpler approximation (e.g., a linear model) may assist in estimating reasonable starting values. It is rare that a wild guess at initial values will lead to convergence. More often, the algorithm will fail to converge. In most algorithms, it is possible to specify boundary values for parameters. For example, one may wish to specify that a parameter must be greater than zero.
3. Select a loss function. In LS, parameter estimates are found that minimize the sum of squares of residuals. This option can be used; however, often a log-likelihood function is chosen to be minimized. This is a function of parameters and data that when minimized yields parameter estimates that are "most likely" to have generated the given data set.
4. Select an estimating algorithm. Many options are available, including the Levenberg–Marquardt method, quasi-Newton method, quadratic programming, and a simplex procedure.
5. Select a step size by which to move: that is, to increment parameters. Typically, optimization algorithms start at the initial values specified and find a direction to move that minimizes the loss function. The question is: How far should it move? If the move is too short, convergence time will increase

significantly; if it is too long, the next point may be far away from the optimum. Many algorithms allow the user to specify the maximum move distance so that this does not happen.

6. Specify the maximum number of iterations for the algorithm. This prevents the algorithm from running forever.
7. Specify a convergence criterion. Typically, when the absolute or relative change in the loss function is within a specified tolerance value, the algorithm is said to *converge* (see the cautionary notes below).

Cautionary notes: No estimation procedure can guarantee convergence to an optimum solution for every data set and nonlinear model. Sometimes a secondary peak will be found. Sometimes convergence will be on a boundary. Occasionally, bifurcation or iterating back and forth without convergence will occur. Sometimes convergence will stop on a saddle point, ridge, or flat surface prior to finding a optimum solution. It is often useful to try different reasonable starting values and algorithms for a complex nonlinear model to ensure that they converge to the same value. As with any output, the investigator should apply a reasonableness check to the results. Bates and Watts (1988) and Seber and Wild (2003) provide a complete discussion of issues relating to nonlinear modeling and estimation.

A simple nonlinear model, called a *discovery process model* (Drew et al., 1995), is used to estimate the cumulative number of undiscovered oil or gas deposits Y for a measure of exploratory effort X. The model is of the form $Y = \beta_0(1 - e^{-\beta_1 X})\varepsilon$. An attempt is made to estimate model parameters from the data shown in Table 4.8. The data represent discovery of oil fields in a U.S. play (play P653, Schuenemeyer and Drew, 1996). Each field contains approximately 50 million barrels of recoverable oil. The data are graphed (Figure 4.19) to see the relationship between the variables and as a guide to

TABLE 4.8 Nonlinear Model Example: Cumulative Number of Oil Discoveries (y) Versus Cumulative Wildcat Wells (x)

x	y	x	y	x	y
153	1	834	13	2473	25
282	2	912	14	2635	26
325	3	972	15	2844	27
437	4	1000	16	3087	28
438	5	1124	17	3167	29
514	6	1263	18	3421	30
541	7	1402	19	3589	31
575	8	1485	20	4422	32
619	9	1613	21	4868	33
681	10	1884	22	5197	34
750	11	2044	23	5725	35
794	12	2419	24	6248	36

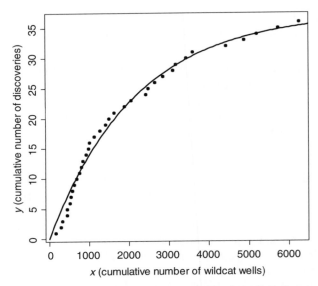

FIGURE 4.19 Nonlinear model example, cumulative number of oil discoveries (y) versus cumulative wildcat wells (x). The curved line is the negative exponential fit.

establishing initial conditions. It appears that the asymptotic y value may be approximately 40, so let $\widehat{\beta}_0^0 = 40$, where the superscript 0 represents the initial starting value. To obtain $\widehat{\beta}_1^0$, the model for $\widehat{\beta}_1 = -(1/X)\ln(1 - Y/\widehat{\beta}_0)$ is evaluated using an arbitrary pair of points on the curve. A reasonable choice is an X and corresponding Y near its midrange. $X = 3087$ and the corresponding $Y = 28$ are chosen. This yields $\widehat{\beta}_1^0 = 0.0004$. The model is run using the R function nls, minimizing the sum of squares of residuals using a Gauss–Newton optimization procedure. Parameter estimates $\widehat{\beta}_0 = 37.23$ and $\widehat{\beta}_1 = 0.00047$ are both highly significant and appear to be reasonable. The fitted model is shown in Figure 4.19. The residual standard error is 1.27 on 34 df. One final point is to be made with respect to this example. There appears to be an inflection point in the data (Figure 4.19) at $X = 500$. This is not captured in the discovery process model and requires the addition of another parameter. In viewing nonlinear processes in many disciplines, it is observed that values at the beginning of a process sometimes differ in form from those that occur after the process becomes established. In this example, a negative exponential process seems to describe the data for X greater than 500.

4.8 SUMMARY

Covariance and correlation are measures of linear association between two random variables with covariance in units squared and correlation, a scaled measure, in dimensionless units. These measures involve computation of sums of squares about means and are influenced by outliers. A nonparametric version of the usual Pearson correlation coefficient is Spearman's correlation coefficient.

Linear regression is a key modeling tool in almost all disciplines. It measures the association between a response variable and one or more explanatory variables. The response variable is a random variable. A quantity that remains after the regression model has been fit is known as a residual. For a model to be at least minimally adequate, the residuals must be random. In the evaluation of a regression model, it is important to use a variety of graphical and statistical procedures. Outliers and influence points can cause least-squares regression fits to be misleading. An important statistical tool in evaluation is the analysis of variance table, which displays the partitioning of the response variability between that accounted for by the regression model and that which is left over. Leave-one-out, K-fold cross-validation, and bootstrap procedures help evaluate the model's ability to predict accurately. When possible, a model should be tested on data not used to estimate model parameters. A valid forecast requires that the assumptions and model fit hold outside the domain of the data. Regression establishes association, not cause and effect.

Other regression procedures are available when the number of explanatory variables is large or standard (normal) assumptions may not hold. Variable selection methods attempt to find an "adequate" subset of explanatory variables to include in the model. Bootstrap, robust, and nonparametric estimation procedures allow a relaxation of standard assumptions.

EXERCISES

4.1 A regression model accounts for 90% of the variability in the response variable. Is this a good model? Explain.

4.2 Suppose that one has the ability to design an experiment that involves obtaining measurements on six explanatory variables. Should the design attempt to make these variables be uncorrelated? Why?

4.3 In SLR, suppose one finds that the slope of the regression line is zero to within sampling error. What does this say about the model?

4.4 In multiple regression, we want dependency between the response variable and the explanatory variables and independence between explanatory variables. Why?

4.5 In the discussion of cross-validation, it is noted that true cross-validation is better than a holdout procedure. In what sense is it better?

4.6 Consider the Pittsburgh coal quality case study (Table 4.1). Estimate three models. The response variable for all three is Btu. For model 1, the explanatory variable is sulfur. For model 2, the explanatory variable is pyretic sulfur (p-sulfur). For model 3, there are two explanatory variables, sulfur and p-sulfur. Compare and contrast the results of these models.

TABLE 4.9

x_1	y_1	y_2	y_3	x_4	y_4
10	8.04	9.14	7.46	8	6.58
8	6.95	8.14	6.77	8	5.76
13	7.58	8.74	12.74	8	7.71
9	8.81	8.77	7.11	8	8.84
11	8.33	9.26	7.81	8	8.47
14	9.96	8.10	8.84	8	7.04
6	7.24	6.13	6.08	8	5.25
4	4.26	3.10	5.39	19	12.5
12	10.84	9.13	8.15	8	5.56
7	4.82	7.26	6.42	8	7.91
5	5.68	4.74	5.73	8	6.89

4.7 Anscombe (1973) constructed the data set shown in Table 4.9. Run the following SLR with y's as the response variable and x's as explanatory variables: (x_1,y_1), (x_1,y_2), (x_1,y_3), and (x_4,y_4). Perform the appropriate diagnostics, including residual plots (residual versus fitted y). Note the R^2 values. Which model(s) are reasonable? Explain.

4.8 Run a robust regression on (x_1,y_3). Contrast the results with the LS results obtained in Exercise 4.7. Which should be reported? Why?

4.9 Explain why in Figure 4.7a there is a significant difference between the Pearson and Spearman correlation coefficients after adding the outlier, whereas no such difference exists after the outlier is added in Figure 4.7b.

4.10 Are influence points bad? Should they be retained or deleted? Why? What assumptions are implied if they are used? Explain.

4.11 In Figure 4.9, $(\widehat{y}_i, e_i)$ is plotted. Suppose that, instead, (x_i, e_i) is plotted. What will the plot look like? Explain. There is no need to replot the data.

4.12 Refer to Figure 4.16. Suppose that $\mathbf{y} = (1,1,0.1)'$ is observed instead of $\mathbf{y} = (1,1,0.5)'$. What will happen to R^2?

4.13 In a permafrost setting, down-hole depth and temperature measurements are obtained (Table 4.10). Compute the correlation between depth and T (temperature). Does depth cause a change in temperature? Explain.

TABLE 4.10

Depth (m)	T (°C)	Depth (m)	T (°C)
0.2	− 3.8	2.4	− 2.6
0.4	− 3.6	3.2	− 1.7
0.8	− 3.1	5.0	− 0.4
1.2	− 3.2	7.0	0.0
1.6	− 3.2	9.8	0.0

4.14 Refer to the data in Table 4.10. Perform a regression analysis assuming that T is the response variable. Perform appropriate diagnostics and comment.

4.15 Refer to Exercise 4.14. Estimate T at 11 m. Compute the error of the estimate. Is the estimate reasonable? Explain.

4.16 Refer to the data in Table 4.10. Do conclusions differ between the use of least squares as a model-fitting procedure versus a robust procedure? Which is best? Comment.

For Exercises 4.17 to 4.19, consider a subset of the 1987 Austfonna ice core chemical data (Table 4.11).

TABLE 4.11

pH	Depth(m)	HCO_3	Cl
5.25	9.32	2.75	0.80
5.53	9.54	3.05	0.31
5.25	11.12	2.90	0.65
5.31	11.45	2.90	0.70
5.27	11.74	2.80	0.79
5.55	15.98	3.21	0.71
5.46	17.18	3.05	0.88
5.46	22.45	3.13	0.71
5.29	24.41	2.90	1.85
5.64	29.96	3.21	0.51
5.55	33.48	3.05	1.17
5.53	35.40	3.05	1.06
5.41	35.76	3.05	1.68
5.46	38.75	2.98	1.22
5.55	40.87	3.05	1.03
5.52	41.50	3.05	1.07
5.59	42.62	3.05	0.81
5.48	43.31	2.98	0.90
5.42	45.31	3.05	0.76
5.43	46.02	2.98	0.86
5.55	47.78	3.05	0.23
5.75	48.53	3.59	0.57

Source: Data from ftp://ftp.ngdc.noaa.gov/paleo/icecore/polar/svalbard/aust87chem.txt.

4.17 The response variable is pH. The explanatory variables are depth, HCO_3, and Cl. Using graphical techniques, select a candidate model. Speculate on the signs of the coefficients.

4.18 Fit the model and comment on the ANOVA output.

4.19 Generate appropriate regression diagnostics and comment.

4.20 Fit the data in Table 4.8 assuming the exponential form discussed in Section 4.7. Examine the residuals. Do the normal assumptions hold? Display the residuals in the original transformed space and comment.

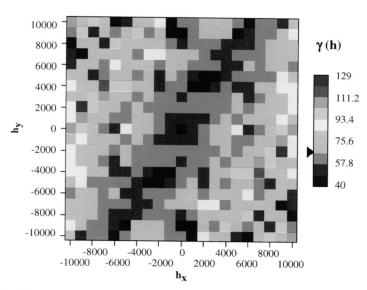

FIGURE 6.20 Semivariogram map for water-well yields in PDmg rocks, Pinardville Quadrangle. [The small discrepancy between this figure and the original in Plate 2C of Drew et al. (2003) was caused by the reclassifying of several wells.]

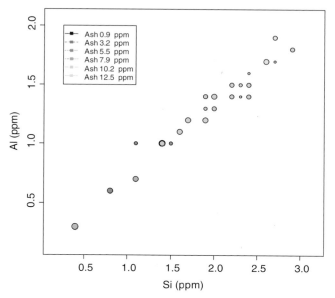

FIGURE 7.3 Bubble plot of Al versus Si with a circle representing the square root of sulfur in ppm. Ash is indicated by color.

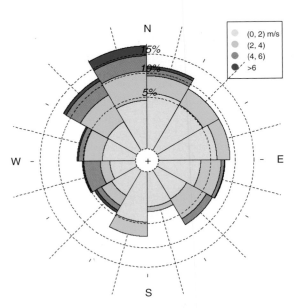

FIGURE 10.3 Wind rose diagram of wind direction and speed (m/s), Mesa Verde National Park, January 2006.

5 Time Series

5.1 INTRODUCTION

The analysis of data taken on a time scale plays a critical role in climatology, geophysics, seismology, geology, and other earth science disciplines. Time series can be temporal, spatial, or spatial–temporal. Time series data are used in short-term prediction, such as near-term weather forecasting or cost of winter heating oil, or long-term prediction, as in global climate change. Studies of climate change involve the examination of temperature, precipitation, and other atmospheric and surface conditions over time. Beyond the Earth, sunspots and other time-dependent activities are monitored extensively. The impact of clean air legislation and other antipollution efforts are evaluated by monitoring SO_2, NO_2, and other air quality emissions over time. Other scales of measurement, such as distance on the Earth's surface and depth below the surface, may be amenable to time series methodology. Geophysicists send signals into the Earth and measure the frequency and return time in order to "see" the undersurface environment and identify potential oil and gas deposits. Geologists, geomorphologists, and biologists often take observations along a transect to detect changes and patterns in soil conditions or plant life. Time series models are omnipresent in economic forecasting. In addition, time series methodology is used in a feedback mode to control such processes as the temperature of a laboratory.

Geophysicists and seismologists tend to work in the frequency domain (defined later). Climatologists usually work in the time domain. Figure 5.1 illustrates that these domains are dual. Sometimes the choice of a domain is arbitrary; for some problems, one domain can yield more insight than another.

In this chapter, methods of analyzing time series data observed at regular intervals over a long period are discussed. In Chapter 6, time series in the spatial and spatial–temporal domains are studied. A different phenomenon, called a *point process*, occurs when an observation of time and magnitude, such as an earthquake, volcanic eruption, or oil discovery contains a random component. The study of point processes requires a different approach from that used to analyze measurements at regularly spaced intervals and is discussed in Chapter 8.

An excellent introduction to time series analysis is presented in a book by Chatfield (2003). Brillinger (2000) has provided a general introduction, including a historical development and references.

Statistics for Earth and Environmental Scientists, By John H. Schuenemeyer and Lawrence J. Drew
Copyright © 2011 John Wiley & Sons, Inc.

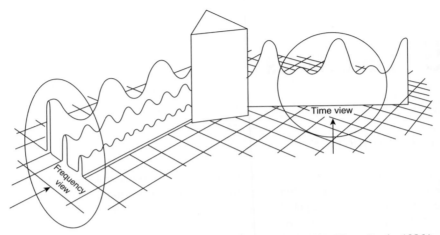

FIGURE 5.1   Duality between the time and frequency domains. (From Davis, 1986.)

5.2   TIME DOMAIN

The reasons for viewing and modeling a time series are similar to those in regression: namely, description, understanding, estimation, and forecasting. Some of the more important patterns in time series are discussed in relation to an example. We depart from the usual procedure of using a real data set so that the various components of a time series can be illustrated. Assume that 101 consecutive responses taken every 0.1 second (time units) are observed (Table 5.1). The time series components are illustrated in Figure 5.2.

In the *time domain*, typically the interest is in understanding and modeling one or more of the following three phenomena:

1. *Long-term trend* (Figure 5.2a). This may comprise small changes that occur over long time periods: for example, global warming. Long-term trends may be linear, exponential, or modeled by a low-order polynomial. Trends may be increasing, such as warming temperatures, or decreasing, as in the thickness of the Greenland ice sheet.
2. *Cyclical effect* (Figure 5.2b). Some cyclical effects are well known, such as annual periodicity in climate data, and may not be of scientific interest. Other cycles are less obvious and more important, such as possible periodicity between times of intense sunspot activity, floods, or droughts. Sunspots, which have an approximate 11-year cycle of activity, affect weather.
3. *Short-term structure* (Figure 5.2c). This may be correlation that occurs within a short time interval or distance relative to a long-term trend or cyclical effects. An example is the event of rain on day t given that it rained on day $t-1$. (Correlation as a measure of linear association between two random variables is

TABLE 5.1   Illustrative Data

Time (discrete units)	y_t	Time t (seconds)	Time (discrete units)	y_t	Time t (seconds)	Time (discrete units)	y_t	Time t (seconds)
0	0.47	0.0	34	0.12	3.4	68	0.81	6.8
1	1.06	0.1	35	0.47	3.5	69	0.00	6.9
2	0.56	0.2	36	0.81	3.6	70	0.68	7.0
3	1.05	0.3	37	0.34	3.7	71	0.33	7.1
4	0.98	0.4	38	0.93	3.8	72	1.10	7.2
5	0.59	0.5	39	1.24	3.9	73	0.40	7.3
6	1.10	0.6	40	0.93	4.0	74	1.19	7.4
7	0.64	0.7	41	0.59	4.1	75	0.34	7.5
8	1.08	0.8	42	0.79	4.2	76	0.73	7.6
9	0.79	0.9	43	0.64	4.3	77	0.20	7.7
10	1.14	1.0	44	0.70	4.4	78	0.46	7.8
11	0.07	1.1	45	0.28	4.5	79	0.79	7.9
12	1.30	1.2	46	0.65	4.6	80	0.19	8.0
13	0.47	1.3	47	0.37	4.7	81	0.34	8.1
14	0.68	1.4	48	0.23	4.8	82	0.42	8.2
15	0.65	1.5	49	0.55	4.9	83	−0.34	8.3
16	0.89	1.6	50	0.26	5.0	84	0.76	8.4
17	−0.16	1.7	51	0.79	5.1	85	0.18	8.5
18	0.59	1.8	52	−0.15	5.2	86	0.70	8.6
19	0.14	1.9	53	0.74	5.3	87	−0.05	8.7
20	0.83	2.0	54	−0.18	5.4	88	0.02	8.8
21	−0.13	2.1	55	−0.10	5.5	89	−0.24	8.9
22	0.53	2.2	56	−0.35	5.6	90	0.21	9.0
23	0.15	2.3	57	0.30	5.7	91	0.35	9.1
24	0.37	2.4	58	−0.17	5.8	92	−0.34	9.2
25	0.57	2.5	59	0.23	5.9	93	0.56	9.3
26	0.30	2.6	60	0.21	6.0	94	0.46	9.4
27	0.59	2.7	61	0.36	6.1	95	−0.62	9.5
28	1.01	2.8	62	1.02	6.2	96	0.34	9.6
29	0.44	2.9	63	−0.11	6.3	97	0.24	9.7
30	0.16	3.0	64	1.07	6.4	98	0.68	9.8
31	0.76	3.1	65	0.34	6.5	99	0.47	9.9
32	0.68	3.2	66	0.74	6.6	100	0.07	10.0
33	0.68	3.3	67	0.09	6.7			

explained in Chapter 4.) Short-term structure may also be referred to as *high-frequency variation*: that is, variation within a short period.

What makes the study of time series challenging is that these long-term trends, cyclical effects, and short-term structures are usually masked at least partially by random variation (noise). Indeed, the remainder (Figure 5.2d) should be random noise if the systematic effects are modeled properly. As in regression analysis, a high random noise component will make it difficult to detect a trend or other structural pattern.

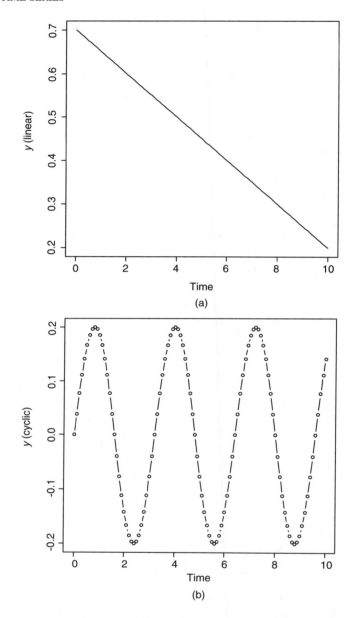

FIGURE 5.2 (a) Long-term trend, illustrative data. (b) Cyclical effect, illustrative data.

Time series in most earth science applications are the result of a continuous process. However, data are most often recorded in discrete, usually equally spaced, intervals. The intervals between measurements Δt may be nanoseconds, days, months, years, or eons, depending on the application. We follow the usual convention

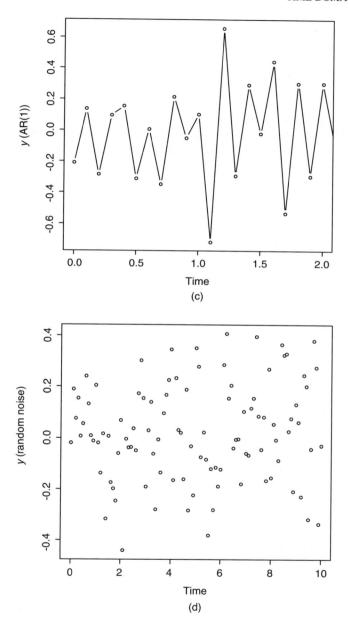

FIGURE 5.2 *(Continued)*   (c) Short-term (high-frequency) structure, illustrative data. (d) Random (white) noise, illustrative data.

in the literature and denote a time series by Y_t, where $t = 0, \pm 1, \pm 2, \ldots$ refers to the discrete time units, not the actual unit of measurement. When a time series is plotted, typically the actual units of measurement on the horizontal axis, such as second, day, month, or year, are used. When differencing of a series is discussed, for example $Y_t - Y_{t-1}$, time t will be an integer (as discussed above), not the original units. Although t is generally used to represent the time domain, in many earth science applications involving responses to distance along a transect, distance is substituted for time. This time series is constructed to contain a linear trend (Figure 5.2a), a cyclical trend (Figure 5.2b), short-term structure (i.e., patterns that occur repeatedly over a relatively short time interval or distance; Figure 5.2c), and random noise (Figure 5.2d).

Long-term effects such as global warming may also be cyclical (Figure 5.2b), repeating over periods of time such as years or centuries. Distinguishing between a long-term trend and cyclical effects is one of the challenges facing scientists who study ozone depletion and other climate phenomena. If only a small portion of a cycle is observed, it may not be possible to distinguish between the two effects. Of course, many time series, including this example, contain both long-term trend and cyclical components.

Climate and economic data often repeat quarterly and/or annually. Other series may have different periods, such as the approximate 11-year cycle of sunspot activity. Consider $Y_t = f(t)$, a function f of t. Y_t has period T if the series repeats itself every T time units. The sine wave shown in Figure 5.2b has a period of $T = 2$ time units. If a series repeats itself every T units, then $f(t + T) = f(t)$. The other term now introduced is *frequency*, which is the reciprocal of the period ($f = 1/T$), where the units are cycles per unit of time. Recall that f is also used as a function, as in $Y_t = f(t)$; however, the meaning will usually be obvious from the context. Cyclical phenomena are not always sinusoidal but may have a sawtooth or other shape.

In many time series, there are more complex and often more interesting structures than linear (long term) or cyclical trends. In this example (Figure 5.2c) there is a tendency for large and small values of y to appear in sequence. This autoregressive (AR) process is discussed in Section 5.2.5. In Figure 5.2c, only a small portion of the series is displayed, so that the short-term structure is more visible.

Figure 5.2d is an example of data generated from a Gaussian (normal) random number generator, in this case with mean 0 and standard deviation 0.20. Sometimes *random noise* is called *white noise*. After the structure is accounted for via a model, a time series plot of residuals should be random noise. This is analogous to what is expected to be seen in residuals after constructing a suitable regression model. Procedures to check for the presence of white (random) noise are discussed in model evaluation in the ozone measurement example.

The combined time series (Figure 5.3) is the sum of the four components shown in Figure 5.2. Although some long-term trends can be discerned from the time series plot in Figure 5.3, other structural effects are less obvious. Therefore, tools are presented that may be used to separate time series into the components shown in Figure 5.2. The modeling approach parallels those used in regression.

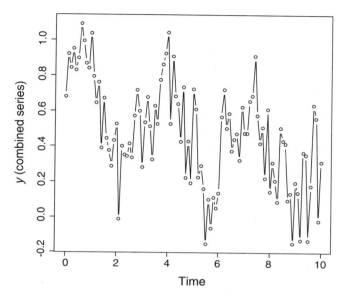

FIGURE 5.3   Combined time series, illustrative data.

5.2.1   Smoothing or Filtering

Smoothing or *filtering* (these terms are used interchangeably in the literature) allows long-term cycles and trends to be viewed by removing or attenuating noise and other high-frequency variation, such as that in Figure 5.2c and d. The type of filter that attenuates high-frequency signals is called a *low-pass filter* because it passes low-frequency signals. Most of the smoothing algorithms discussed in this section are low-pass filters. Analogous to a low-pass filter is a *high-pass filter*, which is used to reduce or eliminate low-frequency variation, such as a long-term trend. This reduction is needed to view and investigate short-term cycles or other structures. High-pass filters are discussed in Section 5.2.2 in the context of trend removal. Additional filters are discussed in Section 5.3.

The structure of a filter is as follows. Suppose that the data are (t, y_t), $t = 1, \ldots, n$. A smoothed estimate of y at the center point t_c, where $c = (m + 1)/2$, with m odd, is of the form

$$\tilde{y}_c = \sum_{i=1}^{m} w_i y_i$$

where there are $m \leq n$ consecutive observations from the original data series and the w_i are weights. Usually, weights are constructed so that they are maximum at the center observation t_c and decline toward the first and mth observations. The weights w_i are nonnegative and must sum to 1. For $m < n$ points, think of a smoother as a window of m points moving across the n points. There are several ways to smooth the endpoints and near endpoints, including using the original data values, y_1 and y_m (Chatfield, 2003). The degree of smoothing is a function of m and the w_i's; a large m

implies that more of the data series is used and produces a smoother series. A more uniform distribution of weights also produces a smoother series. The arithmetic mean results if all the weights are equal. The models described in this section are nonrecursive filters: that is, output is a function of current and previous values. A recursive filter also uses previous output values.

Moving-Average Filters   There are numerous moving-average filters or smoothers. A simple *moving-average filter* is one that results in an arithmetic average of three adjacent points taken across a time series. Four important classes of moving-average filters are kernel smoothers, exponential smoothers, median smoothers, and spline smoothers.

Loess   The loess, a locally weighted regression smoother, was described in Chapter 4. Its smoothness is controlled by a span parameter, which determines the size of the window. It is a low-pass filter. Loess smoothing with spans of 0.75 (Figure 5.4a) and 0.25 (Figure 5.4b) are shown using the data in Table 5.1. Span is the fraction of data used to estimate each point or *m/n* in the notation above. Note the loess curve with the span = 0.75 helps the user identify the longer-term trends, while span = 0.25 accentuates the cyclical effect. Both of these are low-pass smoothers because the signal that is passed (the solid line) has a long period.

Kernel Smoothers   *Kernel smoothers* are a family of statistical procedures used to estimate the real-valued function of the form $Y = f(t; b)$ in the presence of noisy data. This technique is most appropriate for low-dimensional ($p < 3$) data visualization purposes. The kernel smoother represents a set of irregular data points as a smooth line or surface. A *kernel estimator* at point t_c is

$$\widetilde{y}_c = f(t_c; b) = \sum_{j=1}^{m} w(t_c, t_j, b)$$

where the weight function w is of the form

$$w(t_c, t_j; b) = K\left(\frac{t_c - t_j}{b}\right) \Big/ \sum_{j=1}^{m} K\left(\frac{t_c - t_j}{b}\right)$$

and K is the *kernel function* of bandwidth b. *Bandwidth* is a measure of the number of points included in the smoothing process. A larger b results in a smoother time series because it includes more points in the smoother. It is akin to span.

A kernel function may be considered as a class of local average smoothers that include normal, triangular, and quadratic filters. The *Hamming filter* has a bandwidth of three time units and can be represented using the triangular form

$$K_{\text{tri}}(t) = \begin{cases} 1 - |t|/C & |t| \le 1/C \\ 0 & |t| > 1/C \end{cases}$$

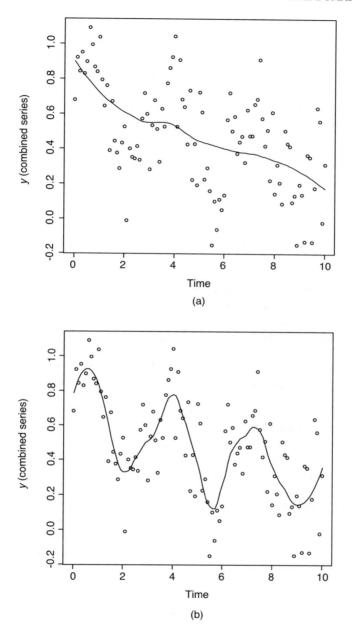

FIGURE 5.4   (a) Loess smoother (solid line) with span = 0.75. (b) Loess smoother (solid line) with span = 0.25.

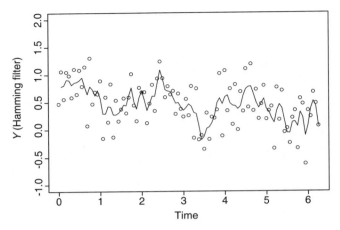

FIGURE 5.5   Hamming filter.

where C is a scaling constant. This triangular function moves across the data, with the center point t_c receiving the most weight. The points on each side receive less weight. Let $b = 2$ and $C = 1$; then three nonzero weights are included in the window with values 0.25, 0.50, and 0.25 for the points t_{c-1}, t_c, t_{c+1}. These are the weights of the Hamming filter. If $b = 1$ in the kernel estimator, no smoothing occurs because only one point will be in the smoothing window at any given time. The t's are integers beginning at zero, not the original time units. The Hamming filter (Figure 5.5) is illustrated using a portion of the data from Table 5.1. Not surprisingly, the resulting smoothed series is not as smooth as that of the previous loess smoother (Figure 5.4). Thus, it passes a higher frequency than does the loess filter with span $= 0.75$.

The *Hanning filter* uses a slightly different computational procedure to obtain weights, but they are close to those of the Hamming filter and the procedure produces similar estimates (Chatfield, 2003). The weights for the Hanning filter are (0.23, 0.54, 0.23).

Exponential Smoothers   Exponential smoothing, suggested by Holt (1957), is a useful procedure for short-term forecasting if a times series does not have a systematic trend and seasonal effects are not present. Suppose that the data (t, y_t), $t = 1, \ldots, n$ are observed and the objective is to make a forecast at y_{n+1}. An *exponentially weighted moving-average* (EWMA) estimate takes the form $\widehat{y}_{t+1} = c_0 y_t + c_1 y_{t-1} + c_2 y_{t-2} + \cdots$. This represents a weighted average of past values $y_t, y_{t-1}, y_{t-2}, \ldots$, where the weights $c_0, c_1, c_2, \ldots$ decline exponentially as in $c_i = \alpha(1 - \alpha)^i$, $i = 0, 1, \ldots$ and $0 < \alpha < 1$, where α is the smoothing parameter. This yields a series of the form

$$\widehat{y}_{t+1} = \alpha y_t + \alpha(1 - \alpha)y_{t-1} + \alpha(1 - \alpha)^2 y_{t-2} + \cdots$$

A large value of α implies that a greater proportion of the weight is on y_t, the most recently observed value. Values of α closer to zero result in more weight being distributed to y_{t-1}, y_{t-2}, ... EWMA is often used in short-term economic, energy supply, and climate forecasting, such as in predicting the demand for motor fuel over the next year.

The *Holt–Winters method* is a modified EWMA which allows estimates to be made in the presence of trend and/or seasonal variation. This and other aspects of EWMA have been discussed more fully by Brockwell and Davis (2002).

Median Smoothers   A *median smoother* uses the median of the points that fall in the window rather than a mean. A simple example is a median smoother of size 3; there are three points in the window: y_{t-1}, y_t, and y_{t+1}. The smoothed value at t is the median of (y_{t-1}, y_t, y_{t+1}). For example, a simple series consisting of the first six points of the illustrative data (Table 5.1) after median smoothing of size 3 is (0.47, 0.56, 1.05, 0.98, 0.98, 0.59). A more complicated median smoother is denoted as 4253H, which means that in succession the series is subjected to median smoothes of size 4, 2, 5, and 3, followed by a Hanning smoother. Variants of this are sometimes called super smoothers. One advantage of this class of smoothers is as protection against outliers. A disadvantage is mathematical intractability. See Tukey (1977) for additional details.

Spline Smoothers   *Splines* are a class of smoothers developed as a subfield of numerical analysis. Unlike other smoothers that have been discussed, spline smoothers are interpolators. An *interpolator* passes through the data points. Consider the time series data $\{(t_i, y_i), i = 0,\ldots, n-1\}$ where $a = t_0 < t_1 < \cdots < t_{n-1} = b$. The n points are called *knots*. A parametric curve S is called a spline of degree k if S is continuous in degree $k - 1$ in the open interval (a, b). In each subinterval or *knot span* (t_i, t_{i+1}), $i = 0,\ldots, n-2$, S is a polynomial of degree k. Thus, a spline is a piecewise polynomial function. The degree of the polynomial governs the degree of smoothness. There are many different types of splines, however, a cubic spline (Figure 5.6) is commonly used and provides reasonable smoothness for many applications. It consists of polynomials up to and including degree three. A cubic spline in the ith subinterval $t \in (t_i, t_{i+1})$ is of the form $y_i(t) = a_i + b_i t + c_i t^2 + d_i t^3$, where model coefficients are found by standard numerical methods. To show the smoothness more clearly in Figure 5.6, only the data up to time 2 are shown. Note that a polynomial of degree 100 also interpolates the data in Table 5.1; however, it oscillates wildly between data points. The cubic spline is much better behaved. Spline functions are used in interpolations of gridded data such as gravity anomalies and rainfall.

Choice of Smoothers   Smoothers have two principal functions. One is to allow a component of the time series that may be of interest to be inspected. A second purpose is to condition the time series data for further analysis. Only a small number of available smoothers have been reviewed, and within each smoother one or more parameters may often be varied. Thus, the choice of a smoother is frequently a function of trial and error and personal preference. However, sometimes it is of interest to pass or eliminate certain classes of frequencies. Another criterion is *leakage*, a by-product of frequency analysis that occurs when small amounts of frequency are observed that are not in the original signal, thus appear to have leaked

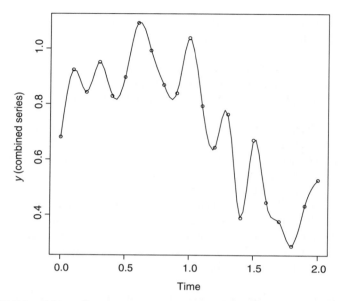

FIGURE 5.6 Cubic spline smoother for illustrative data in Table 5.1, time (0 to 2).

out. For example, the Hanning smoother has less leakage than the Hamming but may not give as fine a resolution of peaks. The immediate past observations typically weigh more heavily with the exponential smoother. For example, in short-term weather forecasting there may be reasons to believe that observations made during the past two weeks are highly relevant while older observations should have little influence. Spline smoothers are useful in applications where there is believed to be minimal measurement error at data points and it is desirable to control the degree of smoothness between observations.

5.2.2 Trends and Stationarity

Trends are of interest because they describe long-term phenomena; examples include the investigation of global warming, ice cap melt, economic growth, and oil depletion. A trend can also be a nuisance factor that interferes with the ability to detect cyclic effects or short-term structure. Models that investigate the latter phenomena generally require that the time series be stationary. *Stationarity* implies that the model parameters do not vary with time. These conditions are of equal concern when space is substituted for time. There are various levels of stationarity. A series is *first-order stationary* if $f(Y_t) = f(Y_{t+k})$, where f is the density function of Y_t, a random variable. Thus, f is independent of k, which implies that the mean $\mu_t = E(Y_t) = \mu$ is independent of t. There is no trend in the data. A more restrictive condition, called *second-order stationary*, is defined as $f(Y_{t_1}, Y_{t_2}) = f(Y_{t_1+k}, Y_{t_2+k})$. This condition implies that $\mu_t = E(Y_t) = \mu$ and that the *autocovariance function (ACVF)*, defined as

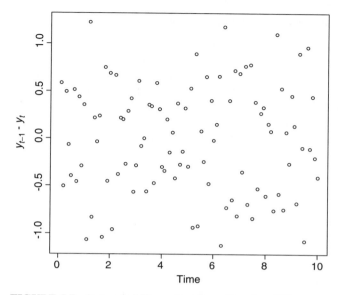

FIGURE 5.7   Detrended illustrative data, using first differences.

$\gamma_t(k) = \text{Cov}(Y_t, Y_{t+k}) = \gamma(k)$, is a function of the lag only where the lag $k = 0, 1, \ldots$ discrete time units. Lag is illustrated in the next section. For $k = 0$, $\gamma_t(0) = \text{Var}(Y_t) = \gamma(0)$. A secondary-order stationary process is also called *weakly stationary*. A *strongly stationary* series occurs if the joint distribution of $Y_{t_1}, \ldots, Y_{t_n}$ is the same as the joint distribution of $Y_{t_1+k}, \ldots, Y_{t_n+k}$. In other words, distribution of the Y's remains the same despite a shift by k units. For most applications, second-order stationarity is a sufficient condition.

Trend removal is also a filtering procedure. A way to remove trend is with a first difference function, defined as $\nabla Y_{i+1} = Y_{i+1} - Y_i$, where the series Y_i lags Y_{i+1} by one time unit ($k = 1$). For the series in Figure 5.7, the first observation in the new series is

$$\nabla y_2 = y_2 - y_1$$
$$= 1.06 - 0.47$$
$$= 0.59$$

The resulting first-differenced series (Figure 5.7) does not show a trend. This is a high-pass filter because the low frequency or long-term trend has been removed, leaving higher-frequency signals. If a trend is still present, differencing needs to be repeated. That is, the first difference operator ∇ is applied to the first-differenced series ∇Y_{i+1}. There are other ways to remove a trend, including regression. To use regression, the series (t, y_t) is fit to a linear model. Its residuals, $e_t = y_t - \hat{y}_t$, become the detrended series. Of course, the usual regression assumptions will not hold, and in general, detrending by differences is recommended.

5.2.3 Autocovariance and Autocorrelation

In Chapter 4, covariance between two random variables is defined. In time series analysis, a similar definition is used except that the two variables are Y_i and Y_{i+k}, where Y_i, $i = 1, \ldots, n$ is the time series and Y_{i+k} is the series lagged by k time units. Y_i is considered to be stationary and, unless stated otherwise, the term stationary refers to first-order stationarity.

The autocovariance between the two series is, as noted previously, expressed as $\mathrm{Cov}(Y_i, Y_{i+k}) = \gamma(k)$ or γ_k. For a discrete data set of n observations, the sample ACVF is

$$\widehat{\gamma}_k = \sum_{t=1}^{n-k} \frac{(Y_t - \overline{Y})(Y_{t+k} - \overline{Y})}{n}$$

The *autocorrelation function* (ACF) is

$$\rho(k) = \frac{\gamma(k)}{\gamma(0)}$$

where $\gamma(0) = \sigma^2$ and $-1 \leq \rho(k) \leq 1$. The ACF is displayed by plotting $\rho(k)$ versus k for the detrended illustrative data (Figure 5.8). This plot is often referred to as a *correlogram*.

The ACF function is computed for at most $k = n/4$ lags because above this too many data are lost. At $k = 0$, the ACF $\rho(0) = 1$. Note the exponential decline in the ACF. The ACF is useful when a time series model is constructed based on diagnostics.

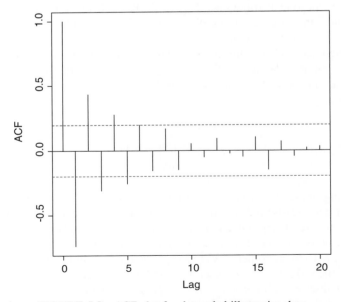

FIGURE 5.8 ACF plot for detrended illustrative data.

5.2.4   Seasonality

A *seasonal effect* is present in many time series, especially climate data. A seasonal effect can be removed with a moving-average window of the form

$$S(Y_t) = \frac{0.5Y_{t-6} + Y_{t-5} + \cdots + Y_{t+5} + 0.5Y_{t+6}}{12}$$

Frequently, a twelfth difference $\nabla Y_{t+12} = Y_{t+12} - Y_t$ is used when the t are monthly intervals. This function is applied to the original or first-differenced series ∇Y_{t+1}, as appropriate.

5.2.5   Stochastic Processes

A *stochastic process* is a collection of random variables; it may be discrete or continuous. The outcome at a given state t is nondeterministic; that is, it is not determined completely by the outcome at step $t-1$. In statistics, the stochastic process most often of interest is a time series. Examples occurring over time and/or space include precipitation, oil futures, radioactive decay, sediment transport, and lineages of fossil records. For a formal definition of a stochastic process and further introduction, see the book by Lawler (1996). Important special cases include the Markov, Poisson (discussed in Chapter 8), autoregressive, and moving-average processes.

Markov Process   A *Markov process* is a stochastic process where the outcome at time t, denoted Y_t, is conditionally independent of the past. The sequence of random variables generated by a Markov process is called a *Markov chain*. Thus,

$$\Pr(Y_t = y | Y_0, Y_1, \ldots, Y_{t-1}) = \Pr(Y_t = y | Y_{t-1})$$

The set of all possible Y's is called the *state space*. It can be discrete or continuous. For example, if the Y's are precipitation, the state space is continuous; if the Y's are geologic eras, the state space is discrete. The key attribute of a Markov process is that it has no memory and thus no trail to allow the user to return to the origin. The probability of moving to a new state at time t depends only on the state at time $t-1$.

The simplest example of a Markov process is a *random walk*, which consists of a sequence of discrete steps of fixed length. Imagine yourself walking through a neighborhood, and each time you come to an intersection you choose randomly from among the available paths. Mathematically, a process $\{Y_t\}$ is a random walk if

$$Y_t = Y_{t-1} + Z_t$$

where the Z_t are mutually independent, identically distributed random variables. For some applications, Z_t is further restricted to be normally distributed. Then $E[Y_t] = t\mu$ and $\text{Var}(Y_t) = t\sigma_Z^2$. Note that $\nabla Y_t = Y_t - Y_{t-1} = Z_t$, where Z_t is referred to as white (random) noise.

When the state space is discrete, a *transition probability* is used: the probability of going from state i to state j. In a Markov chain, the transition probability

$$p_{ij} = \Pr(Y_t = i | Y_{t-1} = j)$$

is independent of t.

A Markov chain combined with certain distributional assumptions is the basis for *Markov chain Monte Carlo* (MCMC) *methods*, which make possible sampling from complex distributions. MCMC methods play an important role in Bayesian statistics.

Autoregressive Process   A process $\{Y_t\}$ is an *autoregressive process* (AR) of order p, denoted AR(p), if

$$Y_t = \alpha_1 Y_{t-1} + \cdots + \alpha_p Y_{t-p} + Z_t$$

where $\{Z_t\}$ is a purely random process with mean zero and variance σ_Z^2 and the $\alpha_1, \ldots, \alpha_p$ are model parameters. As the name suggests, an AR(p) process looks like regression, except that the explanatory variables are past values of Y_t: namely, $Y_{t-1} + \cdots + Y_{t-p}$.

A useful process is AR(1), a first-order process, also known as a Markov process:

$$Y_t = \alpha Y_{t-1} + Z_t$$

with $|\alpha| < 1$. An AR process can be represented by a *backshift operator, B*. Thus, AR(1) process represented in this manner is

$$(1 - \alpha B) Y_t = Z_t$$

where

$$
\begin{aligned}
Y_t &= \frac{Z_t}{1 - \alpha B} \\
&= (1 + \alpha B + \alpha B^2 + \cdots) Z_t \\
&= Z_t + \alpha Z_{t-1} + \alpha^2 Z_{t-2} + \cdots
\end{aligned}
$$

Thus, $B^k Y_t = Y_{t-k}$ for $k \geq 0$. The backshift operator allows the series Y_t to be represented as an infinite sum of current and past realizations of random deviates. It can easily be shown that

$$E[Y_t] = 0 \quad \text{and} \quad \mathrm{Var}(Y_t) = \sigma_Z^2 (1 + \alpha^2 + \alpha^4 + \cdots)$$

For $|\alpha| < 1$, the $\mathrm{Var}(Y_t) = \sigma_Z^2 (1 - \alpha^2)$.

The ACVF for AR(1) is $\gamma(k) = \alpha^k \sigma_Y^2$, also derived from the backshift operator expansion above. The corresponding ACF is $\rho(k) = \alpha^k$. For the ACF in Figure 5.8, $\hat{\alpha} = -0.74$, so the fitted model is $\hat{y}_{t+1} = -0.74 y_t$. What does an AR(1) process look like if $1 > \alpha > 0$? ACFs for two AR(1) processes are shown in Figure 5.9. Figure 5.9a shows the ACF plot with $\alpha = 0.8$ and Figure 5.9b shows the ACF plot with $\alpha = 0.3$. In

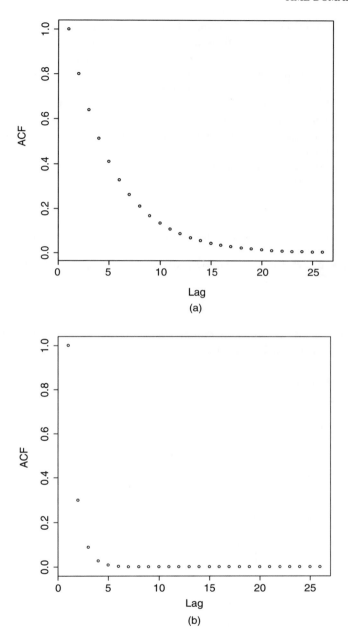

FIGURE 5.9   (a) AR(1) processes, with $\alpha = 0.8$. (b) AR(1) processes, with $\alpha = 0.3$.

the former case the decay is slow, whereas in the latter case it is rapid. Decay refers to $|\rho(k)| > |\rho(k+1)| > 0$, as opposed to cutoff, where $|\rho(k)|$ is significantly greater than 0 and $|\rho(k+1)|$ is at or near zero. When α is small, the decay is rapid (Figure 5.9b). Temperature, precipitation, and other climate data typically are functions of near past values of occurrences, and low-order AR processes should be considered as candidate models. Not surprisingly, the ACF damps out more quickly when $\alpha = 0.3$ as opposed to when $\alpha = 0.8$. Both processes yield ACFs that are positive. If noise is added, some of the values at larger lags may be negative, due to sampling variability.

An AR(1) process is stationarity if $|\alpha| < 1$. For an AR(p) process, stationarity exists if the roots of

$$\phi(B) = 1 - \alpha_1 B - \cdots - \alpha_p B^p = 0$$

lie outside the unit circle. This expression is also useful for forecasting.

Moving-Average Process   The *moving-average process* (MA) for a model of order q, written MA(q), is

$$Y_t = \beta_0 Z_t + \beta_1 Z_{t-1} + \cdots + \beta_q Z_{t-q}$$

where $\beta_0, \beta_1, \ldots, \beta_q$ are model parameters and Z_t is defined as in the AR process; however, the Z's are usually scaled so that $\beta_0 = 1$. An MA model is a linear regression of a series of white noise occurrences, the $Z_t, Z_{t-1}, \ldots, Z_{t-q}$ on Y_t.

The ACVF for MA(q) is

$$\gamma(k) = \begin{cases} \gamma(-k) & k < 0 \\ \sigma_Z^2 \sum_{i=0}^{k-q} \beta_i \beta_{i+k} & k = 0, 1, \ldots, q \\ 0 & k > q \end{cases}$$

and the corresponding ACF is

$$\rho(k) = \begin{cases} \rho(-k) & k < 0 \\ 1 & k = 0 \\ \sum_{i=0}^{k-q} \beta_i \beta_{i+k} \Big/ \sum_{i=0}^{q} \beta_i^2 & k = 1, \ldots, q \\ 0 & k > q \end{cases}$$

The ACVF and ACF for MA(q) become zero after lag q, whereas the AR(p) process declines. Applications of MA processes are discussed in the context of ARIMA models. A MA(q) process via a backshift operator may be represented as

$$Y_t = (\beta_0 + \beta_1 B + \cdots + B^q)Z_t$$
$$= \theta(B)Z_t$$

where $\theta(B)$ is a polynomial of order q.

5.2.6 ARIMA Models

ARIMA is an acronym for the *autoregressive integrated moving-average model*, which is a way of formulating a unified approach to the models discussed previously. When the "integrated" term (I) is omitted, it becomes an *autoregressive moving-average* (ARMA) *model* (Box et al., 1994), which is a combination of the AR and MA models. An ARMA(p,q) process can be written as

$$Y_t = \alpha_1 Y_{t-1} + \cdots + \alpha_p Y_{t-p} + Z_t + \beta_1 Z_{t-1} + \cdots + \beta_q Z_{t-q}$$

where p is on the order of the AR process and q is on the order of the MA process. The advantage of this representation is that most processes can be represented with relatively few terms; generally, $p,q \leq 3$.

An extension of the ARMA model is ARIMA. The ARIMA model is needed because many processes are nonstationary in the mean and the "I" in ARIMA represents the difference function, $\nabla_d Y_t$, defined previously. The process is represented as ARIMA(p,d,q), where there are p AR terms, d differences (the I term), and q MA terms. Usually, only one or two differences are required to achieve stationarity in the mean.

The formulation of an ARIMA model is similar to that of regression except that the explanatory variables are past observations of the time series or random noise. In ARIMA modeling, some terms may represent known effects that can be related to physical or other principles. However, the model is often a best-fit model using available data. Unfortunately, the same level of understanding of the relationships among variables that exists in regression usually is not apparent in time series analysis. However, the principle of parsimony still applies, in that it is desirable to find the simplest model that accounts for the systematic variation in the data. The ARIMA model will help accomplish this.

Model Formulation   The first step in model formulation is to make a time series plot, also called a *run-sequence plot*. The authors recommend visually checking the following:

- Gradual changes in the mean, which indicate nonstationarity in the mean. A difference function needs to be applied.
- Gradual changes in variance, which may require a logarithmic or other transformation of data.

- Seasonality, which needs to be modeled by an appropriate difference function.
- Shifts in the mean or variance, which may occur due to sudden changes in processes observed or to changes in measuring devices or procedures. Shifts are modeled by indicator variables in the model.
- Outliers, which as in regression, can have a significant influence on least-squares procedures and should be examined.

Formal statistical tests are available to investigate these phenomena. Once the series is stationary and other concerns listed above have been addressed, it is necessary to select values of p and q. It is known that the ACF of an AR process declines, so this is not helpful in identifying p. However, the *partial autocorrelation function* (PACF) is a useful tool to help identify p in the AR(p) process. Denote the last (the pth) coefficient in the AR(p) process by π_p (i.e., let $\alpha_p = \pi_p$). Thus, $Y_t = \alpha_1 Y_{t-1} + \cdots + \alpha_{p-1} Y_{t-(p-1)} + \pi_p Y_{t-p} + Z_t$. The coefficient π_p represents the correlation not accounted for by an AR($p-1$) process. The sample PACF is estimated by letting $\hat{\pi}_i = \hat{\alpha}_i$ in AR(i). Note that $\hat{\pi}_i$ cuts off (is statistically insignificant) at $i > p$ because if the process is AR(p), there should not be any statistically significant variability for $i > p$. The PACF is represented as a plot of π_i versus i, $i = 1, \ldots, \max(n/4)$. A PACF corresponding to the AR(1) process (Figure 5.9a) is presented in Figure 5.10. The only nonzero PACF is at lag 1. Of course, when modeling data there will be small values of the PACF beyond lag 1. If the process is an AR(2), there will be large PACF values at lags 1 and 2.

In the determination of q in an MA(q) process, the ACF and PACF display behavior opposite to that of the AR process. The ACF of an MA process cuts off at q (it is

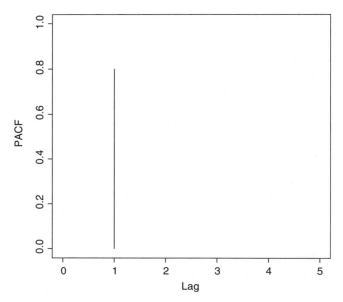

FIGURE 5.10   PACF for an AR(1) process with $\alpha = 0.8$ (Figure 5.9a).

statistically insignificant for $i > q$), and the PACF of an AR process declines exponentially.

Another alternative to model identification is to try various low-order ARIMA processes and evaluate them using *Akaike's information criterion (AIC)*. AIC is a measure of goodness of fit, which minimizes $\ln(\hat{\sigma}_Z^2) + 2(p+q)/n$. As such, it incorporates a penalty function for specifying too many parameters. A related Bayesian procedure, the *Bayesian information criterion* (BIC), is also available (Chatfield, 2003). It is defined to be approximately equal to $-2 \ln(\ell) + k \ln(n)$, where ℓ is the maximized values of the likelihood function for the estimated model, k is the number of free parameters to be estimated, and n is the sample size. The penalty for additional parameters in BIC is at least as strong as AIC. The importance of AIC and BIC measures of fit is in comparing different models. The model with the lowest AIC (BIC) may be considered the best; however, as with regression, goodness-of-fit measures should serve only as guides.

Model fitting is by iteration, unlike linear regression, where an analytic solution is obtained. Usually, nonstationary series will fail to converge. Also, failure to converge may result from specifying too many or too few parameters, the wrong type of process, or failure to account for seasonality (Box et al., 1994).

Model Evaluation   The general procedures to evaluate an ARIMA model follow those used in regression as described in Chapter 4. The significance of parameter estimates is examined and the residuals viewed. As in regression, residuals should be identically, independently distributed. Typically, diagnostic plots include:

- A time series plot of residuals (standardized and/or raw)
- An ACF plot of residuals
- A PACF plot of residuals

If structure exists in the residuals, the model needs to be modified. Model formulation and evaluation are illustrated with the following ozone example.

5.2.7   Example: Ozone Measurements

Consider the daily ozone measurements in parts per million (ppm) from January 1, 1980 through December 31, 2003 from the South Coast Basin, California (California Environmental Protection Agency, 2005) (Appendix III; see the book's Web site). The data are maximum eight-hour averages, which are referred to as "8-hour daily ozone data." A small portion of the data set is shown in Table 5.2. A time series plot of the entire data set is shown in Figure 5.11. In the following sections, various segments of this data set are displayed.

From this graph, a deduction can be made that ozone concentration has declined from 1980 through at least 1995. This decline appears linear but could have some slight degree of curvature. In addition to this "linear" trend, an annual cycle exists. Given that climate exerts a strong influence on ozone concentration, this is not

TABLE 5.2 Subset of 8-Hour Daily Ozone Data from the South Coast Basin, California

Date	Day of Week	OzMax8 (ppm)	Date	Day of Week	OzMax8 (ppm)
12/1/2003	2	0.049	12/17/2003	4	0.045
12/2/2003	3	0.047	12/18/2003	5	0.043
12/3/2003	4	0.050	12/19/2003	6	0.030
12/4/2003	5	0.047	12/20/2003	7	0.037
12/5/2003	6	0.043	12/21/2003	1	0.041
12/6/2003	7	0.049	12/22/2003	2	0.041
12/7/2003	1	0.038	12/23/2003	3	0.043
12/8/2003	2	0.042	12/24/2003	4	0.035
12/9/2003	3	0.041	12/25/2003	5	0.037
12/10/2003	4	0.042	12/26/2003	6	0.040
12/11/2003	5	0.041	12/27/2003	7	0.041
12/12/2003	6	0.038	12/28/2003	1	0.043
12/13/2003	7	0.036	12/29/2003	2	0.044
12/14/2003	1	0.041	12/30/2003	3	0.038
12/15/2003	2	0.040	12/31/2003	4	0.038
12/16/2003	3	0.042			

surprising. In addition, the variability is declining over time. Stabilizing this variability is more difficult than removing a long-term trend because it often requires
a transformation or the use of nonparametric procedures. The long-term trend appears
to have been removed by first differencing (Figure 5.12), however, there still appears
to be a seasonal effect and some decline in variability over time.

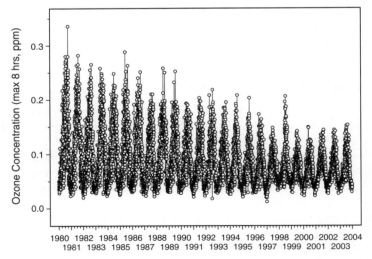

FIGURE 5.11 Time series plot of a subset of 8-hour daily ozone data.

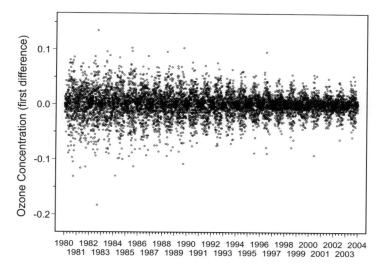

FIGURE 5.12   Detrended (first-difference) 8-hour daily ozone data.

If the main goal is to investigate long-term trends or monthly variability, it may be desirable to work with monthly averages, recognizing that this will mask shorter-term (day-to-day) variability. This is what we have chosen to do, and subsequent analysis will be based on monthly averages (Table 5.3). The monthly time series plot (not

TABLE 5.3   Monthly Ozone Averages (ppm) from January 1, 1980 through December 31, 1981 from the South Coast Basin, California[a]

1980		1981	
Month	Ozone	Month	Ozone
1	0.0436	1	0.047
2	0.0589	2	0.06
3	0.0674	3	0.0635
4	0.102	4	0.104
5	0.0949	5	0.1193
6	0.1576	6	0.1957
7	0.191	7	0.183
8	0.1523	8	0.1869
9	0.165	9	0.155
10	0.1331	10	0.0782
11	0.0842	11	0.0738
12	0.0489	12	0.0583

[a]For a complete data set, see Appendix III (at the book's Web site).

Source: Data from (California Environmental Protection Agency 2005).

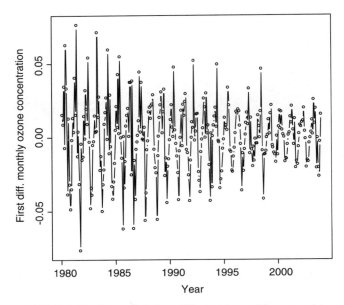

FIGURE 5.13   Detrended (first-difference) monthly ozone data.

shown) reveals the same phenomena as in Figure 5.11, except for the removal of day-to-day variation. The detrended monthly data are shown in Figure 5.13.

Detrending allows some of the other features in a time series to be seen more easily. Figure 5.13 illustrates a decrease in variability of the monthly ozone data from January 1980 through December 2003 (288 observations). An attempt to identify a cause, such as a reduction in amount of motor vehicle traffic, cleaner-burning cars, or climate change, can be made. When this second-order effect of change in variation over time is present, a log or other appropriate transformation may remove it.

The ACF plot (Figure 5.14) neither declines exponentially nor cuts off quickly. Rather, it exhibits a cyclical effect with a period of approximately 12 months. This should not be a surprise, as ozone is subject to annual climatic cycles. This effect needs to be removed from the series by applying a twelfth-differenced function $\nabla Y_{t+12} = Y_{t+12} - Y_t$ in order to investigate a possible shorter-term structure. The ACF on the twelfth-difference of the detrended series (Figure 5.15) still shows some correlation between monthly ozone and monthly ozone lagged by 12 months, which suggests that some of the seasonal effect remains. Climate and anthropogenic factors do not always repeat with a period of exactly 12 months. There appears to be a significant high positive value at lag 11 followed by a significant high negative value at lag 12. The value at lag 24 is not significant, which suggests that the remaining annual periodicity is weak. The two dashed horizontal lines in Figure 5.15 occur at ACF values of approximately -0.1 and $+0.1$ and represent a 95% confidence band. Values of the ACF that are less than -0.1 or greater than $+0.1$ may be statistically significant. This confidence band is similar to that presented in regression (Chapter 4).

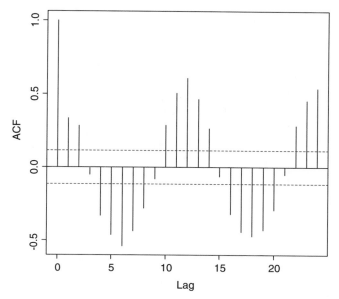

FIGURE 5.14   ACF plot of detrended monthly ozone data (Jan. 1980–Dec. 2003).

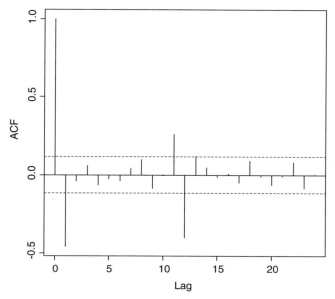

FIGURE 5.15   ACF plot of twelfth-differenced detrended monthly ozone data (Jan. 1980–Dec. 2003).

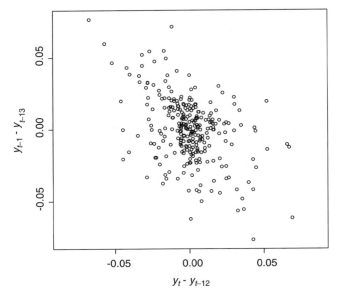

FIGURE 5.16   Lag plot of 1 on the twelfth-differenced monthly ozone data corresponding to ACF (Figure 5.15); observations are from Jan. 1980 to Dec. 2003.

Another way to view this high positive value followed by a high negative value is with a lag plot on the detrended twelfth-differenced ozone data series (Figure 5.16). A *lag plot* of lag k is simply a scatter plot of Y_t versus Y_{t-k}. A lag 12 plot is a plot of ∇Y_{t+12} versus $(\nabla Y_{t+12})_{-1}$ or, equivalently, $(Y_t - Y_{t-12})$ versus $(Y_{t-1} - Y_{t-13})$. The weak negative trend (Figure 5.16) confirms the pattern observed in the ACF (Figure 5.15). An additional step needed to help formulate an ARIMA model is to compute the PACF on this series (Figure 5.17). There is a decline in the PACF with lag, plus what appears to be the remainder of the seasonal effect (at lag 11). The ACF and PACF plots suggest that the detrended-deseasonalized model may be fit with an MA (1) process. Denote this series as V_t. The results of this process are a fitted model

$$\widehat{V}_t = \widehat{\beta}_1 Z_{t-1} + Z_t$$

where $\widehat{\beta}_1 = -1.00$ with standard error $= 0.0001$, L (log likelihood) $= 791$, AIC $= -1576$, and $\widehat{\sigma}^2 = 0.00024$.

The "equivalent model" formulated as a single model is an ARIMA $(0,1,0) \times (0,1,1)$ 12. The first set of numbers, $(0,1,0)$, refer to the detrended series via first differences; the second set of numbers, $(0,1,1)$, refer to an MA(1) process on a seasonal model after seasonal detrending (the middle 1), and 12 is the periodicity of the seasonal component. In the absence of a seasonal effect, the model is ARIMA $(0,1,1)$, an MA(1) process on a detrended series. Unfortunately, the seasonality does not appear to be exactly 12 months, nor does it appear to be 11 or 13 months. Rather, it

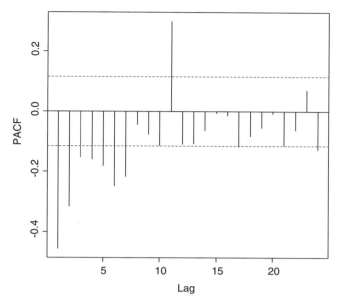

FIGURE 5.17 PACF plot of twelfth-differenced detrended monthly ozone data (Jan. 1980–Dec. 2003).

may occur at some noninteger value. Algorithms exist to model noninteger seasonality, such as the sunspot data (see, e.g., Franses, 1998). In this case, assume a seasonality of 12 months. The candidate model ARIMA(0,1,0) × (0,1,1) 12 is expressed algebraically as $Y_t = Y_{t-12} + (Y_t - Y_{t-13}) + \beta_1 Z_{t-1} + Z_t$. Some differences occur between the estimates obtained previously by proceeding in a sequential manner and those resulting from the single fitted model, which is why "equivalent model" is placed in quotes. For a single fitted model, $\hat{\beta}_1 = -0.83$, AIC $= -1488$, log-likelihood $= 746$, and $\hat{\sigma}^2 = 0.00025$. For $k = 1$ and $n = 288$, BIC $= -2 \times 746 + 1 \times \ln(288) = -1486$. Note that the penalty is slightly more severe (more positive) for BIC than for AIC, although in this case because there is only one estimated parameter, the two measures are essentially equivalent. Consider an alternative model, $Y_t = Y_{t-12} + (Y_t - Y_{t-13}) + \beta_1 Z_{t-1} + \beta_2 Z_{t-2} + Z_t$, in which we have added a second moving-average (MA) term. The AIC for this model is -1487. This is essentially the same as the preceding model that had only one MA term, indicating that the addition of a second MA term is not warranted.

The graphical diagnostic output for the model above consists of a time series residual plot (Figure 5.18a), an ACF plot on residuals (Figure 5.18b), and a PACF plot on residuals (Figure 5.18c). As in regression, the residuals should be distributed randomly. It is important to ensure that no statistically significant autocorrelation patterns exist in the residuals. In the residual plot (Figure 5.18a), possible evidence of a decline is seen in variability over time; however, we judge it insufficient to warrant a transformation. The ACF plot of residuals (Figure 5.18b) shows a negative spike at

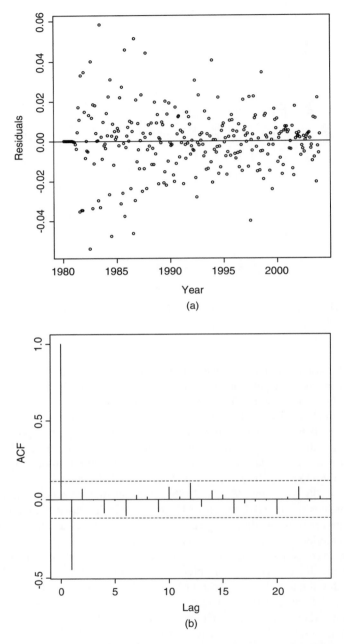

FIGURE 5.18 (a) Residual plot for ARIMA fit to monthly ozone data (Jan. 1980–Dec. 2003).
(b) ACF plot of residuals output for ARIMA fit to monthly ozone data (Jan. 1980–Dec. 2003).

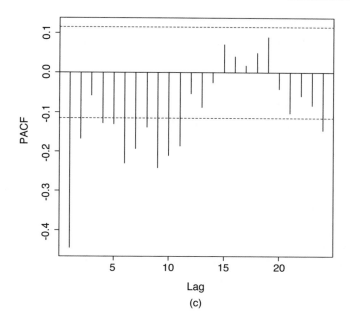

(c)

FIGURE 5.18 (*Continued*)   (c) PACF plot of residuals for ARIMA fit to monthly ozone data (Jan. 1980–Dec. 2003).

a lag of 1, which is a cause for concern. The PACF plot (Figure 5.18c) shows some cyclical effect and several statistically significant negative values. Ideally, both the ACF and PACF plots should have no statistically significant values (except of course at lag 0) or other discernible patterns.

Several other models were attempted to see if a better fit could be obtained; however, none yields a smaller AIC value than the one used. As in regression, judging model adequacy can be a difficult call. Additional terms can ensure a better fit to the data but use up degrees of freedom and result in overfitting the data. As mentioned previously, the AIC criterion is a good measure of fit because it includes a penalty for additional terms. In addition, more complex models are often difficult to interpret. A summary of this modeling process given a time series Y_t ($t = 1, \ldots, n$) is:

1. A time series plot is constructed to investigate long-term trend, seasonal and other cyclical effects, and possibly to help identify outliers.
2. The long-term trend is removed with the first differencing operator, $Y_t - Y_{t-1}$. If a trend still exists, this step will need to be repeated. However, usually only one differencing is required.
3. A twelfth difference $(Y_t - Y_{t-12})$ is applied to attempt to remove annual seasonality effect in the monthly data.
4. The trend removal and twelfth differencing are combined in the following expression: $(Y_t - Y_{t-12}) - (Y_{t-1} - Y_{t-13})$.

5. If over time there are changes in variance which indicate significant second-order nonstationarity, a transformation of the Y's can be applied. However, transformed data can be difficult to interpret and/or to back-transform into the original space. For that reason, it was not done in this example.

6. The resulting model is $Y_t = Y_{t-12} + (Y_t - Y_{t-13}) + \beta_1 Z_{t-1} + Z_t$.

7. The fitted candidate model is $y_t = y_{t-12} + (y_{t-1} - y_{t-13}) + \widehat{\beta}_1 z_{t-1} + z_t$.

8. Evaluation of the model is performed by examining residuals for a statistically significant structure. The three tools used are a time series plot of residuals, an ACF plot of residuals, and a PACF plot of residuals.

5.2.8 Forecasting

One application of time series analysis is to understand the process being studied: to estimate trends, cycles, and short-term structure. Another is to forecast. For example, air quality districts need to predict ozone content months ahead. Energy economists, decision makers, and others need to know the cost and supply of energy resources months and even years ahead.

As in regression modeling, accurate forecasting depends on the model structure being accurate and continuing into the future. Forecasts are based on a weighted average of past observations. Forecasting from an ARIMA model typically uses a Kalman filter to compute weights. [For a discussion of the Kalman filter, see Chatfield's (2003) book.] As noted previously, the estimated ARIMA model for the monthly ozone data is

$$y_t = y_{t-12} + (y_{t-1} - y_{t-13}) + \widehat{\beta}_1 z_{t-1} + z_t$$

then

$$\nabla y_t = y_t - y_{t-1}$$
$$= y_{t-12} - y_{t-13} + \widehat{\beta}_1 z_{t-1} + z_t$$

where $\widehat{\beta}_1 = -0.83$. When the data are discrete, it may be necessary to forecast some number of steps or time units ahead or into the future. A k-step ahead forecast is denoted as $\widehat{y}(t, k)$. A one-step-ahead forecast is written as

$$\widehat{y}(t, 1) = y_t + \nabla y_t$$

Forecasts at future times can be computed recursively. When the value at y_{t+1} becomes known, it is substituted in the forecast equation for $\widehat{y}(t, 1)$.

Other systems of ARIMA forecasting involve the use of ψ weights and π weights (Chatfield, 2003). $\psi(B)$ allows the definition of an ARIMA model to be a purely MA process, while $\pi(B)$ allows it to be a purely AR process where $\psi(B) = \theta(B)/\phi(B)$ and

$\pi(B) = \phi(B)/\theta(B)$. A k-step ahead forecast for a series of length n using ψ weights is represented as

$$y(n, k) = Z_{n+k} + \psi_1 Z_{n+k-1} + \cdots$$

which facilitates computation of forecasting error. The π-weight representation allows formulation of a k-step-ahead forecast in terms of $k-1, k-2, \ldots$, step-ahead forecasts and previous observations as

$$\hat{y}(n, k) = \pi_1(n, k-1) + \pi_2(n, k-2) \cdots + \pi_k y_n + \pi_{k+1} y_{n-1} + \cdots$$

In practice, the πs and ψs are not known exactly and must be estimated from the data. The exponential smoothing discussed previously and a special case, the Holt–Winters method, are especially useful for short-term forecasting.

5.3   FREQUENCY DOMAIN

An alternative way to view a time series is the *frequency domain*. For example, a time series that has a periodicity of 12 months has a frequency of 1/12 cycle per month, the frequency f being the reciprocal of the period T. Electrical engineers, geophysicists, atmospheric scientists, and others in signal processing usually work in the frequency domain.

Previously, time series has been investigated by determining periodicity and other structures using the Box–Jenkins approach. A simple model in the frequency domain is formulated by assuming that a stationary time series with mean zero contains a single sinusoidal component and thus can be represented as

$$Y_t = R \cos(\omega t + \phi) + Z_t$$

where R is the amplitude of the signal, ω is the frequency of the variation, ϕ is the phase angle, and Z_t is the random noise (defined previously). The phase angle ϕ is the amount that Y_t is shifted. It is similar to lag and lead in the time domain. It determines where in the cycle the cosine wave begins. The angle $\omega t + \phi$ is usually expressed in radians ($360° = 2\pi$ radians). Let $Z_t = 0$, $R = 3$, $\phi = \pi/4$ ($45°$) and $\omega = \pi/16$ ($11.25°$). The Y_t is shown in Figure 5.19a. If the time t is in months, $T = 2\pi/(\pi/16) = 32$ months.

More complex signals can be represented as the sum of cosine terms with different phase angles and amplitudes:

$$Y_t = \mu + \sum_{j=1}^{k} R_j \cos(\omega_j t + \phi_j) + Z_t$$

A series with $k = 2$ terms ($R_1 = 3$, $\omega_1 = \pi/16$, and $\phi_j = \pi/4$) and ($R_2 = 2$, $\omega_2 = \pi/6$, and $\phi_2 = 0$), both with $Z_t = 0$, is shown in Figure 5.19b. Figure 5.19c shows the series when

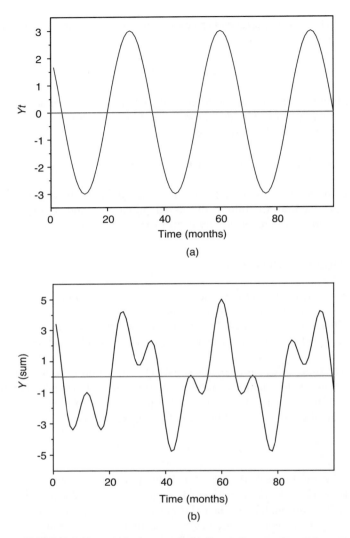

FIGURE 5.19   (a) Cosine wave. (b) Sum of two cosine series.

random noise, $\sigma_Z^2 = 1$, is added. Hence, even a series that appears reasonably complex can be represented by the sum of two cosine terms plus noise.

Since $\cos(\omega t + \phi) = \cos(\omega t)\cos\phi - \sin(\omega t)\sin\phi$, Y_t can be represented as a sum of sines and cosines.

$$Y_t = \mu + \sum_{j=1}^{k}(\alpha_j \cos\omega_j t + \beta_j \sin\omega_j t) + Z_t \qquad (5.1)$$

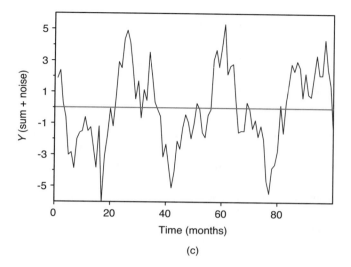

FIGURE 5.19 (*Continued*)   (c) Sum of two cosine series with random noise added.

where $\alpha_j = R_j \cos\phi_j$ and $\beta_j = -R_j \sin\phi_j$ and the $2k + 1$ parameters to be estimated are $\alpha_j, \beta_j, j = 1, \ldots, k$ and μ. Indeed, subject to some mild restrictions, any stationary time series can be represented in this form. This form of the model [equation (5.1)] is similar to the regression model presented in Chapter 4, and a solution can be obtained using least squares.

5.3.1 Nyquist Frequency

When there is the luxury of designing a time series experiment, a key question to be answered is: What range of frequencies is important to the investigator? Clearly, if the desire is to study ice melt and monthly measurements are taken, assessment of day-to-day or week-to-week patterns is not possible. Put another way, assessment of the highest frequency of interest must be determined. If measurements are taken only for a year, it is not possible to measure cyclical changes that occur over longer periods. Thus, the interval between measurements and the time a process is observed are important.

To put this discussion in the context of an equation, the question can be asked: How many frequencies or terms in equation (5.1) are needed? The equivalent statement is: How large should k be? A proof exists that as $k \to \infty$,

$$Y_t = \int_0^\pi \cos(\omega t)\, du(\omega) + \int_0^\pi \sin(\omega t)\, dv(\omega)$$

In the equation above, integration occurs only between 0 and π, as opposed to 0 to ∞, and thus for a discrete process, only frequencies from 0 to π need to be considered.

The justification is quite simple, since

$$\cos[(\omega + k\pi)t] = \begin{cases} \cos\omega t & k \text{ even} \\ \cos(\pi - \omega)t & k \text{ odd} \end{cases}$$

If the t is a unit measure, the highest frequency that can be detected is π. This is called the *Nyquist frequency*, denoted ω_n. If the increment of measurement is Δt, then $\omega_n = \pi/\Delta t$ radians or, equivalently, in cycles per unit of time as $f_n = 1/(2\Delta t)$, where n is the number of (equally spaced) observations. For example, if a time series is measured daily (i.e., $\Delta t = 1$), the Nyquist frequency is $f_n = 1/2$ cycle per day or one cycle every two days. One measurement in a day yields no information about ozone variation within a day. The highest frequency that we can expect to observe is a cycle that repeats itself every two days. When designing a time series study, consideration must be given to the Nyquist frequency.

Another consideration is the lowest frequency to be determined. Clearly, the number of points n must cover a minimum of one cycle. If it is determined that a cycle may be one year and ozone measurements are collected daily, then at an absolute minimum the process needs to be observed for 365 days.

5.3.2   Periodogram Analysis

In the nineteenth and early twentieth centuries, much of the effort in time series analysis attempted to find important specific frequencies. These included studies of planetary orbits and other naturally occurring cyclic phenomena. The formula to accomplish this is a slightly modified equation (5.1), where $k = n/2 - 1$ and $\omega_j = 2\pi j/n, j = 1,\ldots, k$. Then

$$y_t = \widehat{\alpha}_0 + \sum_{j=1}^{k} \left(\widehat{\alpha}_j \cos\frac{2\pi jt}{n} + \widehat{\beta}_j \sin\frac{2\pi jt}{n} \right) + \widehat{\alpha}_{n/2} \cos\pi t$$

where $\widehat{\alpha}_0 = \bar{y}$, $\alpha_{n/2} = \sum_{t=1}^{n} (-1)^t y_t/n$, and

$$\left. \begin{aligned} \widehat{\alpha}_j &= \frac{2\left[\sum_{t=1}^{n} y_t \cos(2\pi jt/n)\right]}{n} \\ \widehat{\beta}_j &= \frac{2\left[\sum_{t=1}^{n} y_t \sin(2\pi jt/n)\right]}{n} \end{aligned} \right\} \quad j = 1,\ldots, (n/2-1)$$

Since there are n observations and n parameters, the data can be fit exactly. This procedure also goes by the names *Fourier analysis* and *harmonic analysis*. The frequency $2\pi/n$ is called the *fundamental frequency* and multiples are called *harmonics*; $2\pi \times 2/n$ is the first harmonic, $2\pi \times 3/n$ is the second harmonic, and so on. The amplitude and phase of the jth frequency components are $R_j = \sqrt{\widehat{\alpha}_j^2 + \widehat{\beta}_j^2}$ and $\phi_j = \tan^{-1}(-\beta_j/\alpha_j)$, respectively.

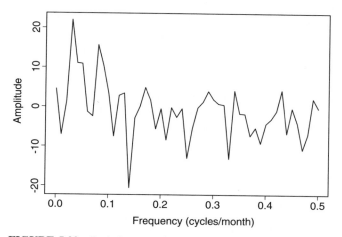

FIGURE 5.20 Periodogram of time series data for Figure 5.19c.

Some algebra shows that

$$\sum_{t=1}^{n}\frac{(y_t-\bar{y})^2}{n}=\sum_{j=1}^{n/2-1}\frac{R_j^2}{2}+\hat{\alpha}_{n/2}^2$$

This is called *Parseval's theorem*. Its importance is that the quantity on the left is a variance and $R_j^2/2$ is the contribution to the variance of the jth frequency component in the interval $\omega_j\pm\pi/n$. In statistical applications, variance is an issue. A high variance can indicate something interesting or, possibly, a problem. In the study of climate dynamics, frequencies associated with a high variance may help identify factors associated with climate change. A histogram plot of $I(\omega_j)=nR_j^2/4\pi$ versus ω_j (or its equivalent expressed in cycles per unit time) is called a *periodogram*. Figure 5.20 is a periodogram for the time series shown in Figure 5.19c. Notice that the two frequencies built into this artificial data set, 0.031 cycle per month ($\pi/16$ radians) and 0.083 cycle per month ($\pi/6$ radians), are seen as large positive amplitudes in Figure 5.20. However, there is considerable noise.

5.3.3 Fast Fourier Transform

Using a *Fourier transform* allows the time series Y_t, $t=0, 1, 2,\ldots$ to be expressed in terms of frequencies as sines and cosines of various amplitudes. It is defined as

$$Y_k=\sum_{j=0}^{n-1}\hat{\alpha}_j e^{-2\pi ijk/n}$$

where $i=\sqrt{-1}$ is a complex number and $k=0,\ldots,n-1$.

Many time series consist of tens of thousands of observations obtained from automated devices. Even with high-speed computers, computational time to obtain parameter estimates can be excessive using the standard Fourier transform. One solution to this problem is the *fast Fourier transform* (FFT), originally discussed by Cooley and Tukey (1965). Instead of summing over n points, the FFT allows summing over $n/2$ points. To use an FFT, n must be able to be factored as a power of 2. The cosine data series contains a 100 points and is not factorable; however, $2^7 = 128$ and the series can be padded by appending 28 zero values. Of course, this will never be done for so few data points. For large data sets, the savings can be significant, reducing the number of computations for n points from $2n^2$ to $2n \log_2(n)$. The expression for the FFT is

$$Y_k = \sum_{j=0}^{n/2-1} \widehat{\alpha}_{2j} e^{-2\pi i (2j)k/n} + \sum_{j=0}^{n/2-1} \widehat{\alpha}_{2j+1} e^{-2\pi i (2j+1)k/n}$$

For a series with $n = 10{,}000$, FFT reduces the number of computations from 2×10^8 to 2.7×10^5.

5.3.4 Spectral Density Function

Of more interest than trying to determine a specific periodicity is an analysis that yields the contribution of a range of frequencies to the signal or time series of interest. In geophysics, for example, the interest may be in determining a range of frequencies that help identify oil-bearing rock. Certain bands of frequency of air quality measurements may help identify anthropogenic-induced climate change.

A tool to help identify these important frequency bands is the *spectral density function*, defined as

$$f(\omega) = \frac{\partial F(\omega)}{\partial \omega}$$

where $F(\omega)$ is the *spectral distribution function*, often abbreviated as the *spectrum* and also called the *power function*. The latter term is from electrical engineering. When the term power is used, think variance. Indeed, *spectral analysis* can be viewed as an analysis of variance technique where $f(\omega) \, \partial\omega$ is the contribution to the variance of the series for frequencies in the range $(\omega, \omega + \partial\omega)$ and $\partial\omega$ is a small increment of frequencies. An estimate of the spectrum $f(\omega) = \partial F(\omega)/\partial\omega$ is

$$\widehat{f}(\omega) = \frac{\lambda_0 \widehat{\gamma}_0 + 2 \sum_{k=1}^{m} \lambda_k \widehat{\gamma}_k \cos\omega k}{\pi}$$

where λ_k are *lag weights*, $\widehat{\gamma}_k$ are the ACVF terms, and $m < n$. Weights can take many forms and result in a smoother function than in a periodogram. Hanning and

Hamming weights are frequently used. Another popular weight is the *Tukey window*,

$$\lambda_k = \frac{1 + \cos(k\pi/m)}{2} \qquad k = 0, 1, \ldots, m$$

Note that weights decline to zero as k increases. Numerous other weight functions are available. Each choice has some statistical implications in terms of the properties of the quantities being estimated. The usual guideline is: The larger the window, the smoother the spectrum. The investigator should try different options and graph the resulting spectrum. Once a choice has been made, the authors recommend doing a sensitivity analysis by changing the weights slightly and seeing the impact on the resulting spectrum.

Smoothing is performed to yield a *consistent estimator*, one that converges to the parameter being estimated as the sample size increases. One may think that the periodogram would be a good estimator of the spectrum; however, it is based on discrete points and is not a consistent estimator of $f(\omega)$.

If $F(\omega)$ is known, the ACVF $\gamma(k)$ is

$$\gamma(k) = \int_0^\pi \cos(\omega k)\, \partial F(\omega)$$

Figure 5.21 shows a power spectrum plot of the cosine data with a span of 10. The unit of the spectrum is the unit of the variance of the time series. A larger span has more points in the lag window and results in a smoother spectrum. It is easier to see that the main contribution to variance comes from frequencies less than 0.1 cycle per month. Contrast this with the periodogram in Figure 5.20. Now consider the spectrum in Figure 5.22 using the 2003 daily detrended ozone data. Notice a sharp drop in frequency at approximately 0.03 cycle/day with period $T \doteq 30$. This may represent a

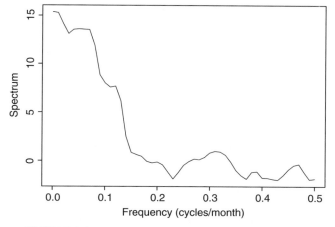

FIGURE 5.21   Spectrum of cosine data with span of 10.

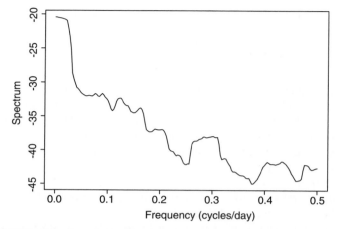

FIGURE 5.22 Spectrum of 2003 daily ozone data in ppm with span = 20.

monthly effect. The spectrum then declines almost linearly until 0.25 cycle/day, where it rises until about 0.31 (a three- to four-day period). It is not obvious what this means, if anything.

As a final example, a spectrum is computed from the AR(1) process used in the illustrative data (Table 5.1). The most power (highest variance) is accounted for by relatively higher frequencies: namely, those around 0.5 cycle/integer unit (Figure 5.23). The ACF of the same series is shown in Figure 5.8.

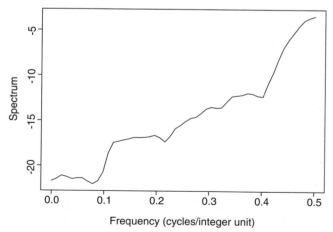

FIGURE 5.23 Spectrum of the AR(1) process versus frequency used in the illustrative data with span = 10.

5.4 WAVELETS

As is noted in the discussion of Fourier analysis, a time series can be represented (modeled) as a sum of sines and cosines. Sometimes this requires a large number of terms and corresponding parameters to estimate. When there are sharp spikes in the data series to be modeled, many terms may be needed. *Wavelet analysis* yields a more compact result by using different sizes of windows. Recall that a Fourier transform uses the same-size window across the domain of the data. However, a window designed to capture a high frequency is inefficient when attempting to capture a low frequency. Wavelet functions are localized in space and thus capture frequency information more efficiently than does the Fourier transform. Wavelet analysis lets the user look at time and frequency components simultaneously. It is applied to earthquake prediction, seismic processing, and the analysis of climate indicators (Burrus et al., 1998).

5.5 SUMMARY

Time series analysis can be performed in the time or frequency domain. The choice of domain is somewhat arbitrary. Many researchers in climate and economics work in the time domain, whereas geophysicists usually work in the frequency domain. In addition to time, many problems in the earth sciences involve distance or space. In time series analysis, the interest is in identifying long-term trends, seasonality, and/or short-term structure. Smoothers (filters) are important diagnostic and modeling tools. Low-pass filters attenuate high-frequency signals and allow one to view long-term trends and cyclic effects. Examples of low-pass filters are Hanning, Hamming, loess, spline, and median. Differencing is an example of a high-pass filter. Exponential smoothers often are useful for short-term forecasting. Accurate forecasts depend on the structure modeled continuing into the future.

Short-term structure can often be modeled in the time domain by autoregressive or moving-average processes or some combination of the two, referred to as an autoregressive moving-average (ARMA) model. A generalization of this model called an autoregressive integrated moving-average (ARIMA) process also models trends and seasonality. An assumption of stationarity, first or second order, is required for many time series models. Nonstationary series in the mean or variance often mask important short-term (high-frequency) variability. Thus, detrending is usually required before proceeding with further analysis. In addition, some forms of seasonality are predictable and well known, such as the yearly cycle of weather data, and should be removed before proceeding. Many of the techniques of model evaluation used in regression, such as viewing residuals for structure, are applicable in the time domain. The autocovariance function and partial autocovariance function plots are important diagnostic tools in the formulation of an ARIMA model.

In the frequency domain, a time series signal can be represented as a sum of sines and cosines. This is called Fourier analysis. The graphical representation of variability versus frequency is called a correlogram or periodogram. This function, when smoothed, yields a spectral density function. Plotting this function gives information about important bands of frequencies in the series: those that account for the most variability and may have physical significance. Wavelet analysis allows a time series to be represented in a more compact form than Fourier analysis by using adaptive windows.

EXERCISES

Exercises 5.1 to 5.13 require the annual precipitation data for Wilmington Delaware for the years 1894 through 2002 (Table 5.4). However, we encourage the reader to substitute or supplement this data set with data from his or her field. There are a multitude of publicly available time series data sets that can be downloaded from the Internet in the areas of climate, water, ice melt, and ocean waves.

5.1 Construct a time series graph. Comment on possible trends.

5.2 Apply several smoothers and contrast the results. Do these help identify trends? Most time series packages have loess, kernel, and median smoothers with the ability to change span and other parameters. Detrend the data if needed.

5.3 Contrast detrending using differences versus regression.

5.4 Is any cyclical/seasonality effect evident from the data? The monthly data can be downloaded from the Web site mentioned above. Using monthly data, does one expect to see any seasonality effect? What is the expected period of the time series?

5.5 Construct an ARIMA model. Generate lag, ACF, PACF, and other graphics as needed. How does one know how many terms to include in the model?

5.6 What is the relationship between what may be observed in a lag plot and an ACF plot?

5.7 Evaluate the fitted model above using appropriate graphics. If the model is reasonable, what does one expect the residual plot to look like?

5.8 Make a forecast for precipitation amount in 2003 using whatever method seems reasonable. The data is available on the Web to check answers.

5.9 Now make a forecast for yearly precipitation in 2023. Comment on the level of confidence in this forecast versus the one made for 2003.

5.10 The time series begins in 1894. How many data are relevant for making a forecast? Explain.

5.11 As discussed in Chapter 4, regression models are often used to make projections beyond the domain of the data. Compare and contrast the use of regression versus time series modeling for making forecasts.

TABLE 5.4

Year	Annual Precipitation (in.)	Year	Annual Precipitation (in.)	Year	Annual Precipitation (in.)
1894	49.9	1931	41.6	1968	31.8
1895	37.1	1932	50.0	1969	40.4
1896	36.7	1933	51.5	1970	38.3
1897	49.2	1934	47.5	1971	52.2
1898	44.2	1935	50.8	1972	48.1
1899	40.2	1936	48.0	1973	47.1
1900	38.6	1937	44.7	1974	39.6
1901	45.9	1938	50.2	1975	49.6
1902	51.1	1939	43.4	1976	33.6
1903	43.4	1940	43.5	1977	40.1
1904	40.8	1941	33.2	1978	51.3
1905	43.9	1942	49.3	1979	53.3
1906	53.5	1943	37.8	1980	33.9
1907	57.0	1944	45.5	1981	35.3
1908	40.6	1945	61.1	1982	41.1
1909	32.8	1946	43.1	1983	54.7
1910	42.8	1947	44.9	1984	41.7
1911	57.1	1948	45.1	1985	33.7
1912	51.4	1949	41.0	1986	42.9
1913	47.0	1950	40.5	1987	36.0
1914	36.7	1951	45.3	1988	37.8
1915	43.7	1952	49.0	1989	49.8
1916	37.1	1953	45.2	1990	44.1
1917	46.4	1954	33.7	1991	39.8
1918	32.7	1955	37.1	1992	37.0
1919	59.5	1956	47.9	1993	46.8
1920	54.3	1957	39.1	1994	45.4
1921	39.8	1958	51.9	1995	40.1
1922	34.3	1959	38.0	1996	52.4
1923	44.4	1960	46.0	1997	28.0
1924	48.2	1961	40.1	1998	36.5
1925	36.0	1962	34.4	1999	47.7
1926	44.0	1963	32.1	2000	45.2
1927	47.2	1964	32.8	2001	33.9
1928	50.9	1965	24.9	2002	39.8
1929	46.3	1966	39.5		
1930	31.6	1967	44.7		

Source: Delaware State Climatologist, http://www.udel.edu/leathers/9595p.txt, accessed Jan. 18, 2004.

5.12 An open question in the study of climate change is long-term trends (say, a linear trend) versus a cycle with a very long period. Do the precipitation data show any effect of cyclic behavior? What is the period of the longest cycle that one can reasonable expect to observe with this data set?

5.13 The converse of Exercise 5.12: What is the shortest period that can be observed? Suppose that the data are monthly as opposed to yearly. What is the shortest period that can be observed? What is the corresponding frequency called?

5.14 Compare and contrast low-pass and high-pass filters. What do these terms mean? When are they to be used?

5.15 Discuss what the phrase stationarity in the mean implies.

5.16 The data from the first few significant earthquakes in the world are listed in Table 5.5. If this series continues such that a reasonable number of observations are obtained, can one analyze it using the methods discussed in this chapter? Why or why not?

TABLE 5.5

Day	Hour	Minute	Second	Latitude	N/S	Longitude	E/W	Magnitude
1	6	25	44.80	5.10	N	92.304	E	6.6
10	18	47	30.10	37.10	N	54.574	E	5.4
10	23	48	50.00	37.02	N	27.804	E	5.4
12	8	40	3.60	0.88	S	21.194	W	6.8
16	20	17	52.70	10.93	N	140.842	E	6.6
19	6	11	36.40	34.06	N	141.491	E	6.5
23	20	10	17.10	1.25	S	119.922	E	6.2
25	16	30	38.90	22.53	N	100.709	E	4.8
25	16	44	16.10	37.62	N	43.703	E	5.9

Source: Data from U.S. Geological Survey Earthquakes Hazard Program, 2005, http://neic.usgs.gov/neis/eqlists/sig_2005.html, accessed Feb. 5, 2007.

5.17 Why is detrending necessary before continuing with further modeling? What happens if there is a failure to detrend when needed?

5.18 When using the Box–Jenkins ARIMA modeling approach, suppose one discovers that the ACF plot declines exponentially and the PACF plot shows only three large (statistically significant) values at lags 1, 2, and 3. What does this suggest about the form of the model?

5.19 Compute a periodogram using the precipitation data in Table 5.4. Now construct a smoothed periodogram (a spectral density plot), trying various smoothing options. Compare the plots. What information do these plots provide?

5.20 Relate the results of the spectral density plot to those of the ACF and PACF made earlier.

6 Spatial Statistics

6.1 INTRODUCTION

Spatial data sets are used to illustrate the methodology of spatial statistics. Graphics play an important role in the understanding of spatial data. We discuss various ways to visualize data in two and three dimensions. Following this, ways to capture spatial continuity are reviewed. The variogram is a traditional tool, similar to the auto-covariance function, used to capture spatial continuity. A general procedure to make estimates in the presence of uncertainty and spatial continuity is called kriging. Various kriging models are presented. Then space–time modeling, used to model changes in precipitation as a function of time and space, is discussed. Other types of spatial models, including lattice structures and point processes, are presented in Chapter 8.

    Geostatistics was defined initially as a branch of geology used to analyze mining processes and estimate mineral resources. The term is sometimes used interchangeably with kriging; however, it is defined more broadly as an analysis of spatial problems that involve correlated variables. It is applied to many other earth science disciplines, including hydrology, climate, oceanography, and ecology.

6.2 DATA

In spatial modeling, the usual preference is for data to be on a grid, thus assuring uniform coverage of the area or volume of interest. An acceptable alternative is for data to be proportional to variability, so that the greatest sampling density is in areas of highest variability. Unfortunately, this rarely happens. Because of convenience or interest, most spatial data are clustered. Three principal data sets used in this chapter are on water-well yield, coal seam thickness, and geochemical analysis.

    In the following sections, a bubble plot will be used to present three-dimensional (x,y,z) data. A *bubble plot* is an extension of a scatter plot (Chapter 1), in which a third dimension (z) such as water-well yield or coal seam thickness is represented by the size of a circle centered at (x,y).

Statistics for Earth and Environmental Scientists, By John H. Schuenemeyer and Lawrence J. Drew
Copyright © 2011 John Wiley & Sons, Inc.

6.2.1   Water-Well Yields in Bedrock Aquifers

A concern in many parts of the world is the availability of an adequate supply of fresh water. Planners and managers want to know how much water is available, scientists want to better understand transport systems and the relationship of water to other geologic phenomena, and homeowners who do not have municipal water need to know where to drill for water on their property. Some of the spatial concepts are illustrated using a water-well yield case study data from Loudoun County, Virginia (Sutphin et al., 2001). A subset of these data is given in Appendix I. The data consists of 754 locations (easting and northing), rock type, and yield in gallons per minute. In the bubble plot of Figure 6.1, the circle size is proportional to the square root of the water-well yield. Questions that an investigator may wish to consider include: "Does yield vary with rock type?", "Is yield spatially correlated?", and "Is there a preferred direction for the spatial correlation?" The last question is suggested because in some studies investigators have found spatial correlation to be present in only one direction. A similar data set from the Pinardville, New Hampshire, Quadrangle (Drew et al., 2003) is also used to illustrate spatial concepts.

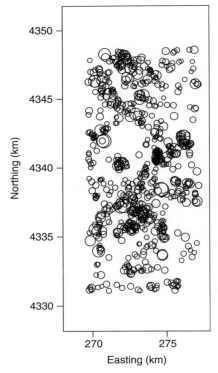

FIGURE 6.1   Bubble plot showing location and size of water-well yields (water-well yield case study).

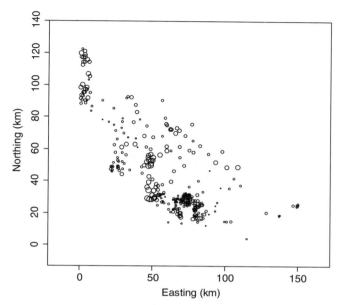

FIGURE 6.2   Bubble plot showing location of drill holes in Harmon coal bed. The circles are proportional to the square root of the coal bed thickness.

6.2.2   Coal

Coal is the primary fuel used in the generation of electrical power. In 1998, the U.S. Geological Survey undertook an assessment of coal resources in the United States. Part of this effort involved estimating spatial correlation in coal bed thickness. One such example is the Harmon coal bed (Fort Union Coal Assessment Team, 1999; Schuenemeyer and Power, 2000). The Harmon coal bed covers an area of 10,464 km^2 in southwestern North Dakota. The data consist of 348 observations of easting, northing, and coal bed thickness. Figure 6.2 shows a bubble plot of the location and size of coal bed thicknesses where the size of the bubble (circle) is proportional to the square root of the thickness. One question to be addressed is: Does coal thickness vary in any systematic way? Another is: Does the spatial correlation allow one to make a better estimate of coal resources than if this information is not used?

6.2.3   Geochemical Data

The geochemical data is from the Cortez Quadrangle (Campbell et al., 1982). This data (Appendix IV; see the book's Web site) is part of the National Uranium Resource Evaluation (NURE) Program. The Hydrogeochemical and Stream Sediment Reconnaissance (HSSR) Program, initiated in 1975, is one of nine components of NURE. The Cortez Quadrangle, a rectangular region, is located in southwestern Colorado and southeastern Utah. A total of 1982 sediment samples and 1706 water samples were collected from 1995 sites and analyzed for uranium. Over 30 other

elements have been analyzed and can be downloaded. Quadrangles from other parts
of the United States are also available.

6.3 THREE-DIMENSIONAL DATA VISUALIZATION

Among procedures to examine three-dimensional data are the previously mentioned
bubble plot, a contour plot, and three-dimensional projections. A *topographic map*,
which shows changes in elevation via contour lines, is an example of a contour plot.
Projections of three-dimensional into two-dimensional space use raw data and/or
fitted regression or other models.

6.3.1 Contouring

Contouring is a way to represent a three-dimensional coordinate system in two
dimensions. The surface is typically a region described by a spatial coordinate
system, such as latitude and longitude. Weather maps showing precipitation or
temperature patterns and elevation maps used by hikers and climbers are examples of
contour maps. Figure 6.3 is a contour plot of a subarea of the water-well yield data in
gallons per minute raised to the fifth root. The purpose of this transformation is to
normalize the skewed data. The contour lines, called *isolines*, represent constant
values of the transformed water-well yield. Circles represent locations of observations. The size of a circle is proportional to the square root of the transformed

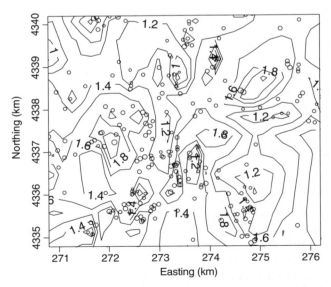

FIGURE 6.3   Contour plot of the fifth root of water-well yields. Circles represent locations of
water-well yield data.

TABLE 6.1 Small Subset of the Water-Well Yield Case Study

Easting	Northing	y (yield)
1	6	6
2	2	4
3	3	15
4	2	8
6	4	5

water-well yield. A contour plot illustrates spatial changes in water-well yield. This plot is examined to see if there are trends, peaks (clustered high values), or discontinuities (sudden jumps) in the data. The compactness of the isolines provides information about how rapidly yield is changing. Sometimes it is possible to identify spatial trends which can be modeled with a linear, quadratic, or cubic surface. The range of the transformed data is 0.428 to 2.512. Several 1.8 isolines appear throughout the region (Figure 6.3). Values are lower above a northing of 4339 km, but in general there is no obvious trend.

How does a contouring algorithm work? A simple contouring procedure is called *triangularization*. Consider a small subset of the water-well yield case study data (Table 6.1). Suppose that one wants a contour line at $y = 10$. This can be achieved by linearly interpolating along the lines connecting the nodes (vertices) to locate the contour for $y = 10$. The interpolated points (Figure 6.4) are connected by dashed lines. Yield values are shown at the nodes (Table 6.1 and Figure 6.4).

Other types of interpolation, such as higher-order polynomials or splines, may give a smoother surface. A common way to create a contour plot is first to "move" data points to a regular grid. This process is called interpolating to a grid, or *gridding*. One approach to gridding is called *nearest neighbor*, which involves taking an average, often an inverse distance weighted average, of points in the vicinity of the grid node. For example, let s_o denote a grid node location, let h_s be the distance (typically, Euclidean) from s_o to data point s, and let r be some arbitrarily chosen radius about s_o. Then an inverse distance-weighted average is

$$Z_{s_o} = \sum_{s \in P} \frac{z_s}{h_s}$$

where P is the set of points within radius r of s_o. Instead of a simple inverse distance weighting, a power of h_s may yield more accurate results. Other procedures involve fitting a surface to the data via polynomial regression, spline functions, or kriging and then interpolating to grid nodes. Kriging is discussed later in the chapter.

There are numerous ways to do interpolation, and there is no unique best way. Interpolated values trim the extremes, and thus estimates, of variance from gridded data are usually biased downward. Interpolation to a grid is illustrated with a small subset of the water-well yield case study data (Figure 6.5). The yield values

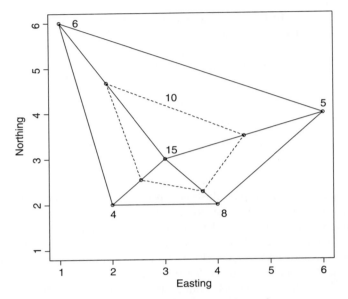

FIGURE 6.4   Constructing a contour plot.

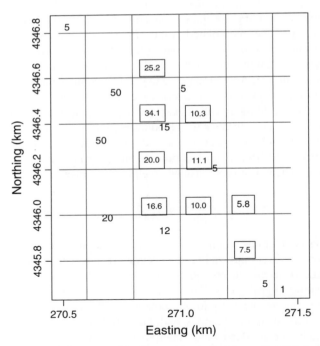

FIGURE 6.5   Interpolation to a grid using a subset of water-well yield case study data.

(the integers) are plotted at their actual locations. The interpolated values are in boxes to the right and above their corresponding grid notes. Linear interpolation is used in the triangles determined by the data points.

Once points are located at the grid nodes, contours lines may be drawn using triangularization or other schemes. Triangularization is the process of forming triangles from a set of points in a plane such that there are no points inside any triangle. For example, four triangles are constructed from the five points in Figure 6.4. Once triangles are constructed, interpolation between points can occur. Some contouring algorithms interpolate through existing data values. Others provide the option of specifying measurement error. A key option is specification of the degree of smoothness of contour lines. Some contour lines follow the data closely, whereas others do not. Some contours bend sharply at data points, whereas others bend more gradually. One note of caution: Contour lines are not extrapolated reliably outside a convex hull enclosing the data set, which is why, as in Figure 6.3, the data point locations should be shown on the contour plot.

There is no right way to contour. Try different features and parameters to gain insight into the data set. Look for trends, anomalies, discontinuities, and other relationships.

6.3.2   Plotting Three-Dimensional Data

Four graphical projection procedures are illustrated using the Harmon coal quality data on a longitude–latitude scale (Figure 6.6). They are a *wire plot* (Figure 6.6a), where the height of the observation, the *z*-variable (coal thickness), is represented as a distance from the *xy*-plane; a *bar plot* (Figure 6.6b), which is similar to a wire plot except that neighboring values of the *z*-variable are summarized in bars, much as in a histogram; a *three-dimensional regression plot* (Figure 6.6c); and a *spline surface* (Figure 6.6d). Each display provides a somewhat different insight into the spatial variability of coal bed thickness. Wire diagrams are sometimes used to identify clusters of data and allow the user to view the raw data. The bar plot, a three-dimensional histogram, smoothes as a function of the length and width of the bar. The regression plot can do more smoothing, depending on the degree of the polynomial used in the fitting (i.e., linear, quadratic, or cubic). The spline surface smoothes but generally follows the data. A major benefit of a three-dimensional plot is that in most software implementations, the user can rotate coordinate systems dynamically to gain different perspectives. Cleveland (1993) has provided further information on three-dimensional graphics.

6.4   SPATIAL ASSOCIATION

How do we characterize the relationships between samples taken at varying distances? Remember that the covariance and correlation coefficients are measures of the degree of linear association between two random variables: for example, the nitrogen (X) and the potassium (Y) content in soil. Instead of estimating the

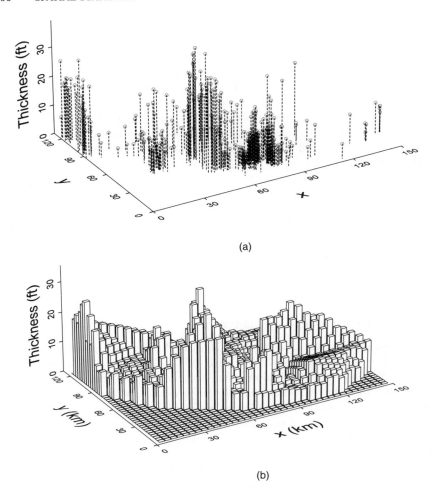

FIGURE 6.6   (a) Wire diagram, Harmon coal bed thickness. (b) Bar plot, Harmon coal bed thickness.

covariance between two random variables X and Y, suppose that we want to estimate the association between a single random variable $Y(\mathbf{s})$ observed at location $\mathbf{s}$, and a second occurrence, $Y(\mathbf{s} + \mathbf{h})$, at a distance $\mathbf{h}$ from $\mathbf{s}$.

The study of spatial association begins with data taken in one dimension, such as down a hole or along a transect. Examples include measurements of dissolved oxygen at points along a stream, sand grain sizes along a beach, or CO_2 measurements in ice cores. Consider the latter example using selected observations from the Vostok ice core data (Ruddiman and Raymo, 2004). A graph of depth and CO_2 in parts per million volume is shown in Figure 6.7.

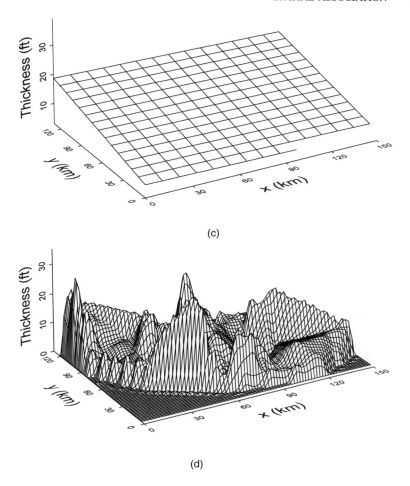

FIGURE 6.6 (*continued*)   (c) Regression plot, Harmon coal bed thickness. (d) Spline surface, Harmon coal bed thickness.

Suppose that our interest is in examining CO_2 between 500 and 1500 m. Clearly, there is an increasing trend in CO_2 with depth. A more interesting study may be the occurrence of structure in the data after this linear trend has been removed. The trend can be removed via linear regression and then an analysis performed on the residuals. However, since the original measurements are not evenly spaced, the data are interpolated to even intervals of approximately 30 m. For illustrative purposes, depth and residual R of Table 6.2 are investigated. The residuals result from a linear fit of CO_2 on the depth in the interval 500 to 1500 m. The columns labeled R1 and R2 are the residuals R lagged by 1 and 2 distance units, respectively. A distance unit in this example is approximately 30 m. The variances (Var), covariances (Cov), and correlations (Cor) are displayed at the bottom of Table 6.2. With each succeeding

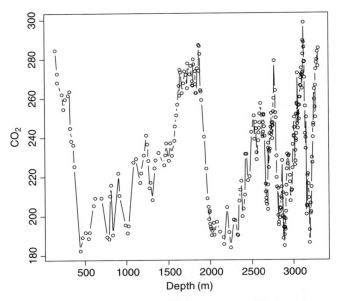

FIGURE 6.7 CO_2 versus down hole depth for a Vostok ice core.

TABLE 6.2 CO_2 Residuals from a Regression on Vostok Ice Core Data, Lagged Values, and Covariances

Depth (m)	R, Residual CO_2 (ppmv)	R1	R2	Depth (m)	R, Residual CO_2 (ppmv)	R1	R2
506.4	0.0	− 6.4	1.8	1021.6	− 18.9	− 4.2	10.1
536.7	− 6.4	1.8	8.9	1051.9	− 4.2	10.1	12.3
567.0	1.8	8.9	12.6	1082.2	10.1	12.3	0.5
597.3	8.9	12.6	13.1	1112.5	12.3	0.5	1.6
627.6	12.6	13.1	10.8	1142.8	0.5	1.6	8.1
657.9	13.1	10.8	6.4	1173.1	1.6	8.1	18.8
688.2	10.8	6.4	− 3.5	1203.4	8.1	18.8	1.5
718.5	6.4	− 3.5	− 20.6	1233.7	18.8	1.5	− 8.7
748.8	− 3.5	− 20.6	12.2	1264.0	1.5	− 8.7	− 11.7
779.1	− 20.6	12.2	− 17.7	1294.3	− 8.7	− 11.7	2.8
809.4	12.2	− 17.7	1.7	1324.6	− 11.7	2.8	4.0
839.7	− 17.7	1.7	15.2	1354.9	2.8	4.0	− 3.2
870.0	1.7	15.2	− 13.2	1385.2	4.0	− 3.2	− 6.3
900.3	15.2	− 13.2	− 16.3	1415.5	− 3.2	− 6.3	4.8
930.6	− 13.2	− 16.3	− 17.4	1445.8	− 6.3	4.8	—
960.9	− 16.3	− 17.4	− 18.9	1476.1	4.8	—	—
991.3	− 17.4	− 18.9	− 4.2				

Var(R)	Cov(R,R1)	Cov(R,R2)	Cor(R,R1)	Cor(R,R2)
122.1	36.6	2.5	0.30	0.02

lag, one observation is lost. As in time series analysis, estimating covariances is not recommended for more than $n/4$ lags, where n is the number of data points. Observe that the covariance between R and R1 is significantly lower than the variance of R. The covariance between R and R2 is even lower. A weak positive correlation (0.30) is observed between R and R1. Visual insight can be gained by plotting the original data set, the R's, against lagged values (see the Exercises). Typically, as the lag increases, the covariance decreases since observations are farther apart.

Many earth science data sets are spatial and have two- or three-dimensional coordinate systems. Thus, instead of the (lag) distance h being a scalar, it will be a vector, denoted as $\mathbf{h}$ (sometimes called the *spatial lag*), and the location will be a vector $\mathbf{s}$. For example, it may be that $\mathbf{h} = (3,5)'$, where 3 and 5 are the distance in the x and y directions, respectively. In addition, most data will not be located on a regular grid.

Three major ways to represent spatial association are (1) the covariance, (2) its standardized version, the correlogram, and (3) the inverse or dual of the covariance, the semivariogram. After presentation of these procedures, we comment on reasons for choosing between a covariogram and a variogram estimator.

6.4.1   Covariogram

Location of water-well yields in the northwestern corner of the water-well yield case study data is the top graph in Figure 6.8. The center of the circle in the graph is

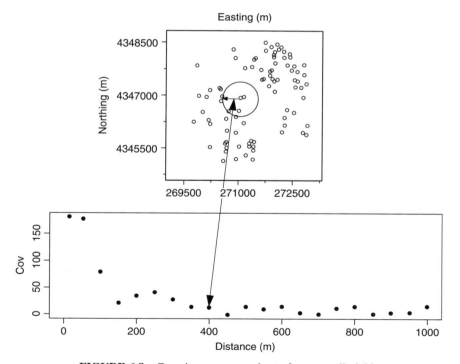

FIGURE 6.8   Covariogram on a subset of water-well yield.

a point $z(\mathbf{s})$. The circumference of the circle represents a locus of points at the head, $z(\mathbf{s} + \mathbf{h})$. The radius of this circle is $h = |\mathbf{h}|$, a vector of length 400 m. The bottom graph is a plot of Cov(h) versus h for the entire 754 observations in the water-well yield case study data. It is called an *empirical covariogram* and is defined in the following paragraph.

Consider the variable $z(\mathbf{s})$ and the variable $z(\mathbf{s} + \mathbf{h})$, which measures the same quantity at a distance $\mathbf{h}$ from $\mathbf{s}$. For example, z can represent water-well yield. The z's at the start of $\mathbf{h}$, the $z(\mathbf{s})$'s, are called *tail values*; those at the end of $\mathbf{h}$, the $z(\mathbf{s} + \mathbf{h})$'s, are called *head values*. Their corresponding means are

$$m_{-\mathbf{h}} = \frac{1}{|N(\mathbf{h})|} \sum_{N(\mathbf{h})} z(\mathbf{s}) \quad \text{and} \quad m_{+\mathbf{h}} = \frac{1}{|N(\mathbf{h})|} \sum_{N(\mathbf{h})} z(\mathbf{s}+\mathbf{h})$$

where $N(\mathbf{h})$ is the set of all pairwise Euclidean distances h and $|N(\mathbf{h})|$ is the number of distinct pairs in $N(\mathbf{h})$. The *covariance function* for random variables z, separated by a distance $\mathbf{h}$, is

$$\text{Cov}(\mathbf{h}) = \frac{1}{|N(\mathbf{h})|} \sum_{N(\mathbf{h})} z(\mathbf{s})z(\mathbf{s}+\mathbf{h}) - m_{-\mathbf{h}}m_{+\mathbf{h}}$$

Cov($\mathbf{h}$) is a shorthand way to write Cov($z(\mathbf{s})$, $z(\mathbf{s} + \mathbf{h})$) and is called a *covariogram*. In time series, it is called an autocovariance function. Because most data are irregularly spaced, data values in the neighborhood of $\mathbf{h}$ are averaged. Note that Cov(0) = Var(z). Most applications require that the covariance be second-order stationary. Recall from Chapter 5 that second-order stationarity requires that $E[z(\mathbf{s})]$ exist and be invariant over all $\mathbf{s}$ and that the Cov($z(\mathbf{s})$, $z(\mathbf{s} + \mathbf{h})$) is only a function of $\mathbf{h}$.

The double arrow connecting the top and bottom graphs in Figure 6.8 shows the covariance corresponding to $h = 400$ averaged over all $z(\mathbf{s})$ and those points that are approximately 400 m away. Does the information in this plot correspond to what is expected? How does it compare with the Vostok ice core data in Table 6.2?

6.4.2   Correlation

Previously, it was noted that correlation is a standardized or unit-free measure of association. Correlation as a function of distance $\mathbf{h}$ is defined as

$$\rho(\mathbf{h}) = \frac{\text{Cov}(\mathbf{h})}{\sqrt{\sigma_{-\mathbf{h}}^2 \sigma_{+\mathbf{h}}^2}} \quad -1 \leq \rho(\mathbf{h}) \leq 1$$

where

$$\sigma_{-\mathbf{h}}^2 = \frac{1}{|N(\mathbf{h})|} \sum_{N(\mathbf{h})} [z(\mathbf{s}) - m_{-\mathbf{h}}]^2 \quad \text{and} \quad \sigma_{+\mathbf{h}}^2 = \frac{1}{|N(\mathbf{h})|} \sum_{N(\mathbf{h})} [z(\mathbf{s}+\mathbf{h}) - m_{+\mathbf{h}}]^2$$

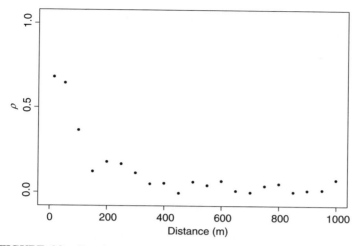

FIGURE 6.9   Correlogram plot on the water-well yield case study data.

are the variances at the tail and head, respectively. The correlogram for the water-well yield case study is shown in Figure 6.9. It is similar in shape to the covariogram in Figure 6.8.

Recall that in time series analysis a correlogram was constructed. The correlation function $\rho(\mathbf{h})$ that has been just constructed is the equivalent of $\rho(k)$ in time series analysis, where k is the lag. Another parallel between time series and spatial analysis is the effect of a trend on correlation. Failure to remove a trend or low-frequency signal in time series analysis may mask important higher-frequency signals. Similarly, trend also needs to be accounted for in spatial modeling.

6.4.3   Semivariogram

The covariance and correlation functions measure association between variables, or in time series analysis, the relationship between a series and the series lagged. When spatial correlation exists, the Cov(**h**) is usually larger for the small **h** (points near each other) and decrease as **h** increases. Another way to view the same phenomenon is with a measure of dissimilarity called an *empirical semivariogram*, which is defined as

$$\gamma(\mathbf{h}) \equiv \frac{1}{2|N(\mathbf{h})|} \sum_{N(\mathbf{h})} [z(\mathbf{s}) - z(\mathbf{s} + \mathbf{h})]^2$$

When nearby observations $z(\mathbf{s})$ and $z(\mathbf{s} + \mathbf{h})$ are similar, the value of $\gamma(\mathbf{h})$ will be small, whereas more distant (large **h**) observations typically are dissimilar, which yields a large $\gamma(\mathbf{h})$. Some authors use the term variogram to describe $\gamma(\mathbf{h})$, but we refer to it as a *semivariogram* and $2\gamma(\mathbf{h})$ as a *variogram*. Cressie (1993) refers to the formulation above as a classic semivariogram estimator.

To estimate Cov(**h**), second-order stationarity is required (see Chapter 5 for a definition), which implies that

$$\gamma(\mathbf{h}) = \text{Cov}(0) - \text{Cov}(\mathbf{h}) \quad \text{and} \quad \rho(\mathbf{h}) = 1 - \frac{\gamma(\mathbf{h})}{\text{Cov}(0)}$$

A less restrictive requirement necessary for a proper semivariogram is called *intrinsic stationarity*, defined as

$$E[z(\mathbf{s}+\mathbf{h})] - E[z(\mathbf{s})] = 0$$

and

$$\text{Var}(z(\mathbf{s}+\mathbf{h}) - z(\mathbf{s})) = 2\gamma(\mathbf{h})$$

where $\text{Var}(z(\mathbf{s}))$ is assumed to be finite. Second-order stationarity implies intrinsic stationarity, but the converse is not true.

Because the classic semivariogram requires a sum of squares, it is sensitive to outliers. An alternative is a *robust semivariogram estimator* (Cressie and Hawkins, 1980), which reduces the effect of outliers and is recommended when the Gaussian assumption is suspect. It is defined as

$$\gamma(\mathbf{h}) \equiv \frac{1}{2} \left[\frac{1}{|N(\mathbf{h})|} \sum_{N(\mathbf{h})} |z(\mathbf{s}) - z(\mathbf{s}+\mathbf{h})|^{1/2} \right]^4 \Bigg/ (0.457 + 0.494/|N(\mathbf{h})|)$$

The robust semivariogram estimator (Figure 6.10a) is below that of the classical estimator because it down-weights the outliers that are present. We recommend

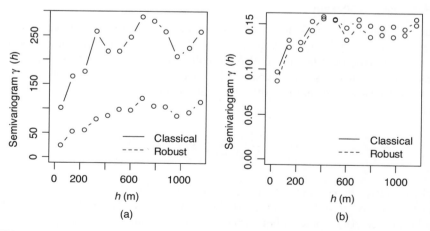

FIGURE 6.10   Classical and robust semivariograms plotted against h for the water-well yield case study data for (a) yield (gpm) and (b) the fifth root of yield.

applying both types of estimators. A significant difference may indicate a skewed distribution, errors in the data, discontinuities, or mixtures of distributions. The semivariogram estimators for the fifth root of yield (Figure 6.10b) are almost congruent because the distribution is symmetric.

Most semivariograms are monotone increasing except for sampling variability. Occasionally, one contains a cyclical pattern. For example, the pattern in the classical estimator (Figure 6.10a, top graph) suggests a possible *hole effect* (Journel and Huijbregts, 1978). A hole effect may indicate important patterns in spatial variability and structure, such as regular bedding in geology or a regular pattern in fractured bedrock in hydrology.

6.4.4 Variogram Clouds

An important diagnostic tool is the *variogram cloud*, which is a plot of a function of pairwise differences of the variable z versus the distance between them. Specifically, if $z(\mathbf{s}_i)$ is the value z at location i (**s** defines the specific coordinate) and $z(\mathbf{s}_j)$ is the value z at location j, their pairwise difference is defined as

$$ZD_{ij} = \frac{(z(\mathbf{s}_i) - z(\mathbf{s}_j))^2}{2}$$

and h_{ij} is the Euclidean distance between points i and j. Variogram clouds can help detect outliers and trends (see Section 6.5 for a discussion of trends).

The example uses the fifth root of the 754 water-well yields in gpm. Figure 6.11a is a variogram cloud for water-well yields. Each open circle on the graph represents the

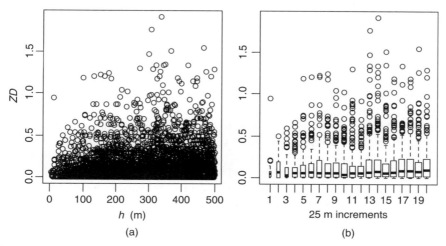

FIGURE 6.11 Variogram cloud on water-well yield data, classical estimator: (a) distance between points are represented by circles; (b) distances are grouped by 25-m increments and shown as boxplots.

squared difference of a pair of points h units apart. In this graph, only pairs of water wells at a distance of 500 m or less are represented. The advantage of this display is that all the data are shown; it is also the disadvantage because patterns may be masked. Figure 6.11b summarizes the cloud distances into boxplots at 25-m intervals. Clearly, distance measurements are highly right-skewed.

6.5 EFFECT OF TREND

Recall that in time series analysis, long-term trends often mask autocorrelation structure. The same is true in spatial statistics. Thus, before computing a semivariogram, any long-term trends need to be removed. Insight into the presence of trends can be gained via data plots, including the use of a smoothing function to filter out high-frequency signals and outliers. Contour plots are also useful tools. For one-dimensional data, a simple plot of the variable versus distance often will suffice. Additional insight can be gained by fitting a nonparametric smoother, such as a loess model, through the data.

Once a trend has been detected, it must be removed. However, only the trend must be removed, not the information on the spatial association between nearby points. In time series analysis, first differences (and occasionally, second) are used to remove a trend. A detrended time series may be represented as

$$Y_t = X_t - X_{t-1}$$

where the X's represent the original series and the Y's the detrended series from first differences. The same procedure can also be used for evenly spaced data along a transect. For spatial data in two or three dimensions, the equivalent of Y_t may be a residual from a polynomial or nonparametric model fit. A useful approach for modeling an additive trend is median polish (Tukey, 1977). Ordinary and generalized linear models (McCullagh and Nelder, 1989) may also be appropriate.

Consider uranium U in the Cortez Quadrangle data set. To investigate the presence of a trend, begin by plotting U versus easting (Figure 6.12a) and northing (Figure 6.12b) distances. What observations can be made about these graphs? Probably not many because the skewness of the data (Figure 6.13a) masks any trend. A highly right-skewed distribution implies that most of the observations are small and only a few are large. The empirical distribution of the U values has a long right-hand tail, with 13 of the 1657 total samples greater than 10 ppm. The maximum value of U is 76.4 ppm. Thus, most of the data are so compressed when plotted as to be unintelligible. A log base 10 transformation results in a more symmetric distribution (Figure 6.13b). A plot $\log_{10}(U)$ versus the easting (Figure 6.14) shows $\log_{10}(U)$ increasing from east to west, whereas there are no apparent trends in $\log_{10}(U)$ from north to south (not shown). The model $\log_{10}U = \beta_0 + \beta_1 X + \varepsilon$ is fit to remove the east–west trend. This model is significant at the 5% level with $R^2 = 0.26$ and a slope, $\widehat{\beta}_1 = 0.001827$.

What happens when a spatial trend is not accounted for in a model? Consider the scatter plot in Figure 6.14 of $\log_{10}(U)$ versus easting. Figure 6.15a shows the

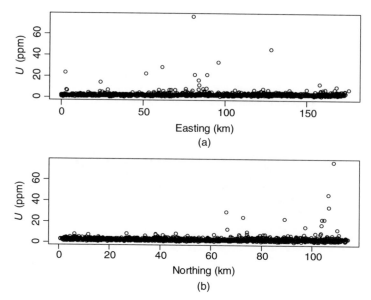

FIGURE 6.12 U versus (a) easting and (b) northing, Cortez Quadrangle data.

semivariogram of $\log_{10}(U)$ of the data shown in Figure 6.14: that is, with the east–west trend present. Note the gradual, almost linear increase in the semivariogram (Figure 6.15a) after 10 km. Contrast this with the semivariogram of the residuals after the trend was removed (Figure 6.15b). Note the increase in the semivariogram to essentially a constant value at 30 km. Consider a third alternative (Figure 6.15c), where the trend (slope) is artificially inflated by an order of magnitude. This semivariogram illustrates the effect of a very strong, unaccounted for linear trend. Figure 6.15b provides the most useful information. Figure 6.15a and c

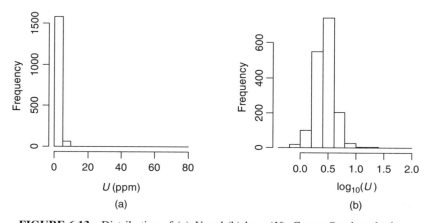

FIGURE 6.13   Distribution of (a) U and (b) $\log_{10}(U)$, Cortez Quadrangle data.

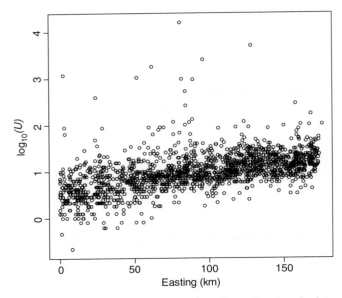

FIGURE 6.14 $\text{Log}_{10}(U)$ versus easting, Cortez Quadrangle data.

violate the intrinsic stationarity requirement, which is $E[Z(\mathbf{s}+\mathbf{h})] - E[Z(\mathbf{s})] = 0$ because of an unaccounted for trend in the data. The horizontal line in these figures is at the approximate asymptotic value of the semivariogram (Figure 6.15b).

6.6   SEMIVARIOGRAM MODELS

When an empirical semivariogram is seen that displays a pattern of spatial associa-tion (e.g., any of the semivariograms in Figure 6.10), the investigator typically wants to fit that pattern with a model. Models are useful to make comparisons between strengths of spatial association among data sets and to provide weights for the method of spatial estimation called kriging, discussed in Section 6.7. Prior to fitting a semivariogram model, it is assumed that there is no trend in the data or that it has been removed. The following models are among the most commonly used to fit an empirical semivariogram.

- *Spherical model:*

$$g(h) = \begin{cases} c\left[\dfrac{1.5h}{a} - 0.5\left(\dfrac{h}{a}\right)^{3}\right] & h \le a \\ c & h > a \end{cases}$$

where c is the covariance contribution or *partial sill* value and a is the range or extent of spatial association. The model is shown in Figure 6.16a.

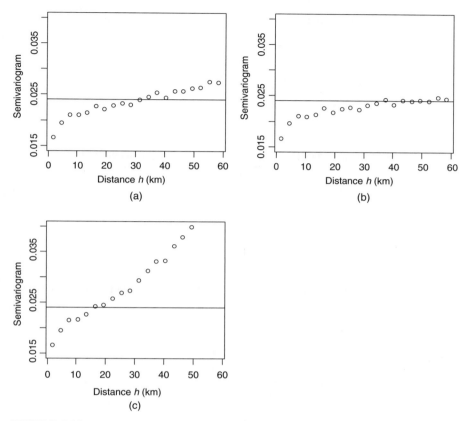

FIGURE 6.15   Semivariogram plots, Cortez Quadrangle data: (a) weak trend; (b) residuals; (c) strong trend. The horizontal line is at the approximate asymptotic value for the detrended (residual) data.

- *Exponential model:*

$$g(h) = c\left[1 - \exp\left(-\frac{3h}{a}\right)\right]$$

This model is shown in Figure 6.16b. Since $g(h) \to c$ as $h \to \infty$, the range a is infinite. A practical range, corresponding to the solution of $g(h) = 0.95c$ for a, is defined arbitrarily.

- *Gaussian model:*

$$g(h) = c\left[1 - \exp\left(-\frac{3h^2}{a^2}\right)\right]$$

This model is shown in Figure 6.16c. As in the exponential model, the range is infinite. A practical range for a is the solution to $g(h) = 0.95c$.

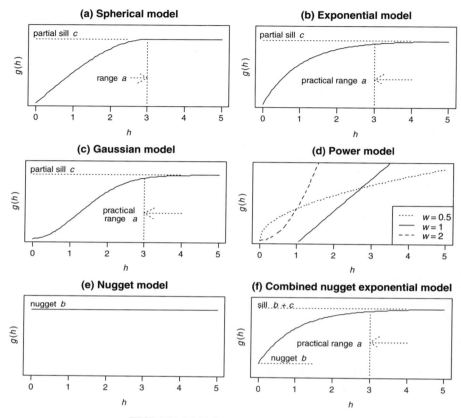

FIGURE 6.16 Semivariogram models.

- *Power model:*

$$g(h) = mh^\omega \qquad 0 < \omega < 2$$

where m is the slope. When $\omega = 1$, this is a linear model. Unlike previous models, $g(h)$ is unbounded at $h = \infty$. The power model for $\omega = 0.5, 1,$ and 2, with $m = 0.5$, is shown in Figure 6.16d. Because it is unbounded, this model has no covariance counterpart.

- *Nugget model:*

$$g(h) = \begin{cases} 0 & h = 0 \\ b & \text{otherwise} \end{cases}$$

A nugget model (Figure 6.16e) implies that there is no covariance component. The nugget is b. For a pure nugget model as shown in Figure 6.16e, regression methods are appropriate; however, an empirical semivariogram often will contain a nugget term plus one or more of the models described previously. For example, in Figure 6.15b the nugget is 0.015 by graphical fit.

- *Combined nugget exponential model:*

$$g(h) = \begin{cases} 0 & h = 0 \\ b + c\left[1 - \exp\left(-\dfrac{3h}{a}\right)\right] & h > 0 \end{cases}$$

This combined model (Figure 6.16f) has a nugget b. The *still* is defined in this book as $b + c$. Some literature refers to c as the *sill*. The nugget is discussed in detail in the next section.

These models are called *permissible*, which means that the variance of any linear combination of $z(\mathbf{s})$, expressed as covariances, must be nonnegative. Any linear combination of these models can be used to fit a semivariogram. For example,

$$g(h) = \begin{cases} b + w_1\left[1 - \exp\left(-\dfrac{3h}{a_1}\right)\right] + w_2\left[1 - \exp\left(-\dfrac{3h^2}{a_2^2}\right)\right] & h \le a_2 \\ b + w_1\left[1 - \exp\left(-\dfrac{3h}{a_1}\right)\right] + w_2 & h > a_2 \end{cases}$$

6.6.1 Model Fitting and Evaluation

How should a model be chosen? It is rarely possible to apply first principles (theoretical considerations) to formulating a semivariogram (or covariance) model as sometimes occurs in regression modeling. The choice of a model is usually based on a best fit, which may be visual or statistical.

When $h \to 0$, $\gamma(h)$ should approach zero and be equal to zero for $h = 0$; however, sometimes, $\gamma(0) > 0$. One reason that $\gamma(0)$ may appear to be greater than zero is that the behavior of $\gamma(h)$ is rarely known near $h = 0$ because of a lack of tightly spaced data. For example, in the water-well yield case study data, the nearest observations are 5.5 m apart and there are relatively few pairs of these. Thus, it is not known from empirical data how $\gamma(h)$ behaves for small h. There may be what is called *microscale variation*, which is natural variability between nearby samples. In addition, there can be *measurement error* resulting from different measured values taken on the same object. Consider the classical empirical semivariogram shown in Figure 6.10a, top graph. The model does not appear to pass through the (0,0) point, and it may be inappropriate to force it. Recall the potential problem in linear regression when one attempts to force a model through (0,0). This is why the data are fit to the model $g(h) + b$, where b is the nugget effect. Later, $b = 0$ can be tested.

To illustrate model fitting, the water-well yield case study data are used. Because the variable of interest, water-well yield in gallons per minute, is highly right skewed,

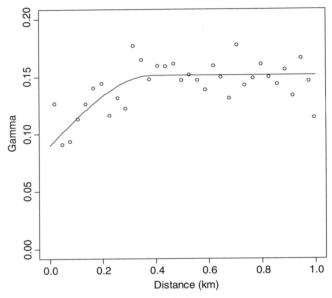

FIGURE 6.17   Spherical model fit to empirical semivariogram, fifth root of water-well yield data; a (range) $= 0.387$, b (nugget) $= 0.090$, $b + c$ (sill) $= 0.151$; water-well yield case study.

the fifth-root transformation is used. Subsequent analysis of the water-well yield case study data is performed on this transformed variable. A spherical model is chosen after viewing the empirical semivariogram (the circles in Figure 6.17). The semivariogram data used for this plot are shown in Table 6.3, where np is the number of observations in $N(h)$, the neighborhood of h. The method used to fit this spherical model is weighted least squares, where the weight is proportional to the number of point pairs $N(h)$. Another useful weighting scheme is for the weight to be proportional to $|N(h)|$ and inversely proportional to h^2. The authors leave it to the reader to consider other fitting options. The semivariogram shows that yields from wells closer to each other are more correlated than those farther apart; that is, nearby wells have more similar yields. When wells are more than 0.387 km apart, the correlation between yields is essentially zero.

Evaluating a semivariogram model is challenging. At first glance, cross-validation might seem like a good approach; however, it has several severe limitations, as described by Goovaerts (1997, Chap. 4). Probably the best approach is to look carefully at the fit graphically and consider the purpose of the study, including the importance of short- versus long-range association.

6.6.2   Preferred Directions: Geometric and Zonal Anisotropy

So far, spatial variability has been considered to exhibit the same properties (values of nugget, sill, range, and form of model) in all directions. This phenomenon is called

TABLE 6.3   Water-Well Yield Case Study Empirical Semivariogram Data

Seq. Num.	np	$h=$Distance (km)	$\gamma(h)=$Gamma	Seq. Num.	np	$h=$Distance (km)	$\gamma(h)=$Gamma
1	42	0.019	0.127	18	327	0.525	0.152
2	89	0.047	0.091	19	297	0.555	0.147
3	140	0.076	0.094	20	335	0.585	0.138
4	190	0.105	0.113	21	308	0.616	0.159
5	195	0.136	0.126	22	339	0.645	0.149
6	236	0.165	0.140	23	340	0.675	0.131
7	268	0.195	0.144	24	396	0.705	0.177
8	268	0.225	0.116	25	375	0.734	0.142
9	259	0.255	0.131	26	419	0.765	0.149
10	305	0.285	0.123	27	359	0.795	0.160
11	298	0.315	0.177	28	373	0.825	0.149
12	344	0.345	0.165	29	424	0.856	0.143
13	330	0.375	0.147	30	436	0.886	0.156
14	288	0.405	0.159	31	408	0.915	0.133
15	303	0.436	0.159	32	471	0.946	0.166
16	307	0.466	0.161	33	514	0.975	0.146
17	319	0.496	0.147	34	159	0.995	0.114

isotropy. Often in nature, characteristics of spatial association vary with direction. This phenomenon is called *anisotropy*. There are two types of anisotropy: geometric anisotropy and zonal anisotropy. *Geometric anisotropy* occurs when the range varies with direction but all other properties of spatial association remain the same (Figure 6.18b). *Zonal anisotropy* occurs when the sill varies with direction.

Isotropy is represented by a circle in Figure 6.18a, indicating that at a given distance r_ϕ the correlation between points on the circumference and the center is the same in all directions. Conversely, in an anisotropic situation, the correlation has a preferred direction, θ in Figure 6.18b. The relative strength or range of association is represented by λ, the ratio of the semiminor r_ϕ-axis to the semimajor r_θ-axis of the ellipse in Figure 6.18b, where $0 \le \lambda = r_\phi/r_\theta \le 1$ and h_x and h_y are the spatial lags in the easting and northing directions, respectively. When geometric isotropy exists, $\lambda = 1$. Geometric anisotropy can be caused by preferred wind or current direction, magnetic fields, or faulting in the Earth's surface. Detecting anisotropy can be difficult. Figure 6.19 is one tool. It represents spatial association as characterized by semivariograms for 566 water wells ranging in value from 0 to less than 40 gpm within the Migmatite Gneiss (PDmg) rocks of the Pinardville, New Hampshire, Quadrangle (Drew et al., 2003). Yield in this range is considered to be low. The columns represent different angles of possible preferred directions (θ in Figure 6.18). In Figure 6.19 they are chosen to be 0°, 45°, 90°, and 135° clockwise, with 0° as north. The pairs of points considered in each of the four directions are within ±22.5° (the *azimuth tolerance*) of the direction preferred. Typically, the azimuth tolerance can be changed by the user. A small tolerance sweeps out fewer points. A larger tolerance

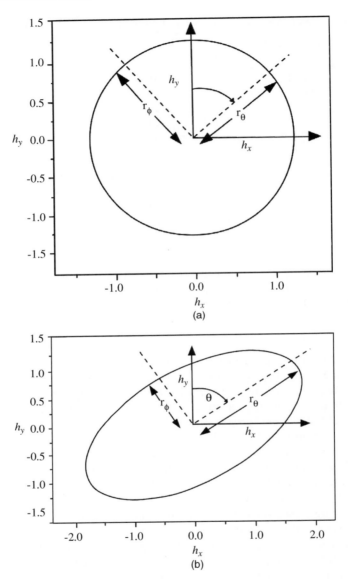

FIGURE 6.18   (a) Representation of isotropy. (b) Representation of geometric anisotropy.

sweeps out more points but may mask a preferred direction. The rows (Figure 6.19) represent geometric ratios of anisotropy. The four rows in Figure 6.19 represent $1/\lambda = 2$, 1.75, 1.5, and 1.25, from top to bottom. The empirical semivariograms appear similar within columns but differ somewhat across columns. Note that at 90° there is a slight dip at a range of approximately 6000 m. No cause can be identified for

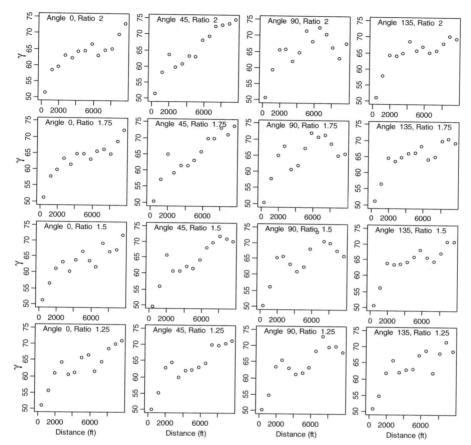

FIGURE 6.19 Display of empirical semivariograms at various azimuth angles of 0°, 45°, 90°, and 135° and ratios of 2, 1.75, 1.5, and 1.25 in the PDmg rocks, Pinardville Quadrangle water-well yield data.

this dip. If there is prior information, one may be able to compute individual empirical semivariograms by specifying a preferred direction (azimuth) and associated toler-ance. The similarity of the empirical semivariograms within columns suggests that there is no zonal anisotropy. Although the empirical semivariograms are noisy, it appears that the best from a directional perspective occur at 135°. This is confirmed by the variogram map (Figure 6.20).

A semivariogram map is another style of display of 566 Pinardville, New Hampshire, Quadrangle low-yield water wells. Each block is the center of lag units in the x and y directions. The block value $\gamma(\mathbf{h})$ is represented by a color scale from dark blue to red (low to high values). The h_x and h_y are the h distances in the x (horizontal) and y (vertical) directions. In this map, the direction of the $\mathbf{h}$'s varies between 0° and 180°. If the spatial pattern is isotropic, one expects to

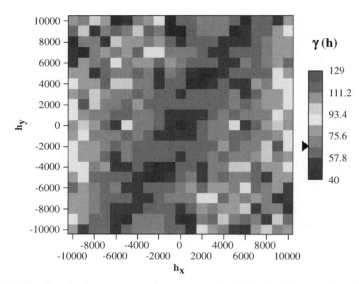

FIGURE 6.20 Semivariogram map for water-well yields in PDmg rocks, Pinardville Quadrangle. [The small discrepancy between this figure and the original in Plate 2C of Drew et al. (2003) was caused by the reclassifying of several wells.] (*See insert for color representation of the figure.*)

see the same $\gamma(\mathbf{h})$ pattern in all directions: namely, blue squares representing lower values of $\gamma(\mathbf{h})$ near the center. The squares should go toward yellow and red as $\mathbf{h}$ increases. This is not the case for the data shown in Figure 6.20. A dark blue line of squares runs southwest to northeast at about 31° clockwise from north. There is no spatial association in this direction, as values of $\gamma(\mathbf{h})$ are similar along this line. However, perpendicular to this line, one sees that $\gamma(\mathbf{h})$ increases for about 4000 ft. Thus, the preferred direction indicated by this semivariogram map is $\theta = 31° + 90° = 121°$ because within 4000 ft (the range of spatial association), the yields of two wells drilled along this azimuth are most dissimilar. If they are drilled along the 31° azimuth, they will be most similar. In practice, when a driller encounters a dry or low-yielding well, the best direction to offset a new well is at 121°, because in this direction the next yield from the next hole drilled is likely to be most dissimilar: that is, higher yielding, whereas when a high yield is drilled, the best direction to encounter a similar well is to offset the new well at 31°.

6.7 KRIGING

Kriging is a generic name for a family of procedures used to make point or block estimates in the presence of spatial association. When there is no spatial association, classical statistical modeling techniques should be used. A South African mining engineer, D. G. Krige (1951), did much of the pioneering work. Kriging is weighted regression with the weights computed from a model of the spatial association. It is a type of interpolation—the surface defined by kriging passes through existing data points. A *kriging estimator* is defined as

$$\widehat{Z}(\mathbf{s}) - m(\mathbf{s}) = \sum_{i=1}^{n(\mathbf{s})} \lambda_i [Z(\mathbf{s}_i) - m(\mathbf{s}_i)]$$

where the $m(\mathbf{s})$ and $m(\mathbf{s}_i)$ are the expected values of $Z(\mathbf{s})$ and $Z(\mathbf{s}_i)$, respectively, and λ_i is the weight assigned to datum $z(\mathbf{s}_i)$. Only the $n(\mathbf{s})$ closest data points are used to estimate the random function $Z(\mathbf{s})$; $n(\mathbf{s})$ may vary with location.

The parameters, the λ_i's, are estimated by minimizing the error variance,

$$\sigma_E^2(\mathbf{s}) = \text{Var}[\widehat{Z}(\mathbf{s}) - Z(\mathbf{s})] \tag{6.1}$$

subject to the constraint that $E[\widehat{Z}(\mathbf{s}) - Z(\mathbf{s})] = 0$. This solution produces a minimum variance unbiased estimator. Estimation of the λ_i's requires a function for the covariance between $\widehat{Z}(\mathbf{s})$ and $Z(\mathbf{s}_i)$ to be a function of distance $h = |\mathbf{s} - \mathbf{s}_i|$. The semivariogram model expressed as a covariance is used for this purpose. As various kriging models are described, we encourage the reader to note the parallels to regression, such as minimizing the error.

Kriging methods require an assumption of stationarity. Estimation of a semivariogram, which is an essential component of kriging, and other parameter estimates require stationarity. In simulation, stochastic realizations from a kriging model are not expected to reproduce a semivariogram exactly but should do so to within sampling error. These discrepancies are called *ergodic fluctuations*.

The general approach to kriging outlined above provides local estimates of the mean and variance at unsampled points. This is often satisfactory for mining and other applications, where the region of interest is relatively small. In regional mineral resource estimation, environmental cleanup, and other applications, an interest is to estimate global uncertainty. A map of variability using local estimates tends to be overly smooth because kriging is a smoother and tends to underestimate highs and overestimate lows, and the variability is only a function of nearby points (Goovaerts, 1997). Simulation should be used to reproduce a more realistic (less biased) global estimate of uncertainty.

Three kriging models are discussed in some detail: simple kriging (SK), ordinary kriging (OK), and kriging with a trend (KT). In addition, indicator kriging (IK), a nonparametric form of kriging; block kriging (BK), a procedure that yields a block estimate as opposed to a point estimate; lognormal kriging; co-kriging; and Bayesian kriging are presented. Equations to show explicitly the form of the models and the relationships between covariogram or semivariogram structure, weights, and estimates of z values and associated error are examined. A detailed understanding of these equations is not required for the reader to obtain a general idea of the applications of kriging.

6.7.1   Simple Kriging

Simple kriging (SK) is a useful building block toward understand more complex models. In SK the mean and second-order stationarity are assumed. Thus, $\mu = m(s_i)\ s_i \in A$, where A is the region of interest. To identify the kriging method, a superscript is placed on a parameter: for example, SK for simple kriging. The SK estimate of the variable of interest or kriging prediction at location s is

$$
\widehat{Z}_{SK}(s) = \sum_{i=1}^{n(s)} \lambda_i^{SK}(s)[Z(s_i) - \mu] + \mu
$$
$$
= \sum_{i=1}^{n(s)} \lambda_i^{SK}(s)Z(s_i) + \left[1 - \sum_{i=1}^{n(s)} \lambda_i^{SK}(s)\right]\mu \tag{6.2}
$$

which is a weighted average of all the data values in region A. The $\lambda_i^{SK}(s)$ weights are obtained by minimizing the error variance $\sigma_E^2(s) = \mathrm{Var}[\widehat{Z}_{SK}(s) - Z(s)]$ under the usual [equation (6.1)] unbiasedness constraint. The quantity $\widehat{Z}_{SK}(s) - Z(s)$ represents the estimation error at location s, which is rewritten as

$$
\widehat{Z}_{SK}(s) - Z(s) = \left[\widehat{Z}_{SK}(s) - \mu\right] - [Z(s) - \mu] \tag{6.3}
$$

the difference between two residuals. Substitute equation (6.2) for $\widehat{Z}_{SK}(s)$ in the right-hand side of equation (6.3) to obtain

$$
\widehat{Z}_{SK}(s) - Z(s) = \sum_{i=1}^{n(s)} \lambda_i^{SK}(s)[Z(s_i) - \mu] - [Z(s) - \mu]
$$
$$
= \sum_{i=1}^{n(s)} \lambda_i^{SK}(s)R(s_i) - R(s)
$$
$$
= \widehat{R}_{SK}(s) - R(s)
$$

where $R(s_i) \equiv \widehat{Z}(s_i) - \mu$, $R(s) \equiv \widehat{Z}(s) - \mu$, and $\widehat{R}_{SK}(s) \equiv \sum_{i=1}^{n(s)} \lambda_i^{SK}R(s_i)$, the weighted sum of residuals $R(s_i)$. Now the error variance $\sigma_E^2(s)$ is able to be expressed in terms of residuals. Thus,

$$
\sigma_E^2(s) = \mathrm{Var}\left[\widehat{R}_{SK}(s) - R(s)\right]
$$
$$
= \mathrm{Var}\left[\widehat{R}_{SK}(s)\right] + \mathrm{Var}[R(s)] - 2\,\mathrm{Cov}\left[\widehat{R}_{SK}(s), R(s)\right]
$$

Because of the assumption of second-order stationarity,

$$
\mathrm{Var}[R(s)] = \mathrm{Cov}_R(0)
$$
$$
\mathrm{Cov}\left[\widehat{R}_{SK}(s), R(s)\right] = \mathrm{Cov}\left[\sum_{i=1}^{n(s)} \lambda_i^{SK}(s)R(s_i), R(s)\right]
$$
$$
= \sum_{i=1}^{n(s)} \lambda_i^{SK}(s)\mathrm{Cov}(s_i - s)
$$

and Cov_R is the covariance of residual R. Therefore,

$$\sigma_E^2(\mathbf{s}) = \text{Var}\left[\sum_{i=1}^{n(\mathbf{s})} \lambda_i^{\text{SK}}(\mathbf{s})R(\mathbf{s}_i)\right] + \text{Cov}_R(0) - 2\sum_{i=1}^{n(\mathbf{s})}\lambda_i^{\text{SK}}\text{Cov}_R(\mathbf{s}_i - \mathbf{s})$$

$$= \sum_{i=1}^{n(\mathbf{s})}\sum_{j=1}^{n(\mathbf{s})}\lambda_i^{\text{SK}}\lambda_j^{\text{SK}}\text{Cov}_R(\mathbf{s}_i - \mathbf{s}_j) + \text{Cov}_R(0) - 2\sum_{i=1}^{n(\mathbf{s})}\lambda_i^{\text{SK}}\text{Cov}_R(\mathbf{s}_i - \mathbf{s}) \qquad (6.4)$$

$$= Q[\lambda_i^{\text{SK}}(\mathbf{s}), i = 1, \ldots n(\mathbf{s})]$$

where Q represents a quadratic form. Next the $\lambda_i^{\text{SK}}(\mathbf{s})$, $i = 1, \ldots, n(\mathbf{s})$ is estimated by taking the derivative

$$\frac{\partial Q(\mathbf{s})}{\partial \lambda_i^{\text{SK}}}$$

for each i. This yields a system of $n(\mathbf{s})$ linear equations

$$\sum_{j=1}^{n(\mathbf{s})}\lambda_j^{\text{SK}}\text{Cov}_R(\mathbf{s}_i - \mathbf{s}_j) - \text{Cov}_R(\mathbf{s}_i - \mathbf{s}) = 0 \qquad i = 1, \ldots, n(\mathbf{s}) \qquad (6.5)$$

Since second-order stationarity is assumed, $\text{Cov}_R(\mathbf{h}) = \text{Cov}(\mathbf{h})$ and therefore the system of equations (6.5) can be expressed as

$$\sum_{j=1}^{n(\mathbf{s})}\lambda_j^{\text{SK}}\text{Cov}(\mathbf{s}_i - \mathbf{s}_j) = \text{Cov}(\mathbf{s}_i - \mathbf{s}) \qquad i = 1, \ldots, n(\mathbf{s}) \qquad (6.6)$$

If equation (6.5) is substituted in equation (6.4), the minimum error variance

$$\sigma_{\text{SK}}^2 = \text{Cov}(0) - \sum_{i=1}^{n(\mathbf{s})}\lambda_i^{\text{SK}}\text{Cov}(\mathbf{s}_i - \mathbf{s})$$

is obtained. Now expand equation (6.6) as follows (recall that a similar operation is used in regression):

$$\lambda_1^{\text{SK}}\text{Cov}(\mathbf{s}_1 - \mathbf{s}_1) + \lambda_2^{\text{SK}}\text{Cov}(\mathbf{s}_1 - \mathbf{s}_2) + \cdots + \lambda_{n(\mathbf{s})}^{\text{SK}}\text{Cov}(\mathbf{s}_1 - \mathbf{s}_{n(\mathbf{s})}) = \text{Cov}(\mathbf{s}_1 - \mathbf{s})$$

$$\lambda_1^{\text{SK}}\text{Cov}(\mathbf{s}_2 - \mathbf{s}_1) + \lambda_2^{\text{SK}}\text{Cov}(\mathbf{s}_2 - \mathbf{s}_2) + \cdots + \lambda_{n(\mathbf{s})}^{\text{SK}}\text{Cov}(\mathbf{s}_2 - \mathbf{s}_{n(\mathbf{s})}) = \text{Cov}(\mathbf{s}_2 - \mathbf{s})$$

$$\vdots$$

$$\lambda_1^{\text{SK}}\text{Cov}(\mathbf{s}_{n(\mathbf{s})} - \mathbf{s}_1) + \lambda_2^{\text{SK}}\text{Cov}(\mathbf{s}_{n(\mathbf{s})} - \mathbf{s}_2) + \cdots + \lambda_{n(\mathbf{s})}^{\text{SK}}\text{Cov}(\mathbf{s}_{n(\mathbf{s})} - \mathbf{s}_{n(\mathbf{s})}) = \text{Cov}(\mathbf{s}_{n(\mathbf{s})} - \mathbf{s})$$

This system is represented in matrix notation $\mathbf{K}_{n(s)}^{\mathrm{SK}} \Lambda_{n(s)}^{\mathrm{SK}} = \kappa_{n(s)}^{\mathrm{SK}}$, where

$$\mathbf{K}_{n(s)}^{\mathrm{SK}} = \begin{pmatrix} \mathrm{Cov}(\mathbf{s}_1 - \mathbf{s}_1) & \cdots & \mathrm{Cov}(\mathbf{s}_1 - \mathbf{s}_{n(s)}) \\ \vdots & \ddots & \vdots \\ \mathrm{Cov}(\mathbf{s}_{n(s)} - \mathbf{s}_1) & \cdots & \mathrm{Cov}(\mathbf{s}_{n(s)} - \mathbf{s}_{n(s)}) \end{pmatrix} \qquad \Lambda_{n(s)}^{\mathrm{SK}} = \begin{bmatrix} \lambda_1^{\mathrm{SK}} \\ \cdots \\ \lambda_{n(s)}^{\mathrm{SK}} \end{bmatrix}$$

$$\kappa_{n(s)}^{\mathrm{SK}} = \begin{bmatrix} \mathrm{Cov}(\mathbf{s}_1 - \mathbf{s}) \\ \cdots \\ \mathrm{Cov}(\mathbf{s}_{n(s)} - \mathbf{s}) \end{bmatrix}$$

An algebraic solution to obtain the weights is

$$\lambda_{n(s)}^{\mathrm{SK}} = \left[\mathbf{K}_{n(s)}^{\mathrm{SK}} \right]^{-1} \kappa_{n(s)}^{\mathrm{SK}} \tag{6.7}$$

What can be inferred from the structure of these equations? First, to obtain an inverse, $\mathbf{K}_{n(s)}^{\mathrm{SK}}$ must be positive definite. If $\mathbf{s}_i = \mathbf{s}_j$, $i \neq j$, this requirement is violated. Thus, there must not be coincident points. In addition, $\mathrm{Cov}(\mathbf{h})$ must be permissible (see our previous discussion on semivariogram models).

Suppose that $\mathbf{s}$ is located at one of the data points, say $\mathbf{s}_1$. Consider the system of equations

$$\sum_{j=1}^{n(s)} \lambda_j^{\mathrm{SK}} \mathrm{Cov}(\mathbf{s}_i - \mathbf{s}_j) = \mathrm{Cov}(\mathbf{s}_i - \mathbf{s}_1) \qquad i = 1, \ldots, n(s)$$

The solution to this system implies that $\lambda_1^{\mathrm{SK}} = 1$ and $\lambda_i^{\mathrm{SK}} = 0$, $i = 2, \ldots, n(s)$. From the SK estimator equation (6.2), one obtains $\widehat{Z}_{\mathrm{SK}}(\mathbf{s}) = Z(\mathbf{s}_1)$. This result, known as the *exactitude property*, means that the estimator $\widehat{Z}_{\mathrm{SK}}(\mathbf{s})$ passes through the data points. The square root of the estimated error $\widehat{\sigma}_E(\mathbf{s}) = \sqrt{\widehat{\sigma}_E^2(\mathbf{s})}$ is the *standard error of prediction* at location $\mathbf{s}$. Because of the exactitude property, when $\mathbf{s} = 0$, the standard error is zero. The standard error of prediction is a function of data, model, and location of $\mathbf{s}$.

To illustrate the computations and their meaning, consider the simple example of estimating $Z(\mathbf{s})$ using four neighboring points of $\mathbf{s}$ (Figure 6.21 and Table 6.4). Assume that $\mu = 3.2$. A known mean is a requirement of SK. Further assume isotropy (Figure 6.18a) and that $\mathrm{Cov}(h) = \exp(-h/3)$, which of course declines with increasing distance h. This yields the covariance matrix shown in Table 6.4. For example, the Euclidean distance between points 1 and 3 is 4.12; thus, the (1,3) element in the model covariance matrix is $\exp(-4.12/3) = 0.25$. $\mathbf{K}_4^{\mathrm{SK}}$ is labeled as a covariance matrix; however, since the diagonal of the Euclidean distance matrix is 0, the diagonal of the covariance matrix is 1, and the off-diagonal elements are less than 1, this can be considered to be a correlation matrix. κ_4^{SK} is the covariance vector between the four data points and point $\mathbf{s}$. As expected, the weight is inversely

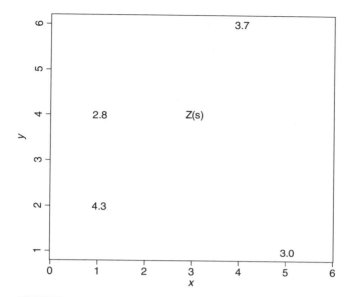

FIGURE 6.21   Data point display for simple kriging example.

proportional to distance of the data point from **s**. The next step is to invert the 4×4 covariance matrix (formed by points 1, 2, 3, and 4), which is denoted as $\left[\mathbf{K}_4^{SK}\right]^{-1}$ (Table 6.4, Inverse). The weights (Table 6.4, Weights) are obtained from equation (6.7). Note that the weights $\lambda_{n(s)}^{SK}$ are a function of the distances between the set of four points and the distance between these points and **s**. Points 2 and 4 are the closest to point **s** and have the largest weights; however, point 2 is slightly closer than point 4 but has a slightly smaller weight (0.32 versus 0.33). The reason is that this is a system of points and all exert some influence on each other. Finally, an estimate of $Z(\mathbf{s})$ can be obtained:

$$
\begin{aligned}
\widehat{Z}_{SK}(\mathbf{s}) &= \sum_{i=1}^{4} \lambda_i^{SK}(\mathbf{s})[Z(\mathbf{s}_i) - \mu] + \mu \\
&= (0.12)(4.30 - 3.2) + (0.32)(2.80 - 3.2) + (0.15)(3.00 - 3.2) \\
&\quad + (0.33)(3.70 - 3.2) + 3.2 \\
&= 3.34
\end{aligned}
$$

One use of SK is in sequential Gaussian simulation, a procedure used to estimate global uncertainty as opposed to an estimate of uncertainty at a given point.

6.7.2   Ordinary Kriging

In *ordinary kriging* (OK), the SK assumption that the mean μ is known and constant over the study area A is relaxed. In OK, μ is unknown and $\mu(\mathbf{s}) = \mu$ is stationary in a

TABLE 6.4   Data Set and Computations for Simple Kriging

Data Set

	Location		Value
Point	x	y	z
1	1.00	2.00	4.30
2	1.00	4.00	2.80
3	5.00	1.00	3.00
4	4.00	6.00	3.70
s	3.00	4.00	Unknown

Distances

	Euclidean Distance (h)				
Point	1	2	3	4	s
1	0.00	2.00	4.12	5.00	2.83
2		0.00	5.00	3.61	2.00
3			0.00	5.10	3.61
4				0.00	2.24

Covariances

	Covariance Matrix, $\mathbf{K}_4^{SK}\|Cov(h) = \exp(-h/3)$				$\kappa_4^{SK}\|Cov(h)$
Point	1	2	3	4	s
1	1.00	0.51	0.25	0.19	0.39
2	0.51	1.00	0.19	0.30	0.51
3	0.25	0.19	1.00	0.18	0.30
4	0.19	0.30	0.18	1.00	0.47

Inverse of Covariance Matrix

	$[\mathbf{K}_4^{SK}]^{-1}$			
Point	1	2	3	4
1	1.398	−0.663	−0.219	−0.027
2	−0.663	1.437	−0.054	−0.295
3	−0.219	−0.054	1.090	−0.138
4	−0.027	−0.295	−0.138	1.119

Weights	
Point	λ^{SK}
1	0.12
2	0.32
3	0.15
4	0.33

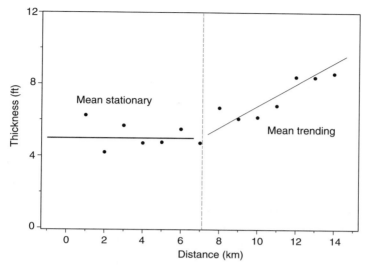

FIGURE 6.22   Stationary mean versus trending mean.

neighborhood centered at **s**, where **s** is the location of the estimate. Figure 6.22 shows the stationary case on the left and the trending case on the right. The trending case is considered in the next section. Ordinary kriging is sometimes called *punctual kriging*.

The OK estimator is

$$\widehat{Z}_{OK}(\mathbf{s}) = \sum_{i=1}^{n(\mathbf{s})} \lambda_i^{OK}(\mathbf{s})Z(\mathbf{s}_i) + \left[1 - \sum_{i=1}^{n(\mathbf{s})} \lambda_i^{OK}(\mathbf{s})\right]\mu$$

The constraint $\sum_{i=1}^{n(\mathbf{s})} \lambda_i^{OK}(\mathbf{s}) = 1$ is imposed to eliminate μ from the equation above. The OK estimator is

$$\widehat{Z}_{OK}(\mathbf{s}) = \sum_{i=1}^{n(\mathbf{s})} \lambda_i^{OK}(\mathbf{s})Z(\mathbf{s}_i)$$

As in SK, an estimate of the weights $\lambda_i^{OK}(\mathbf{s})$, $i = 1, \ldots, n(\mathbf{s})$ results from minimizing the error variance $\sigma_E^2(\mathbf{s})$. Since $E[\widehat{Z}_{OK}(\mathbf{s}) - Z(\mathbf{s})] = 0$, the estimates are unbiased. The solution is slightly more complicated since the constraint on the weights requires the use of a Lagrange parameter, $\mu_{OK}(\mathbf{s})$. The *Lagrange parameter* is a mathematic device used to obtain a solution to the following system of equations:

$$\sum_{j=1}^{n(\mathbf{s})} \lambda_j^{OK} Cov(\mathbf{s}_i - \mathbf{s}_j) + \mu_{OK}(\mathbf{s}) = Cov(\mathbf{s}_i - \mathbf{s}) \qquad i = 1, \ldots, n(\mathbf{s})$$

$$\sum_{j=1}^{n(\mathbf{s})} \lambda_j^{OK} = 1$$

or in terms of the semivariogram model as

$$\sum_{j=1}^{n(s)}\lambda_j^{OK}\gamma(s_i - s_j) + \mu_{OK}(s) = \gamma(s_i - s) \qquad i = 1, \ldots, n(s)$$

$$\sum_{j=1}^{n(s)}\lambda_j^{OK} = 1$$

What issues need be considered? What needs to be done prior to implementing the OK kriging model?

1. Investigate the distribution of the variable of interest and decide on a normalization transformation, if needed.
2. See if the data contain a spatial trend. This can be investigated by:
   a. Plotting the variable of interest, say z, against x and y.
   b. Making a contour plot.
   c. Making a wire diagram or other three-dimensional representation.
   d. Fitting a loess or other regression model.
   e. Testing model significance.
   f. Looking for spatial trends in the residuals.
3. Investigate anisotropy. Construct a variogram map and/or semivariograms at various directions.
4. Model the empirical semivariogram. Choose an exponential, spherical, or other appropriate model.
5. Remove spatial trend as needed.

If OK is appropriate, consider if the interest is in estimating the variability at the data points or estimating the mean value and variability at the nodes on a grid.

The empirical semivariogram and a fitted spherical model for the transformed water-well yield case study data are shown in Figure 6.17. As noted previously, the fifth-root transformed yield in gpm is used as the response variable because the distribution of yield is right-skewed. The spherical semivariogram model serves as input to the OK algorithm (R-project module krige.conv in geoR; Ribeiro and Diggle, 2001). In addition, this kriging algorithm uses point estimates at grid nodes obtained through interpolation. It estimates the standard errors at the nodes. Since kriging is an exact interpolator, the only reason for choosing estimates at the data points is to estimate their standard errors. The usual initial procedure is to specify a dense grid in relation to the number of data points. For this example, a grid of 0.5 km easting by 0.5 km northing is chosen. Since the range of northing is 4331.098 to 4348.591 km, there are 35 cells in the north direction. Similarly, the range of easting cells is 269.7454 to 277.1612, which implies 15 cells in the east direction. The total number of cells is 525 and since there are 754 observations, the density is 1.44 well-water yields per cell, a reasonable number. To make the resulting graph easier to read, only the northern half of water-well yield case study data is shown. Figure 6.23a shows a

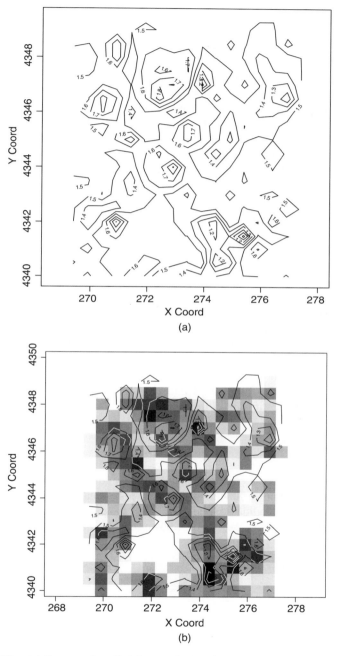

FIGURE 6.23   (a) Contour plot of kriging predictions for the transformed (fifth root) water-well yield (x coordinate is easting in km; y coordinate is northing in km). (b) Image plot of kriging standard errors and contour plot for the transformed (fifth root) water-well yield (x coordinate is easting in km; y coordinate is northing in km).

contour plot of kriging predictions. An image plot of kriging standard errors and contour of the transformed yield is shown in Figure 6.23b. The contour plot (Figure 6.23a) does not show any clear pattern of high values. The highest-valued isoline is 1.8 km in the north-central portion of the graph. The number and spacing of levels is a function of data. Trial and error are usually required to obtain a meaningful plot. In addition, it is possible to plot the data locations (water-well yield in this example). This is particularly useful when data are clustered or otherwise unevenly distributed. Contour lines that are far from any data points may be suspect. This is especially true when the lines are outside a border around the data. A gray-scale image plot is chosen to display the kriging standard errors (Figure 6.23b) because the authors believe that this type of plot makes it easier to spot patterns in the estimated errors. There appear to be a few patterns, including one large white spot centered at $x = 272$ km and $y = 4342$ km indicating low standard errors, and a few areas of darker spots, which may be a function of data coverage.

6.7.3 Kriging with a Trend

In ordinary kriging, it is assumed that there is no spatial trend in the response variable; however, trends exist in many spatial problems. For example, seismic activity typically decreases away from a fault. Precipitation may decrease gradually away from a large body of water. Coal bed thickness often increases gradually with depth below the surface. In these and other situations it cannot reasonably be assumed that the mean μ is stationary, even in the neighborhood of the point being estimated. Thus, there is *kriging with a trend* (KT). In some geostatistics literature, KT is referred to as *universal kriging*. The first step in KT is to model the trend, at least in the neighborhood $W(\mathbf{s})$ of $\mathbf{s}$, the point of interest.

Consider a linear estimator of the form

$$\widehat{\mu}(\mathbf{s}) = \sum_{k=1}^{K} a_k(\mathbf{s}) f_k(\mathbf{s}) \tag{6.8}$$

with $a_k(\mathbf{s})$, the unknown coefficients, approximately equal to a_k in the neighborhood $W(\mathbf{s})$. The $f_k(\mathbf{s})$ are known functions. If one believes that a trend increases linearly from west to east, a candidate model is $\widehat{\mu}(X) = 1 + a_1 X$, where X is the east–west coordinate. By convention, $a_0(\mathbf{s}) = 1$. Substituting equation (6.8) into

$$\widehat{Z}(\mathbf{s}) = \sum_{i=1}^{n(\mathbf{s})} \lambda_i(\mathbf{s}) Z(\mathbf{s}_i) + \left[1 - \sum_{i=1}^{n(\mathbf{s})} \lambda_i(\mathbf{s}) \right] \widehat{\mu}(\mathbf{s})$$

yields

$$\widehat{Z}(\mathbf{s}) = \sum_{i=1}^{n(\mathbf{s})} \lambda_i(\mathbf{s}) \left[Z(\mathbf{s}_i) - \sum_{k=1}^{K} a_k(\mathbf{s}) f_k(\mathbf{s}_i) \right] + \sum_{k=1}^{K} a_k(\mathbf{s}) f_k(\mathbf{s})$$

$$= \sum_{i=1}^{n(\mathbf{s})} \lambda_i(\mathbf{s}) Z(\mathbf{s}_i) + \sum_{k=1}^{K} a_k(\mathbf{s}) \left[f_k(\mathbf{s}) - \sum_{i=1}^{n(\mathbf{s})} \lambda_i(\mathbf{s}) f_k(\mathbf{s}_i) \right]$$

where $\widehat{Z}(\mathbf{s})$ is a weighted average of the data. To obtain a solution, that is, to estimate the λ_i's, a constraint similar to that used in OK is needed. It is $\sum_{i=1}^{n(\mathbf{s})} \lambda_i^{KT}(\mathbf{s}) = 1$. Next, a set of $K+1$ constraints, which result from the need to estimate the trend, is imposed. These are

$$f_k(\mathbf{s}) = \sum_{i=1}^{n(\mathbf{s})} \lambda_i^{KT}(\mathbf{s}) f_k(\mathbf{s}_i) \qquad k = 0, \dots, K$$

This system requires the kth function ($k = 0, \dots, K$) at $\mathbf{s}$ to be a weighted sum of the kth function in the neighborhood of $\mathbf{s}$. After some algebra, the following system of $n(\mathbf{s}) + K + 1$ equations is obtained:

$$\sum_{j=1}^{n(\mathbf{s})} \lambda_j^{KT} \text{Cov}_R(\mathbf{s}_i - \mathbf{s}_j) + \sum_{k=0}^{K} \mu_k^{KT}(\mathbf{s}) f_k(\mathbf{s}_i) = \text{Cov}_R(\mathbf{s}_i - \mathbf{s}) \qquad i = 1, \dots, n(\mathbf{s})$$

$$\sum_{j=1}^{n(\mathbf{s})} \lambda_j^{KT} = 1$$

$$\sum_{j=1}^{n(\mathbf{s})} \lambda_j^{KT}(\mathbf{s}) f_k(\mathbf{s}_j) = f_k(\mathbf{s}) \qquad k = 1, \dots, K$$

The solution to this system of equations yields an estimate of the weights (the λ's), which minimize the variance of prediction. It is assumed that the covariance is known. This system is solved by taking a partial derivative to obtain estimates for the λ's and the corresponding error variance. In the systems of equations above, why does one begin the last system at $k = 1$ as opposed to $k = 0$? (Think about a_0). Note that when $K = 0$, KT reverts to OK.

Many trends are linear or quadratic and can be fit with one of the following models:

- Linear trend in two dimensions ($K = 2$):

$$\mu(\mathbf{s}) = a_0 + a_1 X + a_2 Y \qquad \mathbf{s} = (X, Y)'$$

- Quadratic trend in two dimensions ($K = 5$):

$$\mu(\mathbf{s}) = a_{00} + a_{10} X + a_{01} Y + a_{20} X^2 + a_{11} XY + a_{02} Y^2 \qquad \mathbf{s} = (X, Y)'$$

- Some trends that occur over a large area may be too complicated to be fit by a quadratic model, and higher-order polynomials may not be appropriate. These can be difficult to fit and may be unstable. An alternative is a locally weighted regression model loess (or lowess) of the form $\mu(\mathbf{s}) = g(\mathbf{s}) + e$, where g is a regression function that can be approximated by points in the neighborhood of $\mathbf{s} = (X, Y)'$.

After fitting a trend surface model, compute residuals $Z_R(\mathbf{s}_i) = Z(\mathbf{s}_i) - \widehat{\mu}(\mathbf{s}_i)$. As in regression, examine the pattern of residuals (in this instance, the spatial pattern) and check for randomness.

A semivariogram model must be specified. Now there is a difficulty because the residual semivariogram model $\gamma_R(\mathbf{h})$ will be used for kriging. Since the residuals $Z_R(\mathbf{s}_i)$ are known, the model $\widehat{\gamma}_R(\mathbf{h})$ can be estimated. However, $\widehat{\gamma}_R(\mathbf{h})$ will be dependent on the model used to estimate the trend, and this can lead to biased kriging estimates. There are some alternatives. Frequently, a trend is present in only one direction and $\gamma(\mathbf{h})$ can be estimated using data orthogonal to the trend. To estimate $\gamma(\mathbf{h})$ for small h, the trend is often negligible and can be ignored. Other approaches include the use of filters and SK. Sometimes one simply has no choice but to estimate the semivariogram model from residuals and recognize the possibility of bias.

Example of KT Using the Cortez Quadrangle Data   Figure 6.14 shows clear evidence of a trend in $\log_{10}(U)$ with easting. Fortunately, this trend is only in the easting direction; therefore, an empirical semivariogram orthogonal to this trend is computed: namely, at azimuth $0°$ (north) and tolerance angle $20°$. A semivariogram calculated using residuals from the fitted linear model of the easting trend is shown by the circles in Figure 6.24. The semivariogram model fit to the directional ($0°$ azimuth) semivariogram is

$$\gamma(h) = \begin{cases} 0.0679 + (0.116 - 0.0679)\left[\dfrac{1.5h}{8.224} - 0.5\left(\dfrac{h}{8.224}\right)^3\right] & h \le 8.224 \\ 0.116 & h > 0.8224 \end{cases}$$

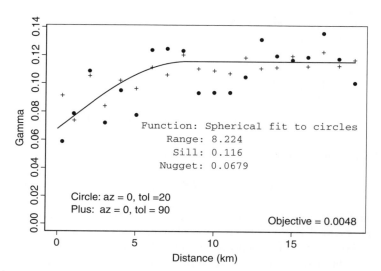

FIGURE 6.24   Empirical semivariograms and spherical model of $\log_{10}(U)$ at an azimuth of $0°$, Cortez Quadrangle data.

The plus signs in Figure 6.24 represent the empirical semivariogram with an azimuth of $0°$ (north) and a tolerance angle of $90°$ (the omnidirectional case). The nugget of a semivariogram fit to the omnidirectional case (not shown) appears to yield a smaller nugget; however, the results are similar. This is not always the case. In the KT model, an investigator specifies

- A data set with location and variable of interest (the z value)
- A grid (prediction locations, if desired)
- A model of the trend
- A semivariogram model

Outputs from KT are predictions of z values and standard errors at the grid points (nodes).

6.7.4 Block Kriging

Block kriging (BK) is useful to practitioners who need to estimate an average over a length, area, or volume. One historical example is from mining, where the interest was in estimating the mean mineral content in a given block. Consider a two-dimensional block $V(\mathbf{s})$, centered on $\mathbf{s}$. This can be a mine face. The block average $z_V(\mathbf{s})$ is obtained by integrating over V as

$$z_V(\mathbf{s}) = \frac{1}{V} \int_{V(\mathbf{s})} z(\mathbf{t})\, d\mathbf{t}$$

In practice, the average of some small finite number of discretizing points N is taken as

$$z_V(\mathbf{s}) \approx \frac{1}{N} \sum_{k=1}^{N} z(\mathbf{t}_k)$$

A BK estimate (Figure 6.25) is $z_V(\mathbf{s}) = \frac{1}{4}\sum_{i=1}^{4} \widehat{z}_{OK}(\mathbf{t}_i)$. A less computationaly intensive system is given by

$$\sum_{j=1}^{n(\mathbf{s})} \lambda_j^{OK}(\mathbf{t}_k) \mathrm{Cov}(\mathbf{s}_i - \mathbf{s}_j) + \mu_{OK}(\mathbf{t}_k) = \mathrm{Cov}(\mathbf{s}_i - \mathbf{t}_k) \qquad i = 1, \ldots, n(\mathbf{s})$$

$$\sum_{j=1}^{n(\mathbf{s})} \lambda_j^{OK}(\mathbf{t}_k) = 1$$

This system is solved at t_k, $k = 1, \ldots, N$. Why this is more computationally efficient than the naive approach is left as an exercise for the reader. The major difference between this equation and that of OK is the right-hand term $\mathrm{Cov}(\mathbf{s}_i - \mathbf{t}_k)$, which is a "point-to-block" covariance instead of a covariance between points.

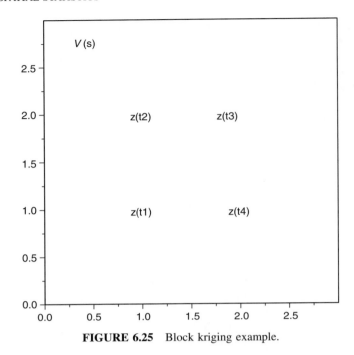

FIGURE 6.25   Block kriging example.

Recall that $z_V(\mathbf{s})$ is a linear average. For highly skewed data it may be desirable to transform the data, say $w(\mathbf{s}) = \log(z(\mathbf{s}))$. Care must be taken in interpretation of results because $\widehat{w}_V(\mathbf{s}) \neq \log(\widehat{z}_V(\mathbf{s}))$. Another consideration is the number of discretizing points, N, to be used. A rule of thumb proposed by Journel and Huijbregts (1978) is to use 16 and 64 for two- and three-dimensional blocks, respectively.

Consider an example using the Harmon coal quality data. The empirical semi-variogram and exponential model are shown in Figure 6.26. For block kriging, the following specifications typically are required.

- A prediction grid (this is also needed in other kriging methods unless it is desired to predict at the data locations).
- A block size (the horizontal and vertical dimensions in Figure 6.25).
- The number of points in a block used to estimate the point-to-block covariance and a resulting block value estimate. As stated previously, this is typically a number between 16 and 64.

6.7.5   Other Kriging Methods

Numerous other kriging approaches carry different assumptions and are useful for non-Gaussian data. Among these are indicator kriging, where interest may be in a variable above or below some threshold value; lognormal kriging for right-skewed

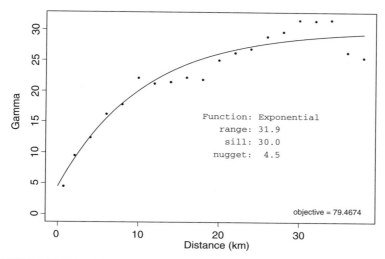

FIGURE 6.26 Empirical semivariogram and exponential model for Harmon coal bed thickness in feet.

data; Bayesian kriging, useful when the spatial covariance structure is unknown; and co-kriging, when relevant secondary information is available.

Indicator Kriging   *Indicator kriging* is a method that allows one to estimate the *continuous cumulative distribution function* (CCDF) F(s;z) through a series of threshold values z_k, $k = 1,\ldots, K$, which can be user specified. Indicator kriging is often appropriate when Gaussian and/or homoscedasticity assumptions are questionable. Threshold values may be those of interest for a particular problem, such as the minimum concentration of mercury in coal allowable under law. Other common choices of threshold values are the first, second, and third quartiles of the distribution of Z. An advantage of indicator kriging is that it can accept different categories of data. Interval or ratio data is usually coded as

$$i(\mathbf{s}_j; z_k) = \begin{cases} 1 & z(\mathbf{s}_j) \le z_k, \quad k = 1,\ldots,K \\ 0 & \text{otherwise} \end{cases}$$

For the water-well yield data, the following cuts were made: $0 \le z \le 0.5$, $0.5 < z \le 1.0$, $1.0 < z \le 10.0$, $10.0 < z \le 40.0$, and $z > 40.0$, because these are demarcation values used in rating the water supply. For thresholds $\{0.5, 1, 10, 40\}$, the resulting indicator vectors are shown in Table 6.5 for five water-well yields $(z(\mathbf{s}_1),\ldots, z(\mathbf{s}_5))$. An example of the coding is $i(j = 2; z_3) = 0$ because $z(\mathbf{s}_2) = 18$, which is greater than 10. Note that each data point becomes a vector of size 4 in this example. An indicator semivariogram $\gamma_I(\mathbf{h}; z_k)$ is computed, which is used as input to the indicator kriging procedure.

TABLE 6.5   Example Coding for Indicator Kriging

Threshold	Indicator Vectors				
$z_4 = 40.0$	1	1	1	0	1
$z_3 = 10.0$	1	0	1	0	1
$z_2 = 1.0$	0	0	1	0	1
$z_1 = 0.5$	0	0	1	0	0
Ordinal Distance j	1	2	3	4	5
$z(\mathbf{s}_j)$	6	18	0.5	50	1

If $z(\mathbf{s}_j)$ is known only to within the interval $a_j < z(\mathbf{s}_j) \le b_j$, the indicator variable is coded as

$$i(\mathbf{s}_j; z_k) = \begin{cases} 1 & z(\mathbf{s}_j) \le z_k \\ \text{uncertain} & z_k \in (a_i, b_i], \quad k = 1, \ldots, K \\ 0 & \text{otherwise} \end{cases}$$

This may occur if there is significant measurement error or if the data are ordinal or categorical. In the expression above, $i(\mathbf{s}_j; z_k)$ is a realization of the indicator random variable $I(\mathbf{s}_j; z_k)$.

Indicator kriging has analogs in SK and OK. The model for *ordinary indicator kriging* (OIK) is presented. Unlike OK, the resulting OIK estimate is a cumulative distribution function

$$\widehat{F}_{\text{OIK}}[(\mathbf{s}; z_k|(n))] = \widehat{I}_{\text{OK}}(\mathbf{s}; z_k)$$

$$= \sum_{i=1}^{n(\mathbf{s})} \lambda_i^{\text{OK}}(\mathbf{s}; z_k) I(\mathbf{s}_i; z_k)$$

This allows an investigation of the probability that $z(\mathbf{s})$ exceeds a threshold value z_k.

Lognormal Kriging   Data generated by many natural processes are right-skewed. *Lognormal kriging* (LK) may be appropriate for such processes. The lognormal process, which can only assume positive values, is defined as $\{Z(\mathbf{s}); \mathbf{s} \in A\}$, where $Y(\mathbf{s}) \equiv \log(Z(\mathbf{s}))$, $\mathbf{s} \in A$ is a Gaussian process, and A is the region of interest. This definition is similar to that used previously to describe a lognormally distributed random variable. The idea behind LK is to transform data from the Z scale to the Y scale to make a prediction of the form

$$\widehat{Y}(\mathbf{s}) = \sum_{i=1}^{n(\mathbf{s})} \lambda_i \log(Z(\mathbf{s}_i)) = \sum_{i=1}^{n(\mathbf{s})} \lambda_i Y(\mathbf{s}_i)$$

where the semivariogram is in the Y scale. The problem with LK is that the back transformed value $\exp(\widehat{Y}(\mathbf{s}))$ is biased (see Cressie, 1993).

Bayesian Kriging   *Bayesian kriging* is an alternative to kriging with a trend (KT) that estimates uncertainty of prediction when the covariance structure is unknown. As noted earlier, the KT predictor $\widehat{Z}(\mathbf{s})$ is the best linear unbiased predictor when the covariance function is known. In the authors' experience with real data, the covariance function is never known and must be estimated from the data. The fitted covariance function (or semivariogram) is often estimated graphically. It is also usually assumed that the user has minimal knowledge about the regression coefficients (the weights). The Bayesian predictive distribution takes into account uncertainty about the covariance structure in the prediction uncertainty. For additional details, see the article by Handcock and Stein (1993). Bayesian kriging can be implemented in WinBugs (2005). Computational examples are given by Banerjee et al. (2004).

Co-kriging   *Co-kriging* (OCK) is a kriging procedure designed to augment the primary variable of interest with secondary information when the primary variable is missing. Co-kriging uses the association between primary and secondary variables and information provided by the secondary variable to produce estimates $Z(\mathbf{s})$ that have lower variance than if the secondary information is ignored. There are co-kriging models that correspond to SK, OK, and KT. Let Z_1 be the primary variable and Z_2 be a single secondary variable. The ordinary co-kriging solution at $\mathbf{s}$ within a neighborhood $W(\mathbf{s})$ is

$$\widehat{Z}_1(\mathbf{s}) = \sum_{i=1}^{n_1(\mathbf{s})} \lambda_{i1}(\mathbf{s})Z_1(\mathbf{s}_{i1}) + \sum_{i=1}^{n_2(\mathbf{s})} \lambda_{i2}(\mathbf{s})Z_2(\mathbf{s}_{i2}) + \lambda_{\mu1}(\mathbf{s})\mu_1(\mathbf{s}) + \lambda_{\mu2}(\mathbf{s})\mu_2(\mathbf{s})$$

where the weights $\lambda_{\mu1}(\mathbf{s}) = 1 - \sum_{i=1}^{n_1(\mathbf{s})} \lambda_{i1}(\mathbf{s})$ and $\lambda_{\mu2}(\mathbf{s}) = 1 - \sum_{i=1}^{n_2(\mathbf{s})} \lambda_{i2}(\mathbf{s})$.

The OCK system is more complicated than OK because it requires the additional estimation of cross covariances between primary and secondary variables. Because of the added mathematical and computational complexity, OCK should be used judiciously. If primary and secondary variables are uncorrelated, there is no advantage to using OCK. In addition, if primary and secondary variables are present at all locations and the two variables are highly correlated, there is little to be gained from using OCK. It is possible in some circumstances to get negative weights from an OCK system, which may necessitate the use of alternative constraints in the solution.

An application of OCK is in the spatial estimation of Cu in the Cortez Quadrangle. This data set contains 1657 observations; however, 82 Cu values (5%) are missing. None of the corresponding Mn observations are missing and the correlation between these variables expressed in logs is 0.46. Thus, Mn can serve as a covariate for Cu. A portion of this data set is shown in Table 6.6.

6.7.6 Comments on Kriging

There are many other kriging models and variations for models, including median-polish kriging, disjunctive kriging, and principal components kriging. Cressie (1993), Goovaerts (1997), and Isaaks and Srivastava (1989) discuss these models in greater

TABLE 6.6   Data Set for Co-kriging, Cortez Quadrangle

Latitude	Longitude	Cu (ppm)	Mn (ppm)	Log Cu	Log Mn
37.05	−109.09	23	429	3.14	6.06
37.05	−109.09	—	289	—	5.67
37.06	−109.11	12	320	2.48	5.77
37.06	−109.11	16	330	2.77	5.80
37.08	−109.10	—	331	—	5.80
37.15	−109.13	15	228	2.71	5.43
37.14	−109.13	17	260	2.83	5.56
37.14	−109.16	12	295	2.48	5.69
37.12	−109.18	14	250	2.64	5.52
37.11	−109.18	—	245	—	5.50
37.09	−109.19	—	231	—	5.44
37.09	−109.19	14	196	2.64	5.28
37.04	−109.19	19	199	2.94	5.29
37.04	−109.20	12	217	2.48	5.38
37.06	−109.22	13	270	2.56	5.60
37.03	−109.22	10	265	2.30	5.58
37.02	−109.22	12	295	2.48	5.69
37.01	−109.21	11	299	2.40	5.70

Source: Data from Campbell et al. (1982).

detail. The journal *Mathematical Geosciences* routinely feature articles on kriging and other aspects of spatial statistics.

Kriging has been demonstrated to work well in many applications. This is especially true in mining applications, where there is hard (drill core) data and the area of interest is typically a relatively small statistically homogeneous stratigraphic unit. In newer applications, such as regional resource analysis, hard and soft data are present and problems are on a different scale. Kriging, which depends on the choice of the semivariogram, is often called *two-point geostatistics* because the semivariogram captures the spatial variability between two points a distance h apart. Fitting the semivariogram is often arbitrary and driven by goodness-of-fit considerations rather than by geology, hydrology, or another science. In addition, different physical situations can result in similar semivariograms.

6.7.7   Geostatistics Simulation

As noted, simulation is required to provide an accurate global assessment of uncertainty. There are numerous approaches to traditional variogram-based simulation. These include sequential Gaussian and indicator simulation, simulated annealing, and p-filed simulation. A relatively new alternative called multipoint geostatistics has been introduced and is discussed below.

Variance-Based Simulation Algorithms   Three widely used procedures (Goovaerts, 1997) are sequential simulation, the p-field approach, and simulated annealing. The sequential simulation process begins with a set of random variables $Z(\mathbf{s}_i^*), i = 1,\ldots, N$ defined by the user at N locations. Note that the asterisk superscript designates locations chosen by the user. These could, for example, be the centers of the cells shown in Figure 6.23b; however, they do not need to be on a grid but should cover the region of interest. The data set is $\{z(\mathbf{s}_j), j = 1, \ldots, n\}$. The objective is to generate K joint realizations of these N random variables conditioned on the data. From this uncertainty, maps can be made and statistics generated.

One procedure begins with a conditional distribution $F(\mathbf{s}_1^*; z|(n))$ being constructed at location $\mathbf{s}_1^*$, where the $z|(n)$ notation means conditioned on the data (Goovaerts, 1997). A draw is made from $F(\mathbf{s}_1^*; z|(n))$. Now suppose that the next node drawn is $\mathbf{s}_2^*$. The distribution at $\mathbf{s}_2^*$ is conditioned on the data and $z(\mathbf{s}_1^*)$. This process continues until all N locations are visited. The result is one set of simulated values. The process is repeated K times with a possibly different path through the cells (nodes).

Multipoint Geostatistics   A technique used to address some of the concerns of traditional kriging methods is called *multipoint geostatistics* (MPS) (Hu and Chugunova, 2008). The major difference between MPS and kriging is that the former uses a training image (TI) while the latter uses a semivariogram. The TI must be representative in a statistical sense of the larger area of interest. MPS is still subject to ergodic and stationarity requirements; however, the strong stationarity requirements may be relaxed. A major advantage of this method is that it allows reproduction of complex multiple-point patterns. An algorithm called *snesim* (Strebelle, 2002) can be used to implement the MPS method.

6.8   SPACE–TIME MODELS

In climate and other disciplines, attributes such as precipitation and temperature vary with space and time. This type of data can be represented by $Z(\mathbf{s},t)$, where $\mathbf{s}$ is the spatial location and t is the time. If stationarity is assumed, the covariance can be represented as

$$\mathrm{Cov}(Z(\mathbf{s}; t), Z(\mathbf{s} + \mathbf{h}; t + u)) = C(\mathbf{h}; u) \quad (\mathbf{h}; u) \in \mathbb{R}^d \times \mathbb{R}$$

where $\mathbf{h}$ and u are the spatial and time series lags, respectively, and $\mathbb{R}^d \times \mathbb{R}$ represents the real space. The superscript d is typically two- or three-dimensional, representing a location in the (x,y) or (x,y,z) coordinate system, respectively. The second $\mathbb{R}$ represents the time domain. $C(\mathbf{h};0)$ and $C(\mathbf{0};u)$ represent purely spatial and time series processes, respectively. A major concern in space–time models is that of separability between the space and time process. Two commonly used assumptions are the separation of space–time processes into a product or sum of purely space and purely time processes, such as $C(\mathbf{h}; u) = C(\mathbf{h}; 0)C(\mathbf{0}; u)$ or $C(\mathbf{h}; u) = C(\mathbf{h}; 0) + C(\mathbf{0}; u)$. Sometimes there is interaction between these processes, which requires a more complex modeling approach (Gneiting, 2001).

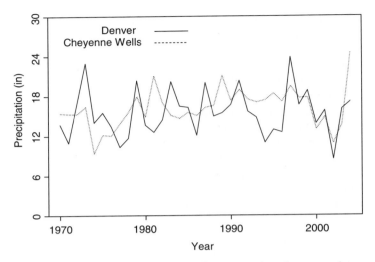

FIGURE 6.27   Colorado precipitation, space–time data example.

Many space–time applications involve a climate variable, which changes over time and is measured at several sites. As an example (Figure 6.27), consider the annual precipitation taken at two sites in Colorado, one in the Denver area and the other on the plains east of Denver in a town called Cheyenne Wells (Western Regional Climate Center, 2005). The Cheyenne Wells site shows possible increasing precipitation from 1970 until about 1989, while precipitation at the Denver site appears to have a constant mean until 2000; however, there is considerable variability in both series. There was a drought in the region from 2000 to 2003, and most stations recorded a significant drop in precipitation. Thus, the spatial distribution of precipitation is misleading unless changes over time are incorporated into the model.

Another example of space–time modeling is hurricane prediction (Jagger et al., 2002). It involves establishing a grid in a hurricane-prone region and then counting the number of cells in the hurricane path by year. Since adjacent cells are not independent, a reasonable model to estimate the hurricane path is an autoregressive process combined with a Poisson process. Thus, instead of the usual observations taken over time at a given site, Jagger et al. present an example of count data on a grid, called a *spatial intensity map* (Figure 6.28). The maps appear to show a slight shift in initial position to the east from years 1900–1943 to years 1944–1993. Also, the final positions have shifted to the east over time and appear to be a bit more spread out in the southwest-to-northeast direction. Such maps can be plotted over time showing space–time trends of hurricanes or other phenomenon. Examples of space–time problems on a lattice include crop yields by county over time.

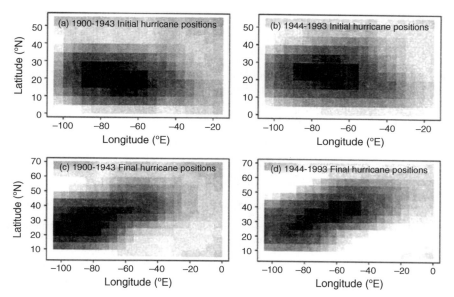

FIGURE 6.28   Spatial intensity maps. (From Jagger et al., 2002.)

6.9 SUMMARY

Models used in spatial statistics are similar to those used in time series analysis. They differ from those used in regression because of the spatial association that occurs in many earth science problems. One of the important and difficult challenges is estimating this spatial association. The estimator takes the form of a covariance function (also called a covariogram). An equivalent representation often used in geostatistical modeling is the semivariogram. Intrinsic stationarity is necessary for a proper semi-variogram. A graph of a semivariogram versus distance exhibits three characteristics: a partial sill (the covariance contribution), a range (the extent of the spatial correlation), and a nugget (local variation or measurement error). If the semivariogram is invariant with respect to direction and sill, it is said to be isotropic; if the range changes with direction, it is said to exhibit geometric anisotropy; and if the sill varies with direction, the semivariogram exhibits zonal anisotropy.

Spatial modeling typically involves predictions at some location (or region) where data do not exist. Kriging is a procedure used to make such an estimate. A kriging estimate is a weighted estimate of known observations. It passes through known points and is optimal under certain assumptions. Kriging requires knowledge of the covariance (or semivariogram) function. This class of models includes simple kriging, ordinary kriging, kriging with a trend, block kriging, indicator kriging, lognormal kriging, Bayesian kriging, and co-kriging. In certain types of applications, particularly climatology, models exist to combine changes over space and time.

EXERCISES

Comment, as appropriate, on the following exercises.

6.1 Consider a simple kriging problem. Assume a spherical model and isotropy. What are the relationships between the weights (the λ_i's) in kriging, distance h between points, and magnitude of the z_i's (the data)? How does the answer change if a nugget model is assumed?

6.2 Can ordinary kriging be used when a trend is present? Explain.

6.3 Should one use kriging as an interpolator when there is a highly clustered set of point locations? Why or why not?

6.4 If the nugget model is appropriate, what can be said about spatial association?

6.5 When there is no spatial correlation, is ordinary least squares equivalent to ordinary kriging? Why or why not?

6.6 Suppose that the response variable of interest is right-skewed. Is a transformation the best or only course of action prior to proceeding with spatial analysis? Why or why not?

6.7 Are the weights assigned by kriging to data points equal in the absence of spatial correlation? Explain.

6.8 Given two data points along a transect with no intervening point, what does one expect the standard error of prediction to look like as one interpolates from one point to the other in a continuous manner?

6.9 Why is ordinary least squares an inappropriate method for modeling data that are spatially correlated?

6.10 Suppose that in a data set, strong geometric anisotropy exists, with the azimuth of greatest range being due north. How does the kriging model handle this anisotropy in terms of any possible transformation of the kriging surface?

6.11 In developing a semivariogram model, is it best to have a good fit to the data at short range? Why or why not?

6.12 How does one know when a kriging model fit is satisfactory?

6.13 Construct lag plots using the data in Table 6.2 for lags 1(R1) and 2(R2). Do these plots correspond to the correlation values?

Exercise 6.14 requires a spatial statistics package and the water-well yield case study data (Appendix I; see the book's Web site). Other spatial data may be used.

6.14 Perform an appropriate transformation on the water-well yield case study data. Repeat the ordinary kriging exercise leading to Figure 6.24. Compare the results. What effect does the transformation have on the results?

Exercises 6.15 to 6.21 require a spatial statistics package and data (Appendix IV; see the book's Web site), represents a random sample of size 197 taken from the Cortez Quadrangle. The complete data set can be found at http://pubs.usgs.gov/of/1997/ofr-97-0492/quad/q_cortez.htm (U.S. Geological Survey, National Geochemical Database, Open-File Report 97-492). Other quadrangles from the NURE HSSR study are available at http://pubs.usgs.gov/of/1997/ofr-97-0492/ (Steven M. Smith, 2001, *Reformatted Data from the National Uranium Resource Evaluation* (*NURE*) *Hydrogeochemical and Stream Sediment Reconnaissance* (*HSSR*) *Program*, ver. 1.30, U.S. Geological Survey, National Geochemical Database, Open-File Report 97-492).

Note to the instructor. You can have your students analyze a subset of these questions and the five elements from Appendix IV (Al, Dy, K, La, and U). You may also choose to substitute another data set and have the students address the following questions.

6.15 Characterize the distributions of the five elements. Which, if any, should be transformed?

6.16 Is there any evidence of a large-scale spatial trend in any of these elements? Include contour plots in the analysis.

6.17 Is there any evidence of spatial anisotropy? Explain.

6.18 Compute empirical semivariograms. Explain the difference between the two types of fit. Verify that each of the lag estimates of gamma is based on at least 30 pairs of points.

6.19 If appropriate software is available, make variogram maps and compare these to the empirical semivariogram computed in Exercise 6.18.

6.20 For each semivariogram, try at least two semivariogram models. Compare estimates of sill, nugget, and range.

6.21 Using what one judges to be the best semivariogram model, estimate the standard errors of prediction using an appropriate kriging method. Generate plots showing the spatial distribution of the errors.

6.22 Are the data points spatially randomly distributed? Use two different methods to address this question.

7 Multivariate Analysis

7.1 INTRODUCTION

Multivariate analysis focuses on developing models to understand relationships among variables. In the earth sciences, multivariate methods are largely exploratory. These methods fall roughly in the following categories:

- *Data reduction.* When a large number of variables are measured, some subsets are almost always a nearly linear combination of others. Data reduction techniques such as principal components analysis (PCA) can help to eliminate this redundancy. In addition, PCA sometimes provides insight into structure and relationships among variables.
- *Grouping of variables.* Grouping of variables that exhibit similar characteristics can yield insight into a process. Factor analysis (FA) is such a tool. In geochemistry, FA may reveal signatures of environmental contamination through the identification of a small set of common factors comprised of similar variables.
- *Clustering of observations.* Grouping observations whose variables have similar characteristics can be an important tool for understanding complex systems, especially when observations have a spatial relationship. An example is a spatial grouping of climate systems.
- *Classification.* Classification of an object into one of several predefined groups may be accomplished with a procedure called discriminant analysis. In archaeology, this procedure may help an investigator assign a pottery sherd to its proper site based on its color, pattern, shape, and other variables (attributes) using statistical comparisons with attributes of the predefined sites.

Other methods include multivariate regression analysis, where there is more than one response variable; multivariate time series analysis, in which multiple measurements are made over time; and multidimensional scaling, a graphical procedure. These are all beyond the scope of this book.

There are numerous textbooks on multivariate analysis, from highly theoretical to purely descriptive. One that we find useful is that of Johnson and Wichern (2002). It contains many examples and presents a sound statistical framework.

Statistics for Earth and Environmental Scientists, By John H. Schuenemeyer and Lawrence J. Drew
Copyright © 2011 John Wiley & Sons, Inc.

7.2 MULTIVARIATE GRAPHICS

Multivariate analysis involves making comparisons, searching for patterns, and making inferences, often across time and/or space. As in univariate analysis, graphics play an important role. Modern computing technology facilitates the use of graphical techniques that involve projections, rotation of coordinate systems, and use of color. A few general-purpose methods are presented first. Then other graphical methods are presented in the context of specific models.

The simplest form of a multivariate graph is the scatter plot. An extension, which can be used to display many variables, is a *scatter plot matrix* or *splom*. It is illustrated using a subset (Figure 7.1) of the Upper Freeport coal bed case study (Appendix V; see the book's Web site). Note that only the upper or lower triangular matrix need be displayed. The data set is discussed in detail in subsequent sections; however, there is:

- An apparent outlier in Ca
- A strong positive correlation between Al and Si

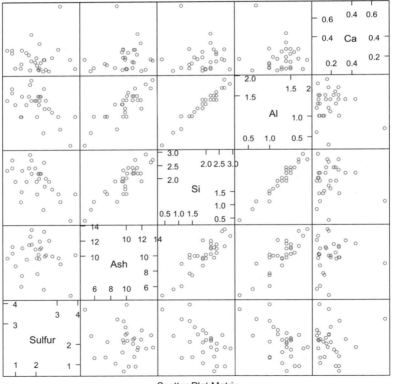

Scatter Plot Matrix

FIGURE 7.1   Scatter plot matrix using a subset of 30 observations and five variables from the Upper Freeport coal bed case study (units are ppm).

- A strong but somewhat weaker correlation between Al and ash and Si and ash
- A weaker negative correlation between Al and sulfur and Si and sulfur

Other patterns may be apparent if the Ca outlier is removed.

Extensions to three or more variables may be made by encoding the third or higher variable in the two-dimensional scatter plot. The relationship among Al, Si, and sulfur is illustrated in the bubble plot of Figure 7.2. A bubble plot is like a scatter plot except that instead of all (x,y) "locations" being represented by a single symbol, a third variable z is represented by a circle whose radius is proportional to z. In Figure 7.2, $z =$ sulfur, the radius, and the size of the circle is proportional to $\sqrt{\pi z^2}$. Thus, the units of size are ppm. To leave the sulfur proportion to the area of the circles places too much visual evidence on larger circles. The three smallest Al and Si pairs have large values of sulfur, as is evidenced by the large circles, but the overall relationship between Al, Si, and sulfur is weak. Ash, a fourth variable, is represented by using color (Figure 7.3). A topological (topo.color module in R-project) palette is used. There is some evidence to suggest that lower sulfur is accompanied by higher ash content.

Another form of plot appropriate for small sample size multivariate data sets is called a *glyph plot*. A glyph plot is an extension of a scatter plot that represents three or more variables. Figures 7.2 and 7.3 can be considered glyph plots because the third variable (sulfur) is coded as bubble size and the fourth (ash) is coded as color. Other examples of glyph plots are the star plot, profile plot, and Chernoff's faces (Thioulouse and Dray, 2007).

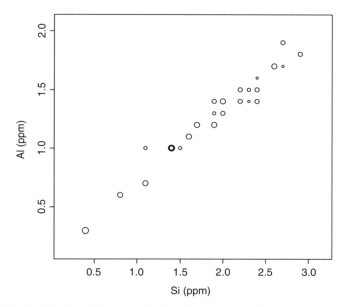

FIGURE 7.2   Bubble plot of Al versus Si with a circle representing the square root of sulfur in ppm.

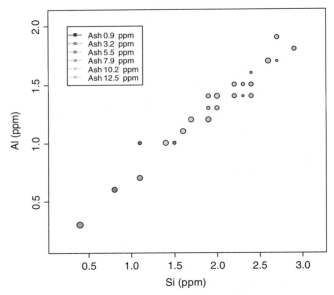

FIGURE 7.3   Bubble plot of Al versus Si with a circle representing the square root of sulfur in ppm. Ash is indicated by color. (*See insert for color representation of the figure.*)

7.3   PRINCIPAL COMPONENTS ANALYSIS

Principal components analysis (PCA) is a mathematical procedure designed to transform a system of correlated variables into one of uncorrelated variables. One purpose of this transformation is to reduce dimensionality. Suppose that p variables ($p > 1$) have been measured and the investigator suspects that the data exist in fewer than p dimensions, which occurs when some variables are a linear combination or a nearly linear combination of other variables. For example, in Table 7.1 variable v2 is a multiple of variable v1, so the data exist in one dimension. Suppose, however, observation 3, variable v2 is 14 instead of 15; then variable v2 is almost a linear combination of v1. This deviation from exact collinearity may be caused by sampling or measurement error. Of course, 14 can be accurate. Variable v2 can be judged to be slightly different than v1 but not significantly so. When many variables are measured,

TABLE 7.1   Simple Example of Collinearity with Three Observations and Two Variables

Obs.	v1	v2
1	2	6
2	1	3
3	5	15

collinearity often occurs. Redundant or almost redundant explanatory variables cause instability problems in regression computation (Chapter 4). The problem of redundant dimensions can be alleviated through the use of PCA, whose main purpose is to reduce the number of variables. PCA results often serve as input to regression, cluster analysis, or other models. Sometimes PCA results lead to a simpler interpretation than when viewing the original data.

7.3.1   Two-Variable Example

A simple illustration of PCA uses two variables, Al and Si from the Upper Freeport coal bed case study (Appendix V; see the book's Web site). Their relationship is shown in a bubble plot (Figure 7.2); however, for the purpose of this example a simple scatter plot will suffice. The basic question to be addressed is: Do Al and Si exist in two dimensions? PCA helps us answer this question. One option in PCA is to use data in original units versus using standardized variables. Since Al and Si are in the same unit, PCA is performed in original (ppm) units. For notational purposes, let $X_1 = $ Si and $X_2 = $ Al. PCA causes X_1 and X_2 to be expressed in a new coordinate system (Y_1, Y_2), which involves a rotation designed to maximize the variance along the Y_1-axis (Figure 7.4). The mechanics of this rotation are explained below.

The variances of Si and Al are 0.347 and 0.125, respectively. Therefore, Si accounts for 73.5% of the total variability and Al accounts for 26.5% (Table 7.2). The scatter plot (Figure 7.4) suggests that these variables are highly correlated. It appears that although the variables are measured in two dimensions (one dimension

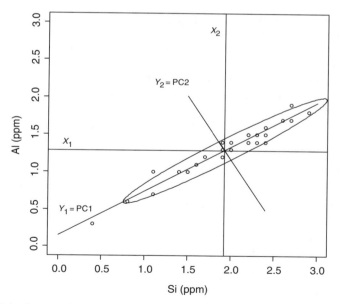

FIGURE 7.4   Scatter plot of Si and Al in original (X_1, X_2) and transformed space (Y_1, Y_2).

TABLE 7.2 PCA Output from the (X_1, X_2) Covariance Matrix

Data	$X_1 = Si$	$X_2 = Al$	Total Variance
Mean	1.923	1.287	
Var(X_i)	0.347	0.125	0.473
Percent of variance (X_i)	73.5	26.5	100.0
Covariance(X_1, X_2)	0.203		

	PC		
	$k=1$	$k=2$	Total Variance
$\lambda_\kappa = $ Var(Y_k) (eigenvalues)	0.468	0.005	0.473
Percent of variance	99.0	1.0	
Cumulative percent	99.0	100.0	

Component Loadings (Eigenvectors)

Variables	$e_1 = $ PC1	$e_2 = $ PC2
$X_1 = Si$	-0.860	-0.510
$X_2 = Al$	-0.510	0.860

Component Loadings (Correlations)

Variables	Y_1	Y_2
$X_1 = Si$	-0.998	-0.061
$X_2 = Al$	-0.986	0.172

for each variable), the variables occupy only one dimension in a statistical sense. That is, variability not accounted for by a linear composite of the two variables may be attributed to sampling or measurement error. Thus, it may be possible to identify a linear composite of Si and Al that accounts for all of the statistically significant variability among Si and Al. PCA can accomplish this. To proceed, define a line $Y_1 = \ell_{11}X_1 + \ell_{21}X_2$. When ℓ_{11} and ℓ_{21} are determined such that the variance of Y_1, Var(Y_1), is maximum, Y_1 (PC1) is called the *first principal component*. This maximization method can be viewed as a trial-and-error process where:

1. A line is drawn in the X_1, X_2 space.
2. Data points are projected normal (perpendicular) to this line.
3. The variance with respect to this projection is computed and saved.
4. The line is then rotated in small steps through 180° and the previous three steps are repeated.
5. After a first principal component is found, the Y_1-axis is held fixed and the second principal component (PC2) is sought.

Iterative computer algorithms are more efficient than the procedure above.

The line that yields maximum variance is the northeast–southwest line (Figure 7.4), labeled Y_1 (PC1). PCA then holds this axis fixed and looks in the remaining $p - 1$

dimensions to find the component that accounts for the second largest amount of the remaining variability. In two dimensions, the second principal component axis, labeled Y_2 (PC2) in Figure 7.4, is fixed when orthogonality is required and there are only two variables. In an orthogonal transformation, such as the one illustrated above, the 90° angle between the axes is maintained, which has the effect of preserving distances. Thus, two points in the (Y_1, Y_2) coordinate system (after the orthogonal transformation) are the same distance apart, as in the original (X_1, X_2) coordinate system. *Oblique transformations*, those in which the resulting coordinate axes are not perpendicular, do not preserve distances and are more difficult to interpret.

The *eigenvalues* (or *characteristic values*) are the variances of the p ($= 2$ in this example) principal components. In the (Y_1, Y_2) space (Table 7.2, heading PC, $k = 1$, $k = 2$), 99.0% of the total variance is accounted for by Y_1 and 1.0% by Y_2. There are two possible conclusions from this example. One is that the data in the (Y_1, Y_2) space represent only one dimension (Y_1), with data in Y_2 attributable to sampling variability and/or measurement error. The other is that the effect of Y_2 is real but small and may be ignored. Without more information, it is not possible to comment on which of these two scenarios is more plausible.

Care must be taken when variables add to a constant. This often happens when units are measured in percent. In this example, that possibility is ignored, since these are only two of many variables; however, it is considered in a subsequent section.

7.3.2   Model

Suppose that there are $p \geq 2$ continuous variables $X_1, X_2, \ldots, X_p$. PCA transforms the X coordinate system into a new coordinate system:

$$Y_1 = \ell_{11}X_1 + \ell_{21}X_2 + \cdots + \ell_{p1}X_p$$
$$Y_2 = \ell_{12}X_1 + \ell_{22}X_2 + \cdots + \ell_{p2}X_p$$
$$\vdots$$
$$Y_p = \ell_{1p}X_1 + \ell_{2p}X_2 + \cdots + \ell_{pp}X_p$$

where the variables, the $Y_1, Y_2, \ldots, Y_p$, are linear combinations of the X's designed to maximize the $\mathrm{Var}(Y_i)$ subject to orthogonality restrictions. It can be shown that if

$$\Sigma = \begin{bmatrix} \sigma_{11} & \cdots & \sigma_{p1} \\ \vdots & \ddots & \vdots \\ \sigma_{p1} & \cdots & \sigma_{pp} \end{bmatrix}$$

is the covariance matrix of the X's, then $\mathrm{Var}(Y_i) = \ell_i' \sum \ell_i$, $i = 1, 2, \ldots, p$ and $\mathrm{Cov}(Y_i, Y_j) = \ell_i' \sum \ell_j$, $i, j = 1, \ldots, p$, $i \neq j$. The covariance matrix (Σ) above is symmetric; that is, $\sigma_{ij} = \sigma_{ji}$. The off-diagonal elements are the covariance terms, and the diagonal elements (also represented as σ_i^2) are the variances.

The PCA algorithm begins by finding the coefficients $\ell_1' = (\ell_{11}, \ell_{21}, \ldots, \ell_{p1})$ such that $\text{Var}(Y_1) = \text{Var}(\ell_1' \mathbf{X})$ is maximized, where $\mathbf{X}' = (X_1, X_2, \ldots, X_p)$. (Bold type represents a vector.) This linear composite $Y_1 = \ell_1' \mathbf{X}$ is the first principal component (PC1) and the ℓ_1 is the *component loading vector* associated with the first principal component. The elements of ℓ_1 are the weights on the X's. The second principal component (PC2) is the linear composite ℓ_2', which maximizes $\text{Var}(Y_2) = \text{Var}(\ell_2' \mathbf{X})$ subject to the constraint that $\text{Cov}(\ell_1' \mathbf{X}, \ell_2' \mathbf{X}) = 0$. The latter constraint ensures that $\ell_2' \mathbf{X}$ is orthogonal to $\ell_1' X$. The algorithm proceeds in this manner through the $p - 1$ variables. The pth linear composite $\ell_p' \mathbf{X}$ is by construction orthogonal to all previous linear composites.

The component loadings are only unique to a nonzero multiplicative constant and thus can be rescaled. For convenience, the ℓ_i often are rescaled such that $\ell_i' \ell_i = 1$. After this rescaling, the notation is changed to $\mathbf{e}_i$ so that $\text{Var}(Y_i) = \mathbf{e}_i' \sum \mathbf{e}_i = \lambda_i$ and $\text{Cov}(Y_i, Y_j) = \mathbf{e}_i' \sum \mathbf{e}_j = 0$, $i \neq j$. Different algorithms may scale differently, but they should be mathematically equivalent.

The $\mathbf{e}_i$ are called *eigenvectors* (or *characteristic vectors*). They determine the direction of the principal component axes. The λ_i are the eigenvalues. The maximization process ensures that $\lambda_1 \geq \lambda_2 \geq \cdots \geq \lambda_p \geq 0$. Because PCA is a mathematical transformation/rotation, $\sum_{j=1}^{p} \lambda_j = \sum_{j=1}^{p} \text{Var}(X_j)$.

An alternative form of the component loading matrix is the representation of the elements as correlations between the Y's and X's. In this formulation, the correlation between Y_k and X_j is $\rho_{Y_k, X_j} = e_{jk} \sqrt{\lambda_k} / \sqrt{\sigma_{jj}}$, where $\sigma_{jj} = \text{Var}(X_j)$. When specifying a covariance matrix, it is convenient to represent the variance as σ_{jj} as opposed to the usual σ_j^2; the covariances are represented as σ_{ij}, $i \neq j$.

The methodology above is illustrated using a two-variable example (Section 7.3.1). The variances and covariance that form the covariance matrix Σ are

$$\Sigma = \begin{bmatrix} 0.347 & 0.203 \\ 0.203 & 0.125 \end{bmatrix}$$

Principal components are extracted from Σ. The linear combination that maximizes $\text{Var}(\ell_1' \mathbf{X})$ is $Y_1 = -0.860X_1 + (-0.510)X_2$. Thus, -0.860 and -0.510 are the component loadings where $(-0.860)^2 + (-0.510)^2 = 1$. Thus, the first eigenvector is $\mathbf{e}_1' = (-0.860, -0.510)$. For this example, once $\mathbf{e}_1$ is determined, no further maximization is required, since there are only two variables and the orthogonality constraint, $\mathbf{e}_1' \mathbf{e}_2 = 0$ implies that $\mathbf{e}_2' = (-0.510, 0.860)$. The corresponding eigenvalues are given in Table 7.2, as are the correlations between the X's and the Y's. For example, the correlation between X_1 and Y_1 is -0.998.

7.3.3   Geometric Interpretation

The direction of Y_1 is determined by the eigenvector $\mathbf{e}_1$ and the direction of Y_2 by the eigenvector $\mathbf{e}_2$. In the two-variable example, the angle of counterclockwise rotation of X_1 into Y_1 is determined by $\theta = \tan^{-1}(e_{21}/e_{11}) = \tan^{-1}(-0.510 / -0.860) = 30.7°$.

The rotation for $p = 2$ with $\boldsymbol{\mu}' = (0,0)$ to simplify notation, after some algebra yields an ellipse of the form

$$\frac{(\mathbf{e}_1'\mathbf{x})^2}{\lambda_1} + \frac{(\mathbf{e}_2'\mathbf{x})^2}{\lambda_2} = c^2$$

Since $Y_i = \mathbf{e}_i'\mathbf{x}$, $i = 1, 2$, $Y_1^2/\lambda_1 + Y_2^2/\lambda_2 = c^2$, the equation of the ellipse in (Y_1, Y_2). This ellipse has semimajor and semiminor axes equal to $c\sqrt{\lambda_1}$ and $c\sqrt{\lambda_2}$, respectively. Thus, for the two-variable example, the semimajor and semiminor axes are proportional to 0.684 and 0.068, respectively.

Since PCA is a mathematical transformation/rotation, it is not necessary to assume a statistical distribution for X_1 and X_2. However, doing so allows a geometric interpretation. If (X_1, X_2) are assumed to be bivariate normal with mean $\boldsymbol{\mu}' = (\mu_1, \mu_2)$ and covariance matrix Σ, then the eigenvalues, eigenvectors, and values of (X_1, X_2) can be represented in the form of an ellipse, which centered on μ is

$$(\mathbf{x} - \boldsymbol{\mu})' \sum{}^{-1}(\mathbf{x} - \boldsymbol{\mu}) = c^2$$

In more than two dimensions the ellipse becomes an ellipsoid.

7.3.4 Standardized Units

PCA is often performed on raw data or, equivalently, on a variance–covariance matrix when variables are measured in the same units, as is the case in the two-variable example in Table 7.2, where Al and Si are in ppm. If, however, one variable is measured in ppm, another in meters, and a third as a percent, comparing variances is meaningless. Therefore, PCA is computed in standardized units by centering and scaling transformation, as in $Z = (X - \overline{X})/\hat{\sigma}_X$. In standardized units, $\sum_{i=1}^{p} \lambda_i = p$, where p is the number of variables and the proportion of variability accounted for by the ith component is λ_i/p. It may also be desirable to standardize variables measured in the same units if they differ considerably in magnitude to give them the same weight: namely, unit variance.

To illustrate the effect of the standardizing–centering transformation, the two-variable example is recalculated and principal components are extracted from the correlation matrix

$$\rho = \begin{bmatrix} 1 & \rho \\ \rho & 1 \end{bmatrix} = \begin{bmatrix} 1 & 0.90 \\ 0.90 & 1 \end{bmatrix}$$

Since each variable now has equal variance (1), the eigenvectors are $\mathbf{e}_1' = (\,0.707 \quad 0.707\,)$ and $\mathbf{e}_2' = (\,0.707 \quad -0.707\,)$, which represent a 45° rotation. Because $\lambda_i = \mathbf{e}_i'\rho\mathbf{e}_i$, simple algebra shows that $\lambda_1 = 1 + \rho$ and $\lambda_2 = 1 - \rho$. For this example, $\lambda_1 = 1.97$ and $\lambda_2 = 0.03$. Note that $\lambda_1 + \lambda_2 = 2$, the total standardized variance. The proportion of variance accounted for by the first principal component is $\lambda_1/p = 0.987$. This differs slightly from the proportion 0.990 accounted for by the first principal components when extracted from the covariance matrix.

TABLE 7.3   Simple Example of the Constant Sum Problem with Two Compounds, *A* and *B*

Observation	Compound	
	$A(\%)$	$B(\%)$
1	20	80
2	25	75
3	35	65
4	27	73
5	11	89

7.3.5   Constant Sum Problem

The Upper Freeport coal bed data (Ruppert et al., 2001) poses a special problem for analysis because the variables add to a constant. See Appendix V in the book's Web site for a description of this data set. Before discussing this more complex example, consider a simple example (Table 7.3) of the measurement as a percentage of a substance consisting of two compounds, *A* and *B*. Clearly, there is an induced negative correlation between compounds *A* and *B* because as one increases, the other must decrease. This same situation exists in data sets containing more variables, although the effect is not always obvious. Because of this induced correlation, special transformations are needed to analyze this type of data properly. In general, when $\sum_{j=1}^{p} y_{ij} = c$, $i = 1, \ldots n$, where p is the number of variables and n is the number of observations, there is an induced correlation because when one variable changes, at least one other variable must change, since the sum is a constant c. Data that add to a constant are also called *compositional data*. To avoid misinterpretation associated with this induced correlation, a transformation in the form of a ratio is often required. There are a number of approaches to this problem, and controversy as to which is best. A key issue is the meaningfulness of the total c. Among the approaches discussed by van den Boogaart (2005) are those of Chayes (1960), who argued that data should be analyzed in real geometry; Aitchison (1986), who argued that the data should be analyzed in relative geometry; and Egozcue et al. (2003), who argued the total is meaningful and data should be analyzed in relative geometry. The R-project program compositions (van den Boogaarta and Tolosana-Delgadob, 2008) can analyze data in these geometries. Their isometric log ratio transformation preserves metric properties. Among the approaches suggested by Aitchison (1986) is to select an unimportant variable. Without loss of generality, call it y_{pi}, $i = 1, \ldots, n$ and divide it into the other variables, forming a set of new variables $y_{i1}/y_{ip}, \ldots, y_{i,p-1}/y_{ip}$ in $p - 1$ dimensions. This method has the advantage of reducing dimensionality and the disadvantage of making results dependent on the choice of the unimportant variable.

The approach used by the authors in the subsequent analysis is the *centered log ratio transformation* (CLR) [see Aitchison (1986) and Pawlowsky and Egozcue (2006)]. This transformation is the log of the observation minus the log of the geometric mean, namely

$$y_{ij}^c = \ln(y_{ij}) - \ln\left(\prod_{j=1}^{p} y_{ij}\right)^{1/p}$$

where y_{ij}^c is the centered log ratio result for the ith observation and jth variable. This approach eliminates the constant-sum problem and preserves symmetry but yields a singular matrix, as the reader may observe by summing each sample over all variables (Table 7.4). A singular matrix is a square matrix that is not invertible. This is not a major problem since many software packages handle singular matrices. The main argument in support of a ratio transformation is that individual variables are influenced by changes in other variables, but ratios are not.

In the Upper Freeport coal bed data (Appendix V in the book's Web site) the sum of rows is not a constant, so why is this data set treated as a constant-sum problem? The basic reason is that there are almost certainly elements missing that cannot be accounted for. To proceed with this analysis, variables expressed in percent are transformed to parts per million by multiplying the variables in percent by 10,000. The rows are summed and then normalized so that each row sums to 1. Following normalization, a CLR transformation is applied (Drew et al., 2008.)

TABLE 7.4   First Nine Observations from the CLR-Transformed Upper Freeport Coal Bed Data Set

Variable	Observation								
	1	2	3	4	5	6	7	8	9
Sulfur	4.50	4.72	5.69	6.09	4.74	5.15	4.48	4.38	5.22
Ash	6.50	6.43	6.54	6.36	6.28	6.44	6.19	6.51	6.24
Si	4.90	4.80	4.55	3.95	4.58	4.75	4.59	4.99	4.15
Al	4.55	4.38	4.29	3.68	4.46	4.36	4.13	4.61	3.67
Ca	2.08	2.08	1.93	3.02	2.01	2.29	2.14	1.88	3.74
Mg	1.28	0.87	0.55	0.37	0.88	1.08	0.57	1.56	0.52
Na	0.58	0.20	−0.28	0.83	0.41	0.50	−0.33	0.73	−0.14
K	2.74	2.33	2.03	0.97	2.19	2.64	2.14	2.92	1.59
Fe	4.24	4.49	5.35	5.38	4.31	4.36	4.18	4.21	4.80
Ti	1.53	1.35	1.09	0.59	1.05	1.53	1.04	1.24	0.44
S	1.70	1.54	1.55	2.49	1.79	1.82	1.47	1.58	2.82
As	−2.21	−2.53	0.47	−0.66	−2.44	−2.32	−1.24	−2.47	−1.72
Cd	−7.89	−6.75	−7.43	−8.76	−8.02	−7.88	−5.89	−8.14	−7.05
Ce	−2.31	−2.07	−1.91	−2.94	−1.83	−2.09	−2.72	−2.30	−3.14
Cu	−2.36	−2.72	−1.50	−3.78	−1.73	−2.24	−3.14	−2.47	−3.67
Hg	−6.08	−6.53	−11.12	−4.79	−5.74	−6.11	−5.68	−6.07	−5.97
Mn	−2.90	−2.32	−2.42	−2.15	−3.02	−2.09	−3.19	−2.38	−1.79
Pb	−3.04	−3.51	−3.03	−3.14	−2.93	−3.37	−2.39	−2.61	−2.67
Se	−4.71	−4.26	−3.86	−3.95	−4.32	−5.51	−4.56	−4.68	−4.75
Sr	−1.10	−0.02	−0.71	−0.64	−0.44	−0.97	−0.79	−1.10	−0.81
Zn	−2.01	−2.47	−1.76	−2.90	−2.24	−2.32	−1.02	−2.38	−1.49

The correlations usually differ between the data set expressed in ppm and the data set after the CLR transformation. For example, the correlation between sulfur (ppm) and ash (ppm) is 0.25 (from Appendix V data; see the book's Web site), while the correlation between sulfur and ash in CLR transformed data is -0.23. Note the change in sign.

A brief description of the coal process follows. The formation of coal begins with the accumulation of a biomass composed of plant and *clastic material* (usually shale). Clastic materials are sediments formed from preexisting rock. *Diagenesis,* a change undergone by sediment after deposition, begins almost immediately and results from chemical changes in the biomass as well as interaction with groundwater. One of the main chemical reactions that occur during the diagenetic process is the production of iron sulfide (pyrite). There are several reactions that produce this sulfide. The first is the reaction of the hydrogen sulfide generated within the biomass reacting with the iron cations (iron with a positive ion) in the biomass and nearby clastic beds. The second is groundwater transporting sulfate material into the coal, where it is reduced to sulfide material and interacts with similarly available iron. The affinity between the iron and sulfide chemical radicals is very strong. To a lesser degree the cations of arsenic, lead, copper, and other chalcophile cations react with the sulfur during the creation of pyrite. *Chalcophile elements* have a preferential tendency to form sulfides. There is also a chemical affinity between the structure of clay minerals and many metal cations, such as copper, the rare earths, magnesium, and titanium. A second process that is active in the formation of a coal bed is the variation of the clastic input to the biomass. It is not unusual to encounter a 5 to 50% variation in the clastic component in a coal bed. A third source of variation in the chemical analysis of a coal is caused by the ashing process used as part of the chemical analysis procedure to determine the amounts of each of the inorganic elements in the ash. Each of these amounts is then converted back to a whole coal basis. A way to analyze coal components is to burn the coal (called *ashing*) and analyze the ash using standard chemical analytic techniques. However, volatile elements can be lost in the ashing process and thus needs to be determined from an analysis of the whole coal. A coal is "ashed" by heating it to 525°C. During the heating process the volatile components (moisture, sulfur, and others) are driven off and the calcium and manganese in the ash and similar cations react with the volatilized sulfur anions to produce calcium sulfate (anhydrite). The total sulfur in the whole coal in Table 7.4 is called "Sulfur," whereas the S variable represents that part of the total sulfur that is captured during the formation of anhydrite in the ashing process.

The purpose here is to determine how well these processes can be isolated from coal chemistry data through the procedure of principal components analysis. We acknowledge the assistance of Frank Delong, Joe Hatch, Curtis Palmer, and Leslie Ruppert of the Coal Branch of the U.S. Geological Survey for their help in isolating a data set and their discussions with us on coal formation.

The results for PC1 through PC7 are shown in Table 7.5. Since there are 21 original variables, there are 21 new variables, the principal components. The Var(PC1) $= 7.97$ and the total variance is the sum of these eigenvalues over all variables; for this example it is 21, since variables have been standardized. Below the eigenvalues are the

TABLE 7.5   First Seven Eigenvalues and the Principal Component Loadings for the CLR Upper Freeport Coal Bed Data

	PC1	PC2	PC3	PC4	PC5	PC6	PC7
Eigenvalue	7.97	3.11	2.32	1.67	1.31	0.96	0.77
Percent	37.97	14.80	11.06	7.94	6.22	4.57	3.69
Cumulative percent	37.97	52.77	63.83	71.77	78.00	82.56	86.25
Variables							
Sulfur	0.24	−0.13	−0.36	−0.18	−0.04	0.07	−0.21
Ash	−0.27	0.11	−0.30	−0.19	0.00	0.08	0.01
Si	−0.34	0.08	−0.01	−0.09	0.09	−0.03	0.04
Al	−0.34	0.01	−0.02	−0.05	−0.03	−0.01	−0.08
Ca	0.22	0.33	0.03	0.21	−0.15	−0.08	0.13
Mg	−0.30	0.20	−0.07	0.01	0.02	−0.03	0.18
Na	−0.17	0.18	−0.33	0.28	0.14	0.14	0.27
K	−0.31	0.16	−0.04	−0.07	−0.02	−0.20	−0.05
Fe	0.22	−0.09	−0.38	−0.35	−0.02	0.05	−0.09
Ti	−0.32	0.04	−0.05	−0.03	0.11	−0.03	0.02
S	0.22	0.36	0.01	0.17	−0.21	−0.09	−0.02
As	0.13	−0.42	−0.17	−0.14	0.04	−0.11	0.10
Cd	0.06	0.08	0.42	−0.38	0.27	0.09	0.04
Ce	−0.28	−0.22	0.01	0.01	−0.13	0.15	−0.16
Cu	−0.20	−0.21	0.10	−0.09	−0.44	−0.26	−0.25
Hg	0.07	−0.04	−0.01	0.37	0.63	−0.07	−0.40
Mn	0.16	0.38	−0.08	−0.12	−0.28	−0.05	0.02
Pb	0.00	−0.23	0.32	0.18	−0.10	−0.52	0.10
Se	0.03	−0.36	−0.01	0.16	−0.05	0.21	0.67
Sr	−0.07	−0.08	0.21	0.30	−0.31	0.65	−0.30
Zn	0.02	0.13	0.39	−0.41	0.13	0.24	0.12

percent of variance and cumulative percent of variance of the principal components. Corresponding to each principal component is the amount of weight or importance assigned to each variable.

The *screeplot* is a graph of the ordered eigenvalues. The screeplot of eigenvalues (variances) are shown on the vertical axis and their order on the horizontal axis (Figure 7.5). Thus, this plot illustrates the amount of variability accounted for by each principal component when ordered from highest to lowest. The screeplot may also be used as a guide to eliminate insignificant principal components, which will be discussed shortly.

A *biplot* is a graphical tool used to represent observations and variables on the same two-dimensional plot (Gower and Hand, 1996). A biplot is useful in the following situation. Suppose that $\mathbf{Y}(n \times p)$ is a data matrix, where n is the number of observations and p the number of variables. Further suppose that $\mathbf{Y} \doteq \mathbf{A}(n \times d)\mathbf{B}'(p \times d)$, where d is 2 or 3. The variables shown as vectors in Figure 7.6 are those in the (scaled) space of the first two principal components, PC1 and PC2 (Table 7.5). Note that there are various implementations of the biplot and that scaling of variables and observations may differ. The angles between vectors represent correlations between the

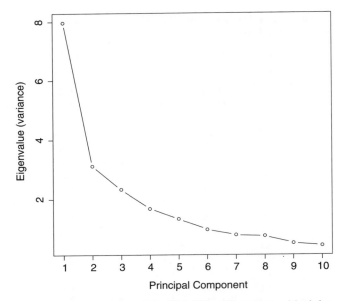

FIGURE 7.5   Screeplot of the CLR Upper Freeport coal bed data.

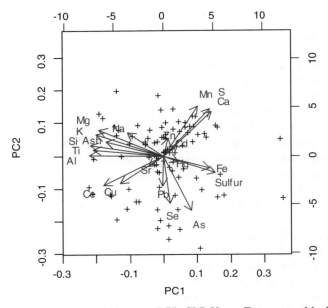

FIGURE 7.6   Biplot of PC1 versus PC2, CLR Upper Freeport coal bed data.

variables. The observations shown by plus signs are actually transformed bivariate observations called *scores*, corresponding to PC1 and PC2. The axes for the scores are on the top and right of Figure 7.6. Scores are discussed in a subsequent section. When the scores are not present, the biplot reduces to a plot of component loadings.

The most striking relationship shown in this biplot is the strong inverse association of the clastic component as measured by the position of the ash content vector (ash) versus the iron pyrite component as measured by the coincident sulfur and Fe content vectors (Table 7.5). The constituents of the ash are shown as the Al, Si, Ti, K, Mg, and Na content vectors, all closely associated with the ash vector. These six constituents are the main components of ash and mineralogically are usually a variety of clay minerals that compose the intercalated shale material. The measure of these materials is per unit volume. The inverse association between the ash and iron pyrite (the ash and Fe vectors are approximately 180° apart) is to be expected because the amount of iron pyrite is a measure of the volume of organic material. As the organic material and shaly materials are associated inversely, the amount of iron pyrite is an inverse proxy for the amount of shaly material in the coal.

The analysis above is successful in showing that the basic deposition and diagenesis processes that produce a coal bed can be isolated from the chemistry of the coal using principal components analysis. As noted previously, not all PCA yields a meaningful interpretation. Many just reduce dimensionality.

7.3.6 Eliminating Insignificant Components

A screeplot (Figure 7.5) provides insight into the number of principal components that the user may wish to retain. It shows the relative importance of the components in terms of their contribution to the total variance.

In the Upper Freeport coal bed case study, 78% of the total variability (Table 7.5) is accounted for by the first five principal components. Thus, a data set with 21 variables may be reduced to one with five variables, resulting in a minimal loss of information. How should an investigator decide how many components to maintain? *Kiser's criterion* suggests omitting components whose variance is less than $K = \sum_{i=1}^{k} \lambda_i/p$. In this study, $p = 21$. When $k = 6$, $K = 0.83$, which is less than $\text{Var(PC6)} = 0.96$; however, when $k = 7$, $K = 0.86$, which is greater than $\text{Var(PC7)} = 0.77$. Thus, Kiser's criterion suggests that the first six principal components be retained. Of course, this is only a guideline and more or fewer components can be retained, based on problem-specific considerations.

7.4 FACTOR ANALYSIS

Factor analysis (FA) is a multivariate procedure used to gain insight into correlation among variables. English psychologist Charles Spearman first proposed what would become factor analysis early in the twentieth century. An early and controversial application was the interpretation of tests of mental ability. Educational psychologists believed that test items (variables) could be condensed into a small number of factors, which identified basic or latent traits, such as spatial ability and reasoning skills. These were called *common factors*. In addition, there was a *unique component* associated with each individual. Harmon (1976) presents issues surrounding this application. Reyment and Jöreskog (1996) discuss applications in the natural sciences. Subsequently, an example from the San Rafael Swell spring and well data area of Utah will be introduced.

Although FA may be considered a formal statistical model in which a number of common factors are posited and tested, most often it is used in an exploratory manner to discover or confirm underlying earth science processes. FA is sometimes confused with PCA, but it is a different procedure. Major differences are:

- In the PCA model, there are $k \leq p$ significant principal components, and the remaining variability, if any, is error (p is the number of variables). In FA, there is an "error" term; however, this term has a unique component and an error component.
- In PCA, the goal is to maximize the variance of the new variables. In FA, the goal is to find correlation structures.

In traditional FA, the rows of data are observations (also called cases or samples). The columns are the variables. A subset of variables may represent a common factor. Factor analysis can either be used to examine the variance–covariance structure among variables or the correlation structure. The latter is often called *R-mode factor analysis*. When units of measurement are the same for all variables, it may be preferable to use the variance–covariance structure. Results will then be weighted by the variances. In a correlation structure, which is appropriate when variables are in different units, variances are all equal to 1.

Another type of factor analysis is *Q-mode* or *inverse factor analysis*, in which the rows are variables and the columns are observations. It has generally been replaced with cluster analysis methods and will not be discussed further.

Factor analysis can be exploratory or confirmatory. In the exploratory mode, the number of common or latent factors is not known in advance and must be discovered. The associated statistical tests on the number of factors should be used as a guideline, since an investigator will have peeked at the data in the process of attempting to identify the number of latent factors. In confirmatory FA, the investigator conjectures, based on theory or experience, that there are a given number of factors and tests accordingly. Most FA studies in the earth sciences are exploratory, and for these studies a range of a number of common factors are often selected. For a given specified number of factors, the hypothesis can do this routinely and produce a p-value. A small p-value will suggest that more factors are needed, however; the result of any statistical test should be used as a guide because interpretability is usually more important in selecting the number of common factors. Indeed, a major concern with factor analysis is that there are almost always multiple plausible interpretations. However, FA can be a useful tool to investigate multivariate structures, but inferences should be made cautiously.

One of the results of PCA is a component loading matrix, which shows association between original variables and principal components. The equivalent matrix in FA is called a *factor loading matrix*, **F**. How is **F** to be found? There are several procedures used to obtain an estimate of the common factors and unique components. The maximum likelihood estimate approach is presented in the next section.

7.4.1   Model

The FA model expressed as a system of linear equations is

$$X_1 - \mu_1 = \ell_{11}F_1 + \ell_{12}F_2 + \cdots + \ell_{1m}F_m + \varepsilon_1$$
$$X_2 - \mu_2 = \ell_{21}F_1 + \ell_{22}F_2 + \cdots + \ell_{2m}F_m + \varepsilon_2$$
$$\vdots$$
$$X_p - \mu_p = \ell_{p1}F_1 + \ell_{p2}F_2 + \cdots + \ell_{pm}F_m + \varepsilon_p$$

where $X_1, \ldots, X_p$ are the observed variables and $\mu_1, \ldots, \mu_p$ are the corresponding means and $\varepsilon_1, \ldots, \varepsilon_p$ are the "errors." $F_1, \ldots, F_m$ is a vector of m common factors. The ℓ_{ij} is the loading (or contribution) of the ith variable ($i = 1, \ldots, p$) on the jth factor ($j = 1, \ldots, m$). For FA to be meaningful, $m \ll p$. (The symbol "$\ll$" means "much less".) If not, the common factors probably will closely correspond to the original variables, and no additional insight is gained with FA. The ε_i's in FA are comprised of a unique or *specific component* and measurement error. However, unless repeated measurements are taken on the observations, it is not possible to separate them. The F's cannot be measured directly. They are known through the X's and ℓ_{ij}'s.

In vector–matrix notation, this system of equations is represented as

$$\underset{(p \times 1)}{\mathbf{X} - \boldsymbol{\mu}} = \underset{(p \times m)}{\mathbf{L}} \quad \underset{(m \times 1)}{\mathbf{F}} + \underset{(p \times 1)}{\mathbf{e}}$$

where the subscripts refer to the number of rows and columns in the vectors and matrices above. The **X** is the vector of observed variables, **μ** the means, **L** the factor loading matrix, **F** the vector of common factors, and **e** the vector of errors or unique components. The unique component $u_{i|m}$ of the ith variable given m common factors can be estimated as the difference between the total variance and the communality.

$$u_{i|m} = \text{Var}(X_i) - \sum_{j=1}^{m} \ell_{ij}^2$$

Two commonly used methods to evaluate model parameters are maximum likelihood estimation and principal factor estimation. Both require iterative solutions. The *maximum likelihood estimate* (MLE) method allows users to perform a test on the adequacy of the number of factors specified. The null hypothesis is that the number of factors specified is sufficient. The alternative is that more factors are needed. A p-value greater than 0.05 may suggest that the number of factors specified is sufficient. MLE finds the parameters that make the data observed most likely.

An alternative, *principal factor estimation*, is a factor analysis method that uses a *communality estimate*, which is an estimate of the proportion of common variance (as opposed to the unique variance) in a variable. From this estimate of common variance, principal factor analysis proceeds in a manner similar to PCA.

7.4.2 Finding a Solution

As noted previously, a goal of factor analysis is to find structure. This is most often accomplished by rotating the factor loading matrix **L** according to some criterion. Numerous algorithms are available to do this, including the following orthogonal rotations: varimax, quartimax, quamax, parsimax, and othomax. A commonly used rotation is *varimax*, which is the default rotation in the R-project language algorithm factanal in the stats library. Varimax is an orthogonal rotation that adjusts elements in the columns of the factor loading matrix **L** by maximizing the variance of their square. This rotation has the effect of making the loadings as close to $+1$, -1, or 0 as possible, which can make interpretation of **F** relatively straightforward because it allows the user to identify variables that are associated with a specific factor (those that have absolute values near 1) and those that do not (those that have values near zero).

The authors recommend the use of orthogonal rotations. Oblique rotations are typically available in most factor analysis algorithms and often can increase the interpretability of individual factors; however, factors resulting from nonorthogonal rotations are correlated and the overall interpretation of the set of common factors becomes muddled.

There are a number of choices in model selection, and in general there is no best procedure. The MLE method is used for estimating model parameters with a varimax rotation in the subsequent example.

7.4.3 Correlation, Covariance, and Robustness

The default option in most factor analysis algorithms is to begin with a correlation matrix, which guarantees equal weighting of the variables. The correlation matrix is the option to use when variables are in different units (e.g., parts per million and centimeters). If all variables are in the same unit and the desire is to have a natural (nonstandardized) weighting, the covariance matrix is recommended. If outliers exist or the variables are highly skewed, there are several options, which include using a robust covariance (or correlation) matrix as input, performing a log or other appropriate transformation, or using a resampling procedure such as jackknifing or bootstrapping. FA can be performed on a correlation or covariance matrix when raw data do not exist.

7.4.4 Example Using the San Rafael Swell Data

In the Upper Freeport coal bed cases, factor analysis is not needed because the component loadings are interpreted directly. In the following example, an FA solution

TABLE 7.6   San Rafael Swell Data

ID	pH	Ca (mg/L)	Mg (mg/L)	Na (mg/L)	K (mg/L)	Cl (mg/L)	SO_4 (mg/L)	alk (mg/L)	C_{13} (sir)[a]	S_{34} (sir)
A	7.12	71.00	66.00	11.00	6.80	3.80	188.00	277.00	−6.00	−22.00
B	6.83	153.00	56.00	420.00	4.10	447.00	663.00	251.00	−2.90	0.90
C	7.00	86.00	30.00	821.00	4.90	846.00	627.00	236.00	−4.10	9.60
D	7.13	113.00	54.00	120.00	4.70	137.00	387.00	200.00	−4.30	7.80
E	7.00	29.00	37.00	8.00	2.50	2.60	62.00	181.00	−7.90	1.00
F	7.00	59.00	31.00	26.00	7.50	15.00	110.00	196.00	−6.00	6.90
G	7.10	160.00	63.00	62.00	15.00	22.00	540.00	210.00	−5.80	5.00
H	7.45	74.00	48.00	15.00	7.40	4.70	225.00	182.00	−6.50	−0.90
I	6.70	81.00	41.00	12.00	4.60	9.50	240.00	140.00	−7.00	5.10
J	8.05	33.00	17.00	5.00	1.80	3.50	24.00	135.00	−6.60	12.10
K	7.20	240.00	95.00	66.00	18.00	30.00	830.00	228.00	−6.90	6.20
L	7.70	33.00	25.00	24.00	4.30	4.20	61.00	179.00	−9.00	0.50
M	7.70	30.00	20.00	35.00	5.90	3.70	53.00	182.00	−9.00	12.00
N	7.60	250.00	65.00	1400.00	6.90	1900.00	1000.00	185.00	−4.10	6.00
O	7.10	85.00	36.00	20.00	8.90	8.30	150.00	274.00	−5.80	−5.90
P	7.00	130.00	53.00	920.00	11.00	1700.00	11.00	147.00	−5.40	6.90
Q	7.00	435.00	152.00	136.00	10.00	134.00	1470.00	312.00	0.40	4.80
R	7.00	106.00	70.00	54.00	3.80	14.00	489.00	200.00	−5.60	11.20
V	7.80	77.00	22.00	33.00	2.10	13.00	70.00	284.00	−10.60	17.10

[a]sir, stable isotope ratio.

with varimax rotation aids in interpretation. The data set used is from springs and flowing wells in the San Rafael Swell area of Utah. It consists of 19 observations and 10 variables (Table 7.6) and was graciously furnished to the authors by Briant Kimball. A description of the data has been given by Kimball (1988). The data are used for purposes of exposition and may differ from results obtained by Kimball. Prior knowledge suggests that two or three common factors may be reasonable. Two factors are used for purposes of illustration. Because variables are in different units, the FA algorithm will operate from a correlation matrix.

Factor 1 in Table 7.7 shows a combination of carbonate dissolution and precipitation (Ca, Mg, alk, C_{13}) with gypsum dissolution (Ca, SO_4). As gypsum continues to dissolve or is added through leakage into the Navajo Sandstone aquifer, it drives the precipitation of carbonate minerals. Since these two processes are clearly not independent, they are not represented by two orthogonal axes. Factor 2 is Na and Cl. The increase in Na and Cl is from a leakage of water from the overlying aquitard (the Carmel Formation, a marine carbonate). It, too, is not independent of the increases that load onto factor 1, but the biplot (Figure 7.7) breaks it out well. pH and S_{34} have a strong unique component; however, K and alk also show significant unique components. S_{34} is unique but is meaningful for water quality interpretation. It does not correlate well with other variables, but its convergence to a single value means that the sulfate in these waters clearly comes from the Carmel Formation and not from within the Navajo Sandstone itself.

TABLE 7.7   Factor Analysis Output for the San Rafael Swell Data[a]

	Factor 1	Factor 2
SS loadings	3.918	2.103
Proportion variability	0.392	0.210
Cumulative variability	0.392	0.602

Factor Loadings

Variable	Factor1	Factor2	Communality	Uniqueness
pH	− 0.310	—	0.096	0.904
Ca	0.971	0.152	0.966	0.034
Mg	0.954	—	0.916	0.084
Na	0.175	0.982	0.995	0.005
K	0.527	—	0.278	0.722
Cl	0.141	0.968	0.957	0.043
SO_4	0.916	0.192	0.876	0.124
alk	0.506	− 0.218	0.303	0.697
C_{13}	0.738	0.254	0.610	0.390
S_{34}		0.150	0.024	0.976

[a]Two-common-factor MLE model solution from the correlation matrix using a varimax rotation; test of the hypothesis that two factors are sufficient. Chi-square statistic $= 79.43$ on 26 df; p-value $= 2.56 \times 10^{-7}$.

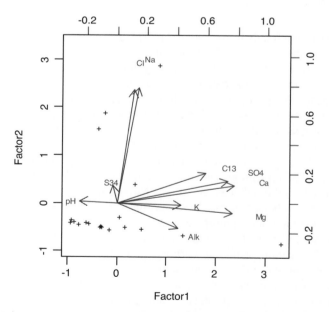

FIGURE 7.7   Biplot of factor 1 versus factor 2 for the San Rafael Swell data. The scales corresponding to the factor loadings are − 1 to 3, shown on the bottom and left. The scales corresponding to the factor scores are on the top (variable 1) and right (variable 2).

The communality and uniqueness components sum to 1 for each variable. *Communality* is the relative proportion of variability accounted for by the common factors and is computed by a sum of squares of factor loadings. For example, the communality of Na accounted for by factors 1 and 2 is $(0.175)^2 + (0.982)^2 = 0.995$ (Table 7.7). *Uniqueness* is the complement. If several variables have a high unique component, an additional common factor may be warranted. Too many common factors may cause a mle FA algorithm to fail to converge.

The test of hypothesis that two common factors are sufficient versus the alternative that they are not is presented in Table 7.7; however, we seldom find it useful and rely on a solution that yields a meaningful interpretation. In the sufficiency chi-square test, the hypothesis that two common factors are sufficient is rejected; the hypothesis that three factors are sufficient (not shown) is also rejected.

The relationship between the original variables and the common factors is illustrated with a biplot (Figure 7.7). The six variables that load heavily on factor 1 point to the east, while the two variables that load heavily on factor 2 point north. The major unique component pH points west. The plus signs are variables 1 and 2 of the factor scores (the transformed observations). Factor scores are described in the next section.

7.4.5 Factor Scores

The primary result in factor analysis is the factor loading matrix, by which the common factors may be interpreted. For diagnostic purposes or when FA output is being used in subsequent analysis, factor scores are needed. *Factor scores* are estimates of values for the unobserved F_i's. They are the data transformed into the space of the common factors. The two most common methods of generating factor scores are *weighted least squares* and regression. Regression scores for the San Rafael Swell data example are shown as plus signs ($+$) in Figure 7.7. Johnson and Wichern (2002) provide computational details that are beyond the scope of this text. The estimates from both methods are usually similar and neither is uniformly superior. Scores for principal component analysis are computed in a similar manner.

7.5 CLUSTER ANALYSIS

Cluster analysis is the name given to a class of methods that seek to group into clusters observations that are most similar. A cluster is typically thought of as a spatially compact group of points. Clustering is a form of stratification that attempts to identify homogeneous subregions.

Among the applications of clustering in the earth sciences are:

- Defining climate zones based on measures of temperature, precipitation, pressure, and other variables.
- Defining coastal zone environments, such as dunes and backshores using sand grain size.

- Grouping metropolitan areas based on quality-of-life measures, such as income, cultural activities, crime, schools, and transportation.
- Defining forest regions based on types and quantity of flora and fauna.
- Identifying areas of similar land use.
- Grouping together molecules that have similar features.
- Identifying hazardous waste sites based on measures of toxicity.
- Identifying archaeological sites based on chemical composition of pottery sherds.
- Partitioning forests into homogeneous subregions for forest and fire management.

Sometimes the purpose of clustering is not to identify spatially compact or connected regions but to identify observations with similar characteristics, such as cities, according to quality-of-life measures.

In some older texts, grouping like variables is considered a cluster analysis procedure, however, this is FA. Everitt et al. (2001) provide an introduction to many popular clustering techniques.

7.5.1 Pottery Classification Study

Archaeologists are interested in the clustering and classification of pottery sherds by site or type, which was illustrated by Hall (2001) using a subset of data. The original data consisted of 92 samples of Early Jomon sherds from fours sites in the Kanto region of Japan. To display the pottery data and results, the authors chose arbitrarily to examine three sites [the Kamikaizuka (K), Narita 60 (N), and Shouninzuka (S)] using 78 observations and five variables (Ti, Mn, Cu, Zn, and Pb) from an x-ray fluorescence analysis (Table 7.8). The data were partitioned further into a training set T and a validation set V for use in discriminant analysis, which is presented later in the chapter. The Kamikaizuka and Shouninzuka sites are within 10 km of each other. The Narita 60 site is more than 40 km from the other sites. Because of differences in data sets and approaches, the results, which are for illustrative purposes, may differ from those of Hall.

7.5.2 Basic Approaches to Finding Clusters

The two basic approaches to clustering observations are hierarchical and partitioning. Hierarchical algorithms are subdivided into agglomerative and divisive methods.

Agglomerative Hierarchical Procedures   An *agglomerate method* is one that forms one cluster from two or more existing clusters. Typically, an agglomerative method begins by letting each observation be its own cluster. Thus, if there are n observations, the method begins with n clusters. Next, the method forms $n - 1$ clusters by grouping the two observations (clusters of size 1) that are most similar according to a prespecified criterion. The method proceeds in this manner until all observations are

TABLE 7.8 Subset of Early Jomon Sherd Pottery Data

Obs.	Training or Validation	Site	Ti (ppm)	Mn (ppm)	Cu (ppm)	Zn (ppm)	Pb (ppm)
1	T	K	9821	805	90	98	38
2	T	K	10304	714	64	76	90
3	T	K	8806	351	19	46	67
4	T	K	9940	453	95	49	30
5	T	K	8751	772	64	62	58
6	T	K	8518	566	53	78	21
7	T	K	9293	433	42	45	31
8	T	K	10990	686	55	81	84
9	T	K	7514	579	43	86	29
10	T	K	7241	462	50	48	48
11	T	K	10112	540	49	68	24
12	T	K	9317	681	42	66	66
13	T	K	10196	570	39	44	58
14	T	K	8399	584	77	55	62
15	T	K	9412	782	83	58	20
16	T	K	6598	368	21	54	56
17	T	K	8595	499	63	74	59
18	T	K	5890	657	49	65	54
19	T	K	5834	587	55	31	48
20	T	K	8832	822	79	81	59
21	T	N	6018	678	64	32	42
22	T	N	5451	533	42	40	60
23	T	N	12033	937	106	52	26
24	T	N	14701	918	81	50	55
25	T	N	14834	1118	86	92	59
26	T	N	9196	633	54	25	11
27	T	N	10865	795	86	63	27
28	T	N	10828	742	103	46	35
29	T	N	10347	1356	106	53	23
30	T	N	9500	965	88	39	56
31	T	N	9754	982	80	33	18
32	T	S	10031	795	92	93	16
33	T	S	8534	1457	91	79	27
34	T	S	9397	625	79	61	41
35	T	S	10994	666	109	91	17
36	T	S	2678	307	19	25	12
37	T	S	8978	856	173	80	58
38	T	S	9958	574	70	70	50
39	T	S	8518	944	84	83	21
40	T	S	10328	1057	128	126	25
41	T	S	9515	1161	125	81	24
42	T	S	11035	1065	131	114	68
43	T	S	8031	979	79	79	15
44	T	S	9736	670	138	64	20
45	T	S	11756	748	77	93	21
46	T	S	11984	572	79	108	26
47	T	S	9348	892	103	87	31
48	T	S	10224	1139	83	78	32

(continued)

TABLE 7.8 (Continued)

Obs.	Training or Validation	Site	Ti (ppm)	Mn (ppm)	Cu (ppm)	Zn (ppm)	Pb (ppm)	Obs.	Training or Validation	Site	Ti (ppm)	Mn (ppm)	Cu (ppm)	Zn (ppm)	Pb (ppm)
49	T	S	10509	1036	207	65	71	64	V	K	8471	762	60	57	38
50	T	S	11843	824	89	101	47	65	V	N	8402	867	102	60	26
51	T	S	9861	885	97	91	34	66	V	N	14815	1140	70	66	18
52	T	S	12054	899	228	81	37	67	V	N	10859	693	69	35	25
53	T	S	10147	851	108	73	32	68	V	N	8817	541	81	49	50
54	T	S	10839	1196	105	140	68	69	V	S	9164	782	146	60	61
55	V	K	7528	532	22	31	46	70	V	S	11302	961	111	83	13
56	V	K	11026	617	69	77	14	71	V	S	10382	1370	85	80	20
57	V	K	10793	1002	79	94	24	72	V	S	10922	921	78	94	33
58	V	K	7597	756	62	65	27	73	V	S	8458	872	104	94	24
59	V	K	8936	788	86	68	88	74	V	S	8272	1124	229	105	28
60	V	K	8507	710	45	68	81	75	V	S	12236	866	115	110	63
61	V	K	7890	720	77	100	18	76	V	S	9773	1416	95	97	78
62	V	K	6914	490	57	51	56	77	V	S	9207	647	96	60	15
63	V	K	8380	3037	72	80	70	78	V	S	8881	1076	138	58	21

Source: Data from Hall (2001).

266

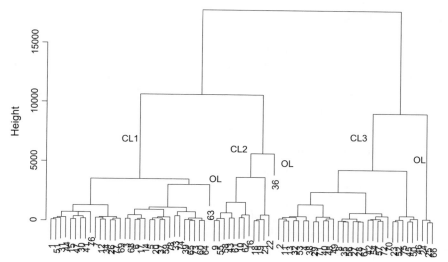

FIGURE 7.8   Dendogram of pottery data using Euclidean distance metric with Ward's method. Horizontal lines link clusters.

combined into a single cluster or a stopping rule is triggered. A stopping rule occurs, for example, when the ratio of the pooled within-cluster variability to the between-cluster variability becomes less than a specified small value. The basic agglomerative procedure is illustrated by a *clustering tree* (also referred to as a *dendogram*) (Figure 7.8). The sample numbers are shown at the bottom of this tree. They are arranged, in so far as possible, according to similarity. Thus, the three rightmost points—24, 25, and 66—are similar using Ward's method of defining the similarity between points and clusters (measures of similarity will be defined shortly). However, the next point in from the right (46) is not as similar. Points 24, 25, and 66 are joined by a vertical line just above a height (vertical axis) of zero. However, the new cluster formed by points 46 and the next five points (from left to right) are not joined to the cluster consisting of points 24, 25, and 66 until a height of approximately 8000 is reached. Indeed, because there are only three points in the rightmost cluster (24, 25, and 66), the algorithm designated them as outliers (OL) instead of as a cluster. The definition of what constitutes an outlier is somewhat arbitrary, and the final decision is left to the investigator. The tree generated by this algorithm suggests three clusters (C1, C2, and C3) and three outliers. Dissimilarity increases with height. At the top height, all observations are in one cluster. Later, guidelines are given to determine the number of clusters and to explain this example further.

Divisive Hierarchical Methods   As the name suggests, divisive methods begin with all observations in one cluster and are then split, most dissimilar first, until each observation is its own cluster, or a stopping rule, is encountered. It is the inverse of the agglomerative procedure.

Partitioning Algorithms   A *partitioning algorithm* begins by assigning observations to k clusters, where k is specified in advance. As an alternative, the procedure may begin with observations assigned to k clusters by another method, such as a hierarchical algorithm. A partitioning algorithm moves observations among the k clusters until some optimal measure of clustering is achieved. For example, the algorithm may terminate when no further reduction of within-cluster to between-cluster variance is possible. The best known of the partitioning methods is k-mean, discussed in Section 7.5.8.

7.5.3   Measures of Dissimilarity Between Observations

Two observations are similar if they are close, but close in what sense? When one thinks about two observations being close spatially, they are typically thought of as being nearby, as in Euclidean distance. However, this is not the only measure of similarity. They may be close in absolute value or in presence–absence. In clustering, it is common to establish a measure of dissimilarity between observations instead of a measure of similarity. The general form of an $n \times n$ dissimilarity matrix is

$$\begin{bmatrix} 0 & & & & \\ d(2,1) & 0 & & & \\ d(3,1) & d(3,2) & 0 & & \\ \vdots & \vdots & \vdots & & \\ d(n,1) & d(n,2) & \cdots & \cdots & 0 \end{bmatrix}$$

where $d(i,j)$ is a measure of the dissimilarity between the ith and jth observations. The measure is usually symmetric: namely, that $d(i,j) = d(j,i)$. Further, $d(i,i) = 0$. Thus, the least dissimilar measure is 0. It increases as dissimilarity increases.

Measures of Dissimilarity for Interval Data   Two common dissimilarity measures of distance for interval data are

$$\text{Euclidean distance } d(i,j) = \sqrt{\sum_{m=1}^{p} (x_{im} - x_{jm})^2}$$

and

$$\text{Manhattan distance } d(i,j) = \sum_{m=1}^{p} |x_{im} - x_{jm}|$$

where p is the number of continuous variables and i and j are the observations.

Measures of Similarity for Binary Data   It is not uncommon for binary data to be included in clustering and classification problems. For example, the presence or

TABLE 7.9 A 2 × 2 Table Showing Presence–Absence[a]

	Observation j		
Observation i	0	1	Total
0	a	b	$a + b$
1	c	d	$c + d$
Total	$a + c$	$b + d$	$a + b + c + d$

[a]0 is absence; 1 is presence.

absence of flora or fauna can help define a biological region. Wildlife biologists may be interested in the presence or absence of fish species in order to cluster streams by assemblages.

Suppose that $\mathbf{x}'_i = (x_{i1}, x_{i2}, \ldots, x_{iq})$ is the ith observation of q binary variables with the presence of x_{ih}, $h = 1, \ldots, q$ indicated by a 1 and its absence indicated by a 0. Table 7.9 shows the presence–absence relationship between observation i and observation j, where the letters, a, b, c, and d represent counts. For example, among the q variables, a are absent in observations i and j. Two common measures of similarity are the *simple matching coefficient* (SMC) $(a + d)/q$, where $q = a + b + c + d$, and *Jaccard's coefficient* or *index* (JI), $d/(b + c + d)$, which ignores the 0 matches. Both measures vary between 0 and 1, with 0 being totally dissimilar (i.e., no species in common) and 1 being an exact match in species between observations i and j.

7.5.4 Standardizing Variables

When the continuous variables are in the same units and it is desired to weight the variables according to the magnitude of the observations measured, no transformation is required. In the pottery classification study, the Ti will have more weight than Pb because the mean and standard deviations of the former, $\widehat{\mu}(Ti) = 9414$ and $\widehat{\sigma}(Ti) = 2201$, are considerably greater than those of the latter, $\widehat{\mu}(Pb) = 39$ and $\widehat{\sigma}(Pb) = 19$. To weight the variables equally or if the variables are in different units, the variables should be standardized. Typically, standardizing will take the form of centering and scaling as discussed for PCA.

7.5.5 Combining Clusters

Deciding on a measure of dissimilarity between observations is relatively straightforward. Deciding on a measure of similarity (or dissimilarity) between clusters, which is needed to agglomerate existing clusters, is not simple because there are many choices. Consider the scatter plot of observations shown in Figure 7.9 plotted in variable space (variable 1 versus variable 2). Three clusters, A, B, and C, have been formed. Can the clusters be agglomerated further? To consider this question, it is necessary to know how far apart they are. Choices include the distance between the two closest points (the solid line), the two most distant points (the dashed line), the cluster centers (the dotted line), or by a

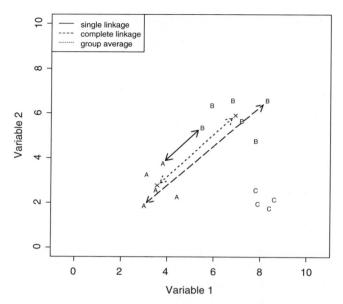

FIGURE 7.9   Measures of cluster similarity.

measure of within-cluster variability. The *cluster center* is not uniquely defined. One example is a vector of p variables whose elements are the means of observations lying within the cluster; this is the measure used to define the dotted line in Figure 7.9. The ×'s in Figure 7.9 define the means of clusters A and B. Without some measure, it is not possible to determine the most similar clusters. For example, if the closest points (minimum distance) are used, clusters A and B will be combined first. However, if the two most distant points are used, clusters A and C will be combined first.

Some common measures of cluster similarity are:

- *Single linkage.* The measure is the minimum distance between clusters, as illustrated by the solid line in Figure 7.9. This method forms long chain clusters.
- *Complete linkage.* The measure is the maximum distance between clusters, as illustrated by the dashed line in Figure 7.9. This method forms tight clusters.
- *Group average linkage.* There are many ways of computing an average. Many algorithms use the arithmetic average of $d(i,j)$, $i \in A$, $j \in B$. Other "averaging" methods include the centroid and the median. See the dotted line in Figure 7.9.
- *Ward's method.* Define a within-cluster error sum of squares for the kth cluster as $E_k = \sum_{m=1}^{p} \sum_{i=1}^{n_k} (x_{im} - \bar{x}_{m(k)})^2$, $i \in k$, where n_k observations are assigned to the kth cluster and $\bar{x}_{m(k)}$ is the mean of the mth variable in the kth cluster; there are p variables. The total within-cluster sum of squares is $E = \sum_{k=1}^{K} E_k$. Clusters

are agglomerated using a minimum increase in the E criterion. Ward's method produces compact clusters but can be influenced by outliers because it uses sums of squares.

There is no best similarity (or dissimilarity) measure between clusters. Each produces different types of clusters. For most applications, tight compact clusters are desired, but exceptions exist, such as when one is interested in defining clusters of molecular chains. In this case, a single linkage method may be preferable. The user should experiment with various measures.

7.5.6   Selecting the Number of Clusters

In the agglomerative hierarchical process, merging the two clusters with the smallest between-cluster dissimilarity forms a new cluster. After the merger, a new dissimilarity matrix is computed and the process is repeated. Most often, the number of clusters is selected by viewing a graphical display or index number. The dendogram (Figure 7.8) is constructed using a Euclidean distance measure and Ward's method of dissimilarity between clusters on the original data. The height axis (Figure 7.8) shows increasing dissimilarity among clusters. Horizontal lines designate a combining of clusters. How many clusters are present in Figure 7.8? As noted previously, it is judged that there are three clusters and a group of three outliers. How does one know that there are three clusters? This is, in part, subjective judgment, and one can make an argument for at least four clusters. Our rationale is that the observations are combined to form clusters at a relatively low height. Look for gaps in height to help determine the number of clusters. In the next section, other methods that yield different results are introduced.

7.5.7   Comparison of Agglomerative Methods

The pottery data is used to illustrate several agglomerative methods.

- Euclidian distance with single linkage and original variables (Figure 7.10a)
- Euclidean distance with single linkage and standardized variables (Figure 7.10b)
- Euclidean distance with group average and original variables (Figure 7.10c)
- Euclidean distance with group average and standardized variables (Figure 7.10d)

There are clear differences between single linkage and group averages. There are more similarities between the original and the standardized variables. The group average method forms clusters that are similar to Ward's method. The single linkage method does not delineate clusters. All of the methods show observation 36 to be an outlier, and all but the single linkage standardized variable method (Figure 7.10b) show observation 63 to be a possible outlier. In addition, observations 24, 25, and 66 appear to be an outlier set for all of the methods except the group average standardized pottery data (Figure 7.10d). The Manhattan distance (not shown)

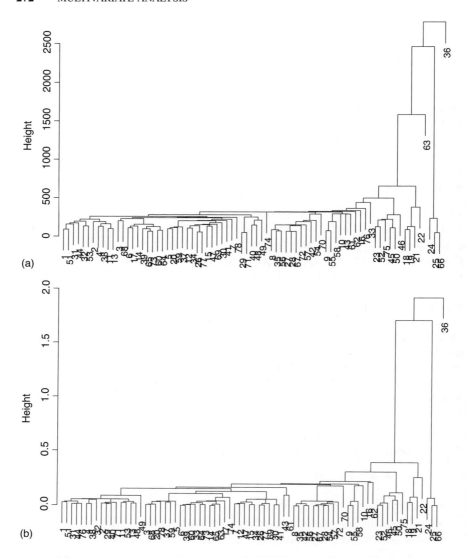

(a)

(b)

FIGURE 7.10 (a) Dendogram using Euclidean distance, single linkage, and original variables for pottery data. (b) Dendogram using Euclidean distance, single linkage, and standardized variables for pottery data.

yields results that are similar to the Euclidean distance measure. Which method yields the correct clusters? There is no right method. The best method is one that provides insight into the investigator's problem, since cluster analysis is an exploratory tool.

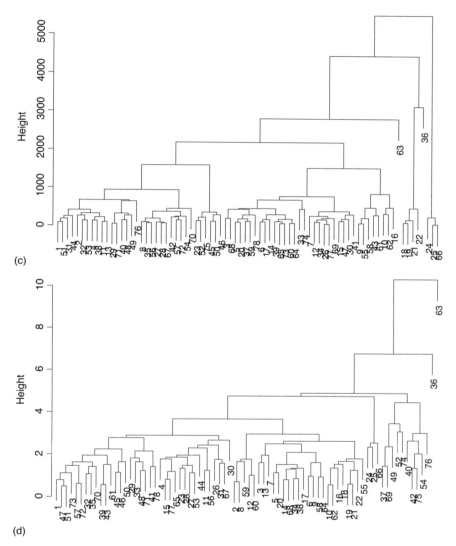

FIGURE 7.10 *(Continued)* (c) Dendogram using Euclidean distance, group average, and original variables for pottery data. (d) Dendogram using Euclidean distance, group average, and standardized variables for pottery data.

7.5.8 Partitioning Algorithms

Partitioning algorithms differ from hierarchical algorithms in two important ways. First, in partitioning algorithms, the number of clusters is specified in advance. Second, partitioning algorithms have the ability to move observations between clusters until some cluster criterion is achieved. Conversely, once an observation is assigned to a hierarchical cluster, it stays in that cluster or some superset or subset of it. A commonly used partitioning algorithm is the k-mean.

k-Mean Algorithm   A k-mean algorithm is a partitioning algorithm in which the number of clusters is specified in advance. Since this number is rarely known, the user typically tries a number of clusters. In the pottery classification study, three and four clusters are chosen. The initial assignment of observations to clusters can be made randomly, from a hierarchical clustering output or from other prior information. Given an initial assignment of observations into K clusters, a centroid is computed for each cluster. The centroid coordinates for the kth cluster are $(\overline{x}_{1(k)}, \overline{x}_{2(k)}, \ldots, \overline{x}_{p(k)})$, where $\overline{x}_{m(k)} = \sum_{i=1}^{n_k} x_{im(k)}/n_k$, $m = 1, \ldots, p$, the number of variables, and n_k is the number of observations in the kth cluster. After the initial assignment is made, the algorithm checks the least-squares Euclidian distance from the ith observation in cluster k to the centroid of the other clusters. If the minimum of the distances is less to another cluster, the observation is moved and the cluster centroids are recomputed. The process iterates through all points in all clusters until no further exchange of points is necessary. The output consists of observation numbers within a cluster. In the pottery classification study (Table 7.10), cluster Ck1 contains a high percentage of observations from the Kamikaizuka and Shouninzuka sites, and Ck2 has a majority of observations in the Shouninzuka site. There is no strong correspondence between the k-mean clusters and sites. The clusters (Table 7.11) are somewhat similar to those produced by Ward's method, with Ck3 and C2 being in close agreement. The

TABLE 7.10   Number of Observations by Cluster and Site for k-Mean Clusters Using the Pottery Data

k-Mean Cluster	Site			Total
	K	N	S	
Ck1	19	6	22	47
Ck2	3	7	10	20
Ck3	8	2	1	11
Total	30	15	33	78

TABLE 7.11   Comparison Between k-Mean Clusters and Ward's Clusters for the Pottery Data

k-Mean Cluster	Ward Cluster			Total
	C1	C2	C3	
Ck1	12	2	30	44
Ck2	18	0	0	18
Ck3	0	11	0	11
Total	30	13	30	73

discrepancy in counts between Table 7.10 and Table 7.11 is because the Ward's method classifies five observations as outliers.

Other Partitioning Algorithms   *Partitioning around medoids* (PAM) is similar to the k-mean algorithm. Instead of the centroid as the cluster center, PAM computes medoids (Kaufman and Rousseeuw, 2005). A *medoid* is the most centrally located object in a cluster. PAM is more robust than the k-mean because it minimizes a sum of dissimilarity distances rather than the sum of squares of Euclidean distances.

Clustering LARge application (CLARA) is for large data sets, such as in data mining applications (Kaufman and Rousseeuw, 2005). Partitioning algorithms described previously hold the entire dissimilarity matrix in memory, which is computationally intensive for a large n. CLARA works with subsets of the data.

Medoid clustering has been shown to be useful for clustering moving objects such as may occur in biological applications (Kriegel and Pfeifle, 2005). The moving object is represented by a density function; samples are taken and clusters formed. The procedure is repeated. Finally, a distance ranking between sample clustering is computed; the smallest ranking value is medoid clustering.

Fuzzy set clustering (also called *fuzzy set partitioning*) is designed to show fractional membership in each cluster, as opposed to methods discussed previously that require an observation to be in a specific cluster. Membership probabilities are determined by minimizing an objective function that consists of unknown memberships between observations and clusters and known distance measures. Three clusters, Cf1 to Cf3, have been chosen using the original pottery data and the Euclidean distance measure. The probability of assignment to a cluster for the first 10 observations is shown in Table 7.12. An advantage of fuzzy set clustering can be seen by viewing observations 6 and 9. Both are to be assigned to Cf3 on the basis of largest probability, however, the strength of the assignments are quite different.

TABLE 7.12   Fuzzy Set Clustering Showing the Probability of Assignment of Observations to Clusters Using the Pottery Data

	Cluster		
Observation	Cf1	Cf2	Cf3
1	0.64	0.23	0.13
2	0.38	0.49	0.13
3	0.49	0.14	0.37
4	0.55	0.30	0.16
5	0.50	0.11	0.39
6	0.37	0.11	0.53
7	0.66	0.14	0.20
8	0.13	0.80	0.07
9	0.18	0.09	0.73
10	0.20	0.10	0.70

7.6   MULTIDIMENSIONAL SCALING

Multidimensional scaling (MDS) is a dimension reducing visualization method. Suppose that the problem of interest is to understand earthquake patterns over time and space. Attributes may include magnitude, location, time, and other properties in p dimensions. One component of an investigation is to look for clusters. However, viewing clusters in $p > 3$ dimensions is difficult. If the problem can be reduced to two or three dimensions, visualization can occur. This is the function of MDS, to reduce the dimensionality of the problem while preserving or approximating the original distances or similarities. Multidimensional scaling and cluster analysis used in an investigation of earthquake patterns has been described by Dzwinel et al. (2005).

MDS is also used as an alternative to FA because it does not require a correlation matrix or the assumption of multivariate normality. A general discussion of MDS may be found in Cox and Cox (2001).

7.7   DISCRIMINANT ANALYSIS

Discriminant analysis (DA) is a multivariate procedure used to:

- Determine separation between known groups and/or
- Classify an observation of unknown origin into one of two or more predefined groups.

In DA, the number of groups and their identifying characteristics are in theory known in advance, as opposed to being determined from the data, as in cluster analysis. In the pottery example, the groups are assumed to be sites K, N, and S. The variables used to investigate separability are the chemical elements Ti, Mn, Cu, Zn, and Pb. Since the population characteristics of those elements are never known in practice, they must be estimated from data. A key assumption is that the investigator knows with certainty which observations are associated with each of the three sites. The usual assumption, at least for hypothesis testing, is that the observations are multivariate normal. Each group is characterized by a mean and variance–covariance. For example, if there are $k = 1, \ldots, K$ groups, the kth group will have a mean $\boldsymbol{\mu}_k = (\mu_{1(k)}, \ldots, \mu_{p(k)})'$, where p is the number of variables and $\sum_k$ is a $p \times p$ variance–covariance matrix in which the diagonal elements are variances and the off-diagonal elements are covariances. It is often important to determine the amount of separability between groups by some distance measure, to see which variables contribute most to determining the separation. An alternative use of DA occurs when the investigator has an observation of unknown origin but suspects that it is a member of one of his predefined groups. For example, a pottery shred may be found outside sites K, N, or S, and the challenge is to see if it belongs to one of these sites.

7.7.1   Discriminant Analysis for Prediction of Separability

DA is analogous to multiple regression, except that the response variable in DA is a nominal variable indicating membership in a group, and the predictor variables usually are assumed to be multivariate normal. The discriminant function computes "regressionlike" coefficients that find as much separability between the groups as possible according to some defined criterion.

For two groups (π_1 and π_2) characterized by two variables (X_1 and X_2), DA is as illustrated in Figure 7.11. These two groups are impossible to separate using only X_1 or X_2, because a projection onto neither the X_1- nor the X_2-axis results in significant separability. However, when groups π_1 and π_2 are projected onto the discriminant function (the 45° line in the top left-hand corner of Figure 7.11), considerable separability is achieved. The ellipses (Figure 7.11) can be viewed as isolines from two bivariate normal distributions. It is often of interest to know how much separability can be achieved by a discriminant function and which variables contribute most to this separability.

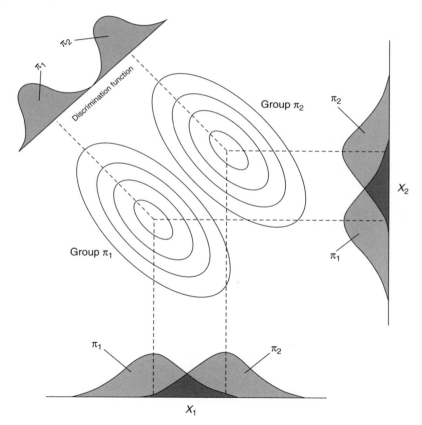

FIGURE 7.11   Two-group, two-variable discriminant analysis graphical representation.

MULTIVARIATE ANALYSIS

7.7.2 Variance–Covariance Assumptions

The form of the discriminant function depends in part on assumptions that can be made about the $\sum_k$'s, $k = 1, \ldots, K$. They include the following models:

- *Homoscedastic model.* It is assumed that $\sum_k = \sum$ for all $k = 1, \ldots, K$. This assumption implies that Fisher's *linear discriminant function* (LDF) model should be used. We refer to the procedure as *linear discriminant analysis* (LDA), although the terms are used interchangeably in the literature.
- *Heteroscedastic model.* This is the most general form of a DA model. For $K = 2$, $\sum_1 \neq \sum_2$. For K groups this model has $Kp(p + 1)/2$ parameters and thus has the greatest number of parameters. This assumption implies that a quadratic discriminant function (QDF) model should be used.
- *Spherical model.* The assumption is that the covariances are zero. Thus, $\sum_k = \text{diag}(\sigma_{k1}^2, \ldots, \sigma_{kp}^2)$. This equation can be simplified further if it is assumed that $\Sigma_k = \Sigma$ for all k.

Other modeling assumptions include assuming that covariances are proportional and that correlations are equal but variances are unequal. These assumptions can be tested.

7.7.3 Test of Equality of Variance–Covariance Matrices

To test the equality of variance–covariance matrices, consider the following null and alternative hypotheses:

$$H_0: \sum_1 = \cdots = \sum_K$$
$$H_a: \sum_i \neq \sum_j \qquad \text{for some } i \neq j$$

Assume that the observations are drawn randomly from multivariate normal populations. If C_i is the estimator of $\sum_i$, the likelihood ratio test statistic LR is

$$\text{LR} = h(n - K)\ln|C| - \sum_{i=1}^{K}(n_i - 1)\ln|C_i|$$

where

$$C = \frac{\sum_{i=1}^{K}(n_i - 1)C_i}{n - K}$$

is the pooled variance estimate and

$$h = 1 - \frac{(2p^2 + 3p - 1)(K + 1)}{6(p + 1)(K - 1)}$$

When H_0 is true, this statistic has a chi-square distribution with $p(p + 1)(K - 1)/2$ df. A high p-value implies that the null hypothesis cannot be rejected, and thus it may be reasonable to use a pooled estimate of the covariance matrices from the K groups. Therefore, the LDA is appropriate. Should this not be the case, quadratic discriminant analysis is used.

7.7.4 Discriminant Functions

Suppose that $\sum_1 = \cdots = \sum_K$ and the data are multivariate normal. When there are two groups ($K = 2$), only one LDF is computed. When $K > 2$, at most $(K - 1)$ LDFs will be computed. Consider the ratio

$$S(\mathbf{w}) = \frac{\mathbf{w}'\mathbf{S}_B\mathbf{w}}{\mathbf{w}'\mathbf{S}_W\mathbf{w}}$$

where $\mathbf{S}_B$ is the estimated between (among) groups variance–covariance, $\mathbf{S}_W$ is the estimated pooled within-group variance–covariance, and $\mathbf{w} = (w_1, \ldots, w_p)'$ is a vector of weights. $S(\mathbf{w})$ is essentially a signal-to-noise ratio. For $K = 2$ this ratio becomes

$$S(\mathbf{w}) = \frac{\mathbf{w}'(\widehat{\boldsymbol{\mu}}_1 - \widehat{\boldsymbol{\mu}}_2)(\widehat{\boldsymbol{\mu}}_1 - \widehat{\boldsymbol{\mu}}_2)'\mathbf{w}}{\mathbf{w}'\mathbf{S}_W\mathbf{w}}$$

where

$$\mathbf{S}_W = \mathbf{S}_{\text{pooled}} = \frac{(n_1 - 1)\widehat{\boldsymbol{\Sigma}}_1 + (n_2 - 1)\widehat{\boldsymbol{\Sigma}}_2}{n_1 + n_2 - 1}$$

The $\mathbf{w}$ is determined by optimization to yield the maximum separability between groups, which for $K = 2$ occurs when $\mathbf{w} = (\widehat{\boldsymbol{\mu}}_1 - \widehat{\boldsymbol{\mu}}_2)'\mathbf{S}_W^{-1}$. The resulting estimated LDF is

$$y = \mathbf{w}'\mathbf{x}$$

For $K = 2$,

$$\max_{\mathbf{w}} S(\mathbf{w}) = (\widehat{\boldsymbol{\mu}}_1 - \widehat{\boldsymbol{\mu}}_2)'\mathbf{S}_W^{-1}(\widehat{\boldsymbol{\mu}}_1 - \widehat{\boldsymbol{\mu}}_2)$$

The square root of this quantity is known as the *Mahalanobis distance*. It is the maximum distance between $\widehat{\boldsymbol{\mu}}_1$ and $\widehat{\boldsymbol{\mu}}_2$ adjusted for $\mathbf{S}_W$. When $\mathbf{S}_W = \mathbf{I}$, the identity, the Mahalanobis distance is equivalent to the Euclidian distance. The LDA model uses only $p(p + 1)/2$ parameters (independent of the size of K) because the $\mathbf{S}_W$ is a pooled estimate of the Σ_K.

Fisher's LDF can also be used to classify a new observation $\mathbf{x}_0$. Let $\bar{\mathbf{x}} = (\hat{\boldsymbol{\mu}}_1 + \hat{\boldsymbol{\mu}}_2)/2$. Let the mean distance between $\hat{\boldsymbol{\mu}}_1$ and $\hat{\boldsymbol{\mu}}_2$ along the LDF be

$$m = (\hat{\boldsymbol{\mu}}_1 - \hat{\boldsymbol{\mu}}_2)' \mathbf{S}_W^{-1} \bar{\mathbf{x}}$$

Let $y_0 = (\hat{\boldsymbol{\mu}}_1 - \hat{\boldsymbol{\mu}}_2)' \mathbf{S}_W^{-1} \mathbf{x}_0$. Now allocate to $\mathbf{x}_0$ to π_1 if $y_0 \geq m$; if not, allocate $\mathbf{x}_0$ to π_2.

The unequal covariance assumption results in an estimated *quadratic discriminant function* (QDF) of the form

$$g_i(\mathbf{x}) = -\frac{1}{2}(\mathbf{x} - \hat{\boldsymbol{\mu}}_i)' \hat{\boldsymbol{\Sigma}}_i (\mathbf{x} - \hat{\boldsymbol{\mu}}_i) - \frac{1}{2}\ln\left(\left|\hat{\boldsymbol{\Sigma}}_i\right|\right) + \ln(p_i)$$

where p_i is the prior probability that $\mathbf{x}$ belongs to group π_i. If in the two-group case $g_1(\mathbf{x}) > g_2(\mathbf{x})$, $\mathbf{x}$ is assigned to group π_2. As in the LDF case, the densities are assumed to be normally distributed. The associated procedure is referred to as *quadratic discriminant analysis* (QDA).

7.7.5   Expected Minimum Cost

A more general allocation rule than the one based on Fisher's LDF is called the *expected minimum cost* (ECM) rule. For $K = 2$, when $\sum_1 = \sum_2$ this rule calls for an allocation of $\mathbf{x}_0$ to π_1 if

$$(\hat{\boldsymbol{\mu}}_1 - \hat{\boldsymbol{\mu}}_2)' \mathbf{S}_W^{-1} \mathbf{x}_0 - \frac{1}{2}(\hat{\boldsymbol{\mu}}_1 - \hat{\boldsymbol{\mu}}_2)' \mathbf{S}_W^{-1}(\hat{\boldsymbol{\mu}}_1 + \hat{\boldsymbol{\mu}}_2) \geq \ln\left(\frac{c(1|2)\, p_2}{c(2|1)\, p_1}\right)$$

Otherwise, allocate $\mathbf{x}_0$ to π_2. This rule has the advantage of incorporating costs of misclassification, the $c(1|2)$ and $c(2|1)$, and prior probabilities, p_1 and p_2, of an observation being from π_1 or π_1, respectively. The cost of misclassifying an observation into π_1 given it is from π_2 is $c(1|2)$, and the cost of misclassifying an observation into π_2 given it is from π_1 is $c(2|1)$. In some cases these costs may not be equal. Since each new observation will be classified as belonging to π_1 or π_2, it may be desirable to specify a prior probability that an observation belongs to π_k, $k = 1, 2$. For example, if the investigator believes that the data represent a random sample from the region containing π_1 and π_2, he or she may choose to specify a proportional prior, namely $p_1 = n_1/(n_1 + n_2)$, and in this case, $p_2 = 1 - p_1$. An alternative is to specify a uniform prior, namely $p_k = 1/2$ or, in general, $p_k = 1/K$. If $c(1|2) = c(2|1)$ and $p_1 = p_2$, the ECM rule is the same as Fisher's LDF allocation.

7.7.6   Discriminant Analysis Example Using Pottery Data

To illustrate LDA, the pottery data (Table 7.8) are used. Observations have been split randomly into two groups: a training set (T) and a validation set (V). Only the K and N sites and variables Cu and Pb are used in this example. The training set is used to

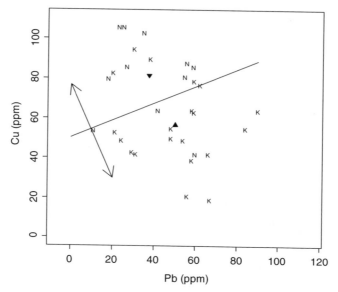

FIGURE 7.12 Scatter plot of the K and N sites plotted in the Cu and Pb variable space.

initialize (train) model parameters because in real problems these are not known in advance. A scatter plot of the observations (the N's and K's), the LDF (the line with the arrows), and a line orthogonal to the LDF are shown in Figure 7.12. The latter line is halfway between the sample means. The upside-down triangle represents the sample mean of the N sites, and the triangle represents the sample mean of the K sites. This line achieves the greatest degree of separability between K and N sites. Below and to the right is the K domain; above the line and to its left is the N domain.

Group means are shown in Table 7.13. Each prior is 0.5, implying that a uniform prior has been chosen. The covariance structure is a pooled estimate, which requires a homoscedasticity of variance–covariance matrix assumption: namely, that $\Sigma_1 = \Sigma_2$. Given that the p-value for the likelihood ratio test for homogeneity of covariances (Table 7.13) is 0.99, the hypothesis of equality cannot be rejected, and thus one can conclude that it is reasonable to use LDA. If this hypothesis is rejected, QDA is used. The constants and linear coefficients specify the linear discriminant function. For LDA (Table 7.13), let $f(N) = f(K)$ and solve the resulting equation to obtain $0.042047Cu - 0.01759Pb = 2.116411$ (the discriminant function line shown in Figure 7.12).

Given covariance homoscedasticity, a test for equality of means is conducted. Numerous tests are available with slightly different properties; however, they usually yield similar results. The one presented here is *Wilks' lambda*, which is a multivariate test of the equality of means. The null hypothesis is that the group means are equal versus the alternative that at least two groups means differ. In this example (Table 7.13) there are only two groups, K and N. The p-value is 0.0094, implying a difference between the group K and group N population means. The Mahalanobis distance is

TABLE 7.13   LDA Output, Cu and Pb Pottery Data

Group Means

	Cu	Pb	n	Priors
K	56.60	50.10	20	0.5
N	81.45	37.45	11	0.5

Covariance Structure Assuming Homogeneity of Covariances

	Cu	Pb
Cu	431.23	-78.33
Pb	—	370.23

Likelihood Ratio Test for Homogeneity of Covariances

Chi-square	df	p-value
0.15	3	0.99

Linear Discriminant Function

$-0.01759Pb + 0.042047Cu = 2.116411$

Equation Orthogonal to LDF

$Cu = -2.34189(Pb - 20) + 30$

Tests for the Equality of Means: Group Variable: SC[a]

	Statistics	F	df 1	df 2	p-value
Wilks' lambda	0.72	5.54	2	28	0.0094

[a]Tests assume covariance homoscedasticity.

Mahalanobis Distance

	K	N
K	0	1.62
N		0

Kolmogorov–Smirnov Test for Normality

	Statistic	p-value
Cu	0.06	1.00
Pb	0.11	0.77

1.62. As noted in the discussion of DA, it is a measure of mean distance between group means adjusted for variance–covariance. Finally, results for the Kolmogorov–Smirnov goodness-of-fit test of normality for Cu and Pb are presented (Table 7.13); both are judged to be normally distributed.

7.7.7 Errors in Classification

When the parent population parameters are known, estimation of misclassification errors is relatively straightforward; however, this rarely occurs in real problems. We investigate next how well the DA model performs when classifying observations used to estimate model parameters (i.e., the training set). For two groups, the *total probability of misclassification* (TPM) is defined as

$$\text{TPM} = \text{Pr(observation is from } \pi_1 \text{ and misclassified as belonging to } \pi_2)$$
$$+ \text{Pr(observation is from } \pi_2 \text{ and misclassified as belonging to } \pi_1)$$

In DA, one goal is to obtain the smallest value of TPM through the choice of classification regions R_1 and R_2. This value is called the *optimum error rate* (OER). Both the TPM and OER depend on knowledge of the density functions of π_1 and π_2, which in real problems are unknown. When density functions have to be estimated, the situation is more complicated and various options are available. A desirable result is to obtain the *actual error rate* (AER) for a given DA model; however, this computation also requires knowledge of unknown density functions. We look next at some alternatives.

Table 7.14 shows two classification tables from the pottery study. The top one is based on classifying observations in the training set that are used to estimate model parameters. The number of observations misclassified is shown in the off-diagonal elements. Row 1 shows that 16 of the 20 observations belonging to site K are classified correctly. Row 2 shows that 8 of the 11 observations from site N are classified correctly. The column on the right (Table 7.14, first two rows) displays the *apparent error rate* (APER$_i$) for groups 1 and 2 ($i = 1, 2$), defined as

$$\text{APER}_i = \frac{n_{iM}}{n_i}$$

TABLE 7.14   Misclassification Tables, Cu and Pb Pottery Data

Classification Table Using Training Set Observations

Actual Membership	Predicted Membership K	N	APER
K	16	4	0.200
N	3	8	0.273
Overall			0.226

Cross-Validation Table

Actual Membership	Predicted Membership K	N	EstE[AER]
K	15	5	0.250
N	3	8	0.273
Overall			0.258

where n_{iM} is the number of observations belonging to group i that are classified in another group and n_i is the total number of observations (actual membership) in group i. It is biased in favor of correct classification because the data are used to estimate model parameters. The overall APER is

$$\text{APER} = \frac{\sum_{i=1}^{2} n_{iM}}{\sum_{i=1}^{2} n_i}$$

$$= \frac{4+3}{31}$$

$$= 0.226$$

for two groups.

A less biased procedure is a cross-validation or "holdout" procedure (see the cross-validation table in Table 7.14), which omits the observation to be classified from estimation of the classification function. The result is an estimate of AER (EstE [AER]), which has a higher error rate and is less biased than APER.

Another approach to determining an error rate is to split the data into training and validation sets. The model parameters are estimated with the training set and the resulting model is used to classify observations in the validation data set. This requires a reasonably large data set for results to be meaningful. A portion of the validation set (Table 7.8) is shown in Table 7.15 and the model parameters established from the pottery data training set. The K and N columns list the probabilities of classification

TABLE 7.15   Results of Applying Observations from the Validation Set into the Model Estimated by the Training Set for the Pottery Data

Observation	Actual Membership	Classification K	N
1	K	0.93	0.07
2	K	0.53	0.47
3	K	0.89	0.11
4	K	0.72	0.28
5	K	0.61	0.39
6	K	0.59	0.41
7	N	0.34	0.66
8	N	0.27	0.73
9	N	0.50	0.50
10	N	0.27	0.73
11	N	0.10	0.90
12	N	0.35	0.65
13	N	0.39	0.61
14	N	0.38	0.62

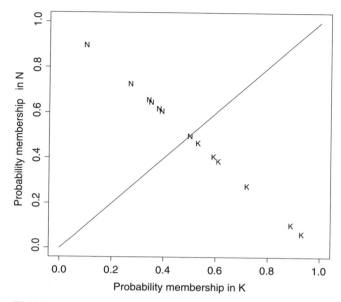

FIGURE 7.13   Graph of the classification results in Table 7.15.

into groups (sites) K and N. In this example, all observations are classified correctly, although the strength of the classification varied. (Since observation 9 has a 50–50 chance of being classified as a K or an N site, it can be argued that one observation is misclassified.) In general, there will be misclassifications. Since there are only two sites, the results (Table 7.15) plot as a straight line (Figure 7.13).

An important and often incorrect assumption is that observations in the training set are classified correctly or misclassified at random. In the pottery classification study it is assumed that all the observations listed as belonging to site K or N belong to those sites. Frequently, this is not the case. When misclassification occurs, the observations that are closest to the border between two groups are most likely to be misclassified. Should this occur, group means will be biased in the direction of more separability than occurs if observations are classified correctly. Thus, the result of classification appears overly optimistic, and some observations at or near the boundary between the two regions may be classified incorrectly.

7.7.8   Comments

In many earth science applications, data are not multivariate normally distributed but often tend to be right-skewed. This may leave the user in a quandary. There are several options:

- If the data are only mildly nonnormal, normal theory models can be used with minimal problems, especially if the use is exploratory.
- If the data are highly skewed, there are three options:

1. Transform the data using a log, square root, Box–Cox, or other appropriate transformation.
2. Use a nonnormal distribution model. Most univariate distributions, including the beta, gamma, and lognormal, have multivariate counter parts and have been well studied.
3. Use a nonparametric procedure. Converting the data to rank order and then using a multivariate normal procedure is a type of nonparametric procedure.

Other approaches described by Flury (1988), McLachlan (1992), and Ripley (1996) include principal components and canonical methods.

7.8 TREE-BASED MODELING

Tree-based modeling is an increasingly popular method that may be viewed as a nonparametric alternative to LDA for classification and also as an alternative to regression. It is a very general procedure in which results can be viewed as binary trees. Consider a general model where a response can be expressed as a function of a set of explanatory variables. The response variable can be continuous, counts, or categorical. Assume that an area has been partitioned into $R_1, \ldots, R_m$ regions. If the response variable is continuous, the tree model may take the form

$$f(x) = \sum_{m=1}^{M} c_m I(x \in R_m)$$

where the c_m, $i = 1, \ldots, m$ are parameters to be estimated and I is an index function, which is equal to 1 when $x \in R_m$ and is 0 otherwise. The c_m are determined by some criterion, such as minimizing sum of squares of residuals, as in parametric regression (Chapter 4).

In the case of classification, the response variable is categorical, and in the context of the simple pottery example (Table 7.8) it is sites, K, N, and S; however, only sites K and S are used in this example . The five elements (predictor variables) are Ti, Mn, Cu, Zn, and Pb. The classification process is first discussed generally (Figure 7.14) and then in the context of the specific pottery example. A tree procedure first splits at the root node on the variable that best minimizes a classification rule. The rule used in the pottery example is to make the terminal nodes as pure as possible. The terminal nodes are the rectangles labeled T1 through T4. An impure node is one that in the pottery example contains both K and N sites. A terminal node is reached when some stopping rule criterion is reached. One example is that no greater level of purity can be achieved. The top node, where the first split is made (split 1 in Figure 7.14), is called the *root node*. It is made after reviewing all of the predictor variables to see which yields the best split. This leaves the data set partitioned into two parts. After checking the stopping rule, it is determined that the right split yields the terminal node T1. Split 2 is made on some other variable than is used in split 1, again partitioning the original

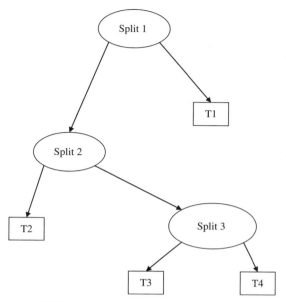

FIGURE 7.14   Tree-based classification.

left part of the data set into two parts. The left data set cannot be split further and yields terminal node T2. The right portion of the data set can again be split. Split 3 can use the same variable used in split 1 or another variable but not the one used in split 2. The result is four terminal nodes as pure as the data will allow. In the pottery example, the first split is at Cu $=79.5$. The branch is to the left if Cu < 79.5 and to the right if Cu $\geq$ 79.5 (Figure 7.15). This split results in two nodes. Node T1 (the K node) contains 20 observations; 17 are from site K and 3 are from site N. Node T2 (the N node) contains 11 observations, 3 from site K and 8 from site N. Splitting continues until a stopping rule is encountered. In this example only one split is necessary. To see why the split at Cu $=79.5$ is optimal, view the Cu data sorted in ascending order (Table 7.16).

As noted, the impurity of a node refers to mixed sites. For example (Table 7.17), node T1 (corresponding to region R_1) would be pure if it contained all observations

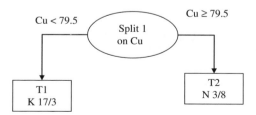

FIGURE 7.15   Classification from tree-based method using simple pottery data; K is the Kamikaizuka site and N is the Narita 60 site.

TABLE 7.16   Pottery Data Sorted by Cu (ppm)

Site	Cu	Site	Cu
K	19	K	64
K	21	N	64
K	39	K	77
K	42	K	79
K	42	N	80
N	42	N	81
K	43	K	83
K	49	N	86
K	49	N	86
K	50	N	88
K	53	K	90
N	54	K	95
K	55	N	103
K	55	N	106
K	63	N	106
K	64		

from site K. Node T_2 corresponds to region R_2. More formally, an *impurity index* at node m can be defined as

$$I(m) = \sum_{i=1}^{C} f(p_{im})$$

where C is the number of groups (in the pottery example, it is two: sites K and N) and p_{im} is the proportion of observations in node m (region R_m) that belong to group i. The function f is:

- An *information index*, $f(p) = -p \log(p)$, or
- The *Gini index*, $f(p) = p(1 - p)$

For this example, the recursive partitioning R-project algorithm rpart is used, including the Gini index. In addition to assigning a splitting function (information or Gini index), the user can typically specify:

TABLE 7.17   Classification Matrix Resulting from Partitioning at Cu = 79.5

Site	Assignment		Total
	R_1	R_2	
K	17	3	20
N	3	8	11

- Prior probabilities of assignment of observations to regions (similar to those available in LDA)
- A loss function, which may be important if losses of misclassified assignments are not equal
- Pruning and stopping rules

An important question is: Where should the tree be pruned? In the simple pottery example, the decision is easy because no further improvement is made after the split at the root node. Splitting on Cu < 79.5 yields only six misclassifications and a misclassification error of 0.194, which is less than the APER of 0.226 using LDA (Table 7.14).

The advantages of tree-based methods include:

- Speed.
- Interpretability—no transforms are required.
- Data can be of any type (nominal, ordinal, interval, or ratio).
- Missing data are easy to handle.

There are various methods of splitting and various pruning and stopping rules. For further details of tree-based methods, see the book by Breiman et al. (1984).

7.9 SUMMARY

The multivariate techniques discussed in this chapter are useful for gaining insight into high-dimensional data. For large sample sizes and many variables, the information often lies in a small but not always easily detectable subset of the data. Principal components analysis, factor analysis, cluster analysis, multidimensional scaling, and tree-based methods are tools to gain insight into the structure of data. They are generally used as exploratory tools and in data-mining applications. Often, there is no single best approach. Successful exploratory data analysis involves the use of a combination of methods, including graphical tools.

PCA helps determine dimensionality via the use of orthogonal rotations to account for successively smaller components of variance. FA uses rotation as a means to group (cluster) similar variables. Cluster analysis helps define groups of observations (samples) based on similar attributes. Two major clustering procedures are hierarchical and k-mean. Clustering involves defining a measure of similarity, a way of grouping, and a stopping rule. A dendogram is a useful graphic for visualizing the hierarchical clustering process. DA allows one to examine the degree of separability between defined groups and to classify an observation of unknown origin into a group. Tree-based modeling procedures, which use data splitting based on rules, are an alternative to DA as classification tools.

EXERCISES

7.1   Can PCA and FA methods be used interchangeably? Explain.

7.2   Suppose that a set of variables are pairwise independent and uncorrelated. What does one expect a resulting screeplot from a PCA to look like? Does it matter if the variables are in original units or standardized? Explain.

7.3   In a factor loading matrix, what can one learn by viewing the uniqueness component?

7.4   A DA model is used to establish two climate zones. On the boundary where two zones meet, there may be errors in the training sets (sample values used to estimate means and variances). How may this affect a description of separability between the climate zones?

7.5   What are the advantages and disadvantages of hierarchical clustering techniques? Why?

7.6   Give an example of when to use a single linkage method to form a cluster. Justify.

7.7   Cite a major advantage and disadvantage to Ward's method of forming clusters. Explain.

7.8   Hierarchical clustering methods can be either agglomerative or divisive. Describe reasons to choose one over the other.

7.9   How does one select the number of clusters when using hierarchical clustering?

7.10 Explain why observation 36 in Table 7.8 is an outlier for all the methods presented in this chapter.

7.11 Explain why observation 66 in Table 7.8 is not identified as a possible outlier in the Euclidean distance, single linkage, and standardized variable dendogram (Figure 7.10b).

7.12 In Table 7.8, five variables are used in the cluster methods. For Ward's method (Figure 7.8), determine which variables are most important in forming clusters.

7.13 Run fuzzy set clustering using the data in Table 7.8. Assume three clusters and find or construct a graphical method that illustrates the probabilistic nature of this method.

7.14 Instead of using the K and N sites (Table 7.8) as in the LDA pottery example, perform an LDA using the K and S sites, again with variables Cu and Pb. What are the conclusions?

7.15 Refer to Exercise 7.14. Use all three sites (K, N, and S) and perform an LDA using all five variables. State the conclusions.

7.16 Refer to Exercise 7.15. Are there any concerns about using all five variables? What if one has 15 variables instead of five? Explain.

7.17 Refer to Exercise 7.15. Determine the relative importance of the five variables in the LDA. Explain.

7.18 Refer to Exercise 7.15. Instead of using LDA, should QDA be used? Justify the answer.

7.19 Contrast the use of tree-based methods with the LDA procedures.

7.20 In clustering methods, what is the major concern with outliers?

7.21 Verify that in the pottery example (Figure 7.15) with a tree model, using Cu for the first split is better than using Pb.

7.22 Verify that in the pottery example (Figure 7.15) with a tree model, no improvement in classification is possible after the first split.

8 Discrete Data Analysis and Point Processes

8.1 INTRODUCTION

Discrete data often are generated by counting processes. The binomial process, for example, describes the number of successes that occur in n experimental trials. Other common processes are the Poisson, often used to describe a random outcome such as a distribution of counts of anomalous points within cells, and the negative binomial, which can be used to describe spatial clustering of points.

A proportion is an alternative representation for some forms of discrete data. Instead of x successes in n trials, one may consider the ratio x/n and see if it changes over time or differs between populations.

An important application in discrete data analysis is the study of dependencies between two or more discrete variables. This may be accomplished by forming a contingency table. An example is to study the relationship between houses with and without energy-saving refrigerators (one category) versus household income (a second category).

Many applications examined in this chapter are special cases of the generalized linear model (McCullagh and Nelder, 1989). One such case is logistic regression, where the response variable is discrete.

8.2 DISCRETE PROCESS AND DISTRIBUTIONS

Three important processes resulting in discrete outcomes are the binomial, the Poisson, and the negative binomial. They are characterized by distributions of the same name. These processes and associated distributions are reviewed and then illustrated with examples. All of these processes involve counting outcomes of a random event. In earth science applications such processes often occur in two dimensions.

The population probability of success on a single trial is denoted as π. The corresponding sample statistic (estimate) is denoted as $\hat{\pi}$ or, equivalently, p.

Statistics for Earth and Environmental Scientists, By John H. Schuenemeyer and Lawrence J. Drew
Copyright © 2011 John Wiley & Sons, Inc.

8.2.1   Binomial Process and Distribution

A *binomial process* is a random process generated by an experiment that is characterized by the following properties:

- The experiment results in one of two possible outcomes: for example, success or failure.
- The experiment is conducted for a fixed number of trials, n, determined in advance.
- The outcome of each trial, called a *Bernoulli trial*, is independent of the outcome of any other trial.
- The probability of success π on a single trial remains constant across trials.

A simple example of the binomial process is to toss a six-sided die (the sides are marked with one to six dots) 10 times, where a success is defined to be a 1 (i.e., a single dot) and failure is any other outcome. The probability of success on a single trial is $\pi = 1/6$ and remains constant for all trials. Clearly, the die has no memory, so trials are independent. This process is described by the *binomial probability function*

$$\Pr(X = x) = \binom{n}{x}\pi^x(1-\pi)^{n-x} \qquad x = 0,1,\ldots,n$$

where n is the number of trials (fixed in advance of conducting an experiment), π is the probability of success on a given trial, and x is the number of successes. One question to be asked is: What is the probability that in 10 trials (tosses of the die), exactly three successes will result? The solution is

$$\Pr(X = 3) = \binom{10}{3}\left(\frac{1}{6}\right)^3\left(\frac{5}{6}\right)^7$$
$$= 0.16$$

The probability is 0.16 that in 10 tosses of the die, three 1's will appear and the other seven outcomes will not be 1's. The order of their occurrence is unimportant.

Often, the interest is in the probability of x or fewer successes, in which case the *cumulative distribution* is used: namely,

$$\Pr(X \le x) = \sum_{k=0}^{x}\binom{n}{k}\pi^k(1-\pi)^{n-k}$$

For the die example, with 10 tosses,

$$\Pr(X \le 3) = \Pr(X = 0) + \Pr(X = 1) + \Pr(X = 2) + \Pr(X = 3) = 0.93$$

The $\Pr(X \ge 4) = 0.07$ because $\Pr(X \le 3) + \Pr(X \ge 4) = 1$.

Consider a second example, in which a well driller is searching for water. He has a budget sufficient to drill five wells, and from past experience he believes that the probability of success on any given well is 0.2. A reasonable question to ask is: What is the probability that the driller will have at least one success from drilling five wells? First, it needs to be determined if this is a binomial process, by asking the following questions:

- Are there a fixed number of trials? If the driller is going to drill all five wells regardless of outcomes, the answer is yes.
- Are there only two possible outcomes? The answer is clearly yes—the drilling results in success or failure.
- Is the probability of success constant from trial to trial, or expressed another way, are the trials independent? This question is more difficult to answer. If the driller obtains information from trial 1 that helps him site the next well (trial 2), the answer is no because learning occurs and the probability of success may improve as drilling progresses. However, if no information is gained, trials may be assumed to be independent and therefore the binomial model is appropriate.

The answer to the question about the probability of at least one success, assuming the binomial model, is

$$P(X \geq 1) = \sum_{k=1}^{5} \binom{5}{k} 0.2^k (0.8)^{5-k}$$

which is equivalent to

$$1 - P(X = 0) = \binom{5}{0} (0.2)^0 (0.8)^5$$

$$= 0.67$$

The probability function and cumulative distribution function for this example are tabulated in Table 8.1. The probability function graphed in Figure 8.1 is a right

TABLE 8.1   Probability Function and Cumulative Distribution Function for a Binomial with $n = 5$ and $\pi = 0.2$

k	$Pr(X = k)$	$Pr(X \leq k)$
0	0.3277	0.3277
1	0.4096	0.7373
2	0.2048	0.9421
3	0.0512	0.9933
4	0.0064	0.9997
5	0.0003	1.0000

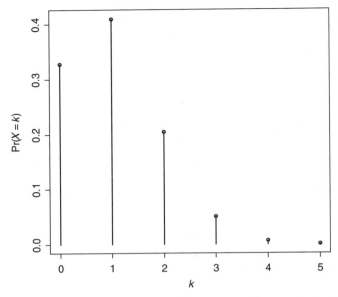

FIGURE 8.1   Display of probability function of a binomial distribution with $n = 5$ and $\pi = 0.2$.

(positively)-skewed distribution. In the exercises the reader is asked to examine other combinations of n and p and investigate the resulting distributions.

From the binomial model, a prospective purchaser of the property wants to determine how many wells may need to be drilled to have a reasonable chance of success. If five wells are drilled, the chance of at least one success is 0.67 given $\pi = 0.2$.

If the random variable X is binomial, it is easy to show that the expected number of successes $E[X] = n\pi$ and the variance of X, $\text{Var}[X] = n\pi(1 - \pi)$. For the drilling example, these are $E[X] = 1$ and $\text{Var}[X] = 0.8$.

When more than two outcomes are possible—for example, if the well driller's outcome is a success, a possible success, or a failure—the distribution is multinomial (Agresti, 2002). In contingency table analysis, the multinomial is often approximated by a chi-square distribution.

8.2.2   Poisson Process and Distribution

A *Poisson process* is important in earth science applications because it is character-ized by independent outcomes over time and/or space. The following are examples of processes that can be Poisson. They include the number of:

- Earthquakes along the San Andreas fault occurring in a unit of time
- Floods in the Upper Mississippi River Valley in a unit of time

- Anomalous values of mercury in a mining region
- Illnesses attributed to the West Nile virus in a neighborhood
- Stars in a three-dimensional volume of space
- Lightning strikes per unit of time and/or area

Consider an example involving the frequency of lightning strikes in a county. One question is: Do lightning strikes occur independently and with exponentially distributed interarrival times?" The response may provide insight into the processes governing occurrences of lightning. Examples of such a distribution will be given shortly. Two alternatives to a Poisson process are that lightning strikes occur uniformly over time or that they are clustered (occur) within a small number of days. A more complex question is: Can a Poisson process be used as an appropriate model for lightning strikes over time and space?

There are different types of Poisson processes. A homogeneous Poisson process must satisfy the following properties:

- The probability of exactly one arrival in a sufficiently small interval h is $\Pr(1,h) = \lambda h + o(h)$, where λ is the *intensity function* (the number of occurrences per unit of time) and $o(h)$ is a small number that goes to zero as $h \to 0$. In the lightning strike example, this implies that in a short time interval h, the probability of exactly one strike is approximately proportional to h. The intensity λ is the mean number of strikes per day. It is assumed that λ does not change systematically over the time duration of the study. If it does, the process is inhomogeneous.
- The probability of two or more arrivals in a sufficiently small interval h is essentially 0: namely, $\Pr(\text{no. of arrivals} \geq 2, h) = o(h)$.
- The numbers of occurrences in nonoverlapping intervals (units of time) are independent for all intervals. For example, if three lightning strikes are observed in a given day, this does not influence the number of strikes that are likely to occur the next day. The process has no memory.

In the notation above, h is a time interval; however, it can be a distance, area, volume, or space–time unit.

Difference equations are derived from the postulates above. Let $\Pr_n(t) = \Pr(n)$, the probability of n arrivals in time 0 to t. Then

$$\Pr_n(t+h) = \Pr_n(t)(1 - \lambda h) + \Pr_{n-1}(t)\lambda h + o(h) \qquad n > 0$$
$$\Pr_0(t+h) = \Pr_0(t)(1 - \lambda h) + o(h)$$

where the $1 - \lambda h$ and λh are weights reflecting the large chance of n events occurring in the interval 0 to $t + h$ and the small chance of $n - 1$ events occurring in the interval 0 to t. If each of these equations is divided by h and the limit taken as $h \to 0$, the Poisson distribution results. The *Poisson distribution* is typically the first distribution

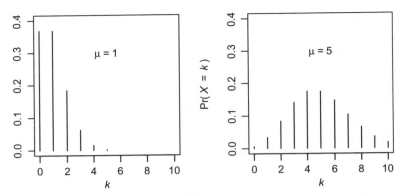

FIGURE 8.2   Poisson probability function for (a) $\mu = 1$ and (b) $\mu = 5$.

or model that is applied to time or spatial point data. If the data are Poisson distributed, no further investigation may be needed. Conversely, if the distribution observed is not Poisson, further investigation may be required to help determine causal factors.

The form of the probability function for the Poisson distribution is

$$\Pr(X = x) = \frac{\lambda^x e^{-\lambda}}{x!} \qquad x = 0, 1, \ldots$$

where λ is the intensity or mean number of occurrences in some unit of time or space. This function is also interesting because $E[X] = \mathrm{Var}[X] = \lambda$. The cumulative distribution function is

$$\Pr(X \leq x) = \sum_{i=0}^{x} \frac{\lambda^i e^{-\lambda}}{i!}$$

The symbol λ is commonly used to represent the mean μ as the Poisson distribution parameter; the two symbols will be used interchangeably. Figure 8.2a shows a Poisson probability function with $\mu = 1$; Figure 8.2b shows a Poisson probability function with $\mu = 5$. As μ becomes larger, the Poisson distribution becomes more normal in appearance.

Relationship Between a Poisson Process and an Exponential Distribution   Take an arbitrary starting time and call it 0. Now suppose that the Poisson events arrive at times $t_1 < t_2, \ldots$. Let $\tau_1 = t_1 - 0$, $\tau_2 = t_2 - t_1$, $\ldots$ be the interarrival times of Poisson events (Figure 8.3). It is easy to show that $\tau_1, \tau_2, \ldots$ are independent of each other and each has an *exponential distribution* with mean $1/\lambda$. Thus, if the counts are Poisson, the interarrival times are exponentially distributed. Recall that an exponentially distributed random variable is continuous.

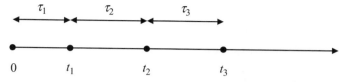

FIGURE 8.3   Relationship between the Poisson events and exponential interarrival times.

8.2.3 Negative Binomial Process and Distribution

A *negative binomial process* is characterized by the number of failures before the rth success in a sequence of Bernoulli trials. A negative binomial process can be viewed as a generator of clustered observations. It is often the first alternative model when an attempt to fit a Poisson distribution fails. The negative binomial process has the following characteristics:

- The experiment results in one of two possible outcomes, such as success or failure.
- The outcome of each trial is independent of the outcome of any other trial.
- The probability of success π on a single trial remains constant.
- The process stops when the rth success is achieved.

Note the similarity to the binomial process. The difference is that in the binomial process, the number of trials is fixed; in the negative binomial process it is a random variable. The usual way to characterize the negative binomial probability function is

$$\Pr(X = x) = \binom{x + r - 1}{r - 1} \pi^r (1 - \pi)^x$$

where x is the number of failures prior to the rth success. Trials continue until the rth success is achieved after x failures. Therefore, the random variable is x, the number of failures. This model may be appropriate to use to answer such questions as: What is the probability that an archaeologist needs to observe 10 sites in order to find three sites that contain decorated pottery if the probability of success at each site is estimated to be 0.1 and trials are assumed to be independent. The desire is to know the probability of exactly seven failures prior to the third success. Sequences may take the form FFSFFFSFFS, SSFFFFFFFS, and so on. The process must stop with a success (S). The answer to the question is

$$\Pr(X = 7) = \binom{7 + 3 - 1}{3 - 1} (0.1)^3 (1 - 0.1)^7 = 0.017$$

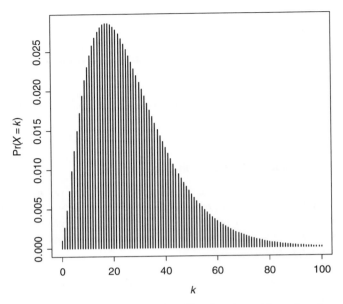

FIGURE 8.4   Negative binomial probability function with $r = 3$ and $\pi = 0.1$.

The probability function for $\pi = 0.1$ and $r = 3$ is shown in Figure 8.4. It is often more informative to view the cumulative negative binomial distribution

$$\Pr(X \le x) = \sum_{i=0}^{r} \binom{x+i-1}{x-1} \pi^x (1-\pi)^i$$

In the previous problem, $\Pr(X \le 7) = 0.070$.

For the negative binomial distribution, $E[X] = r(1-\pi)/\pi$ and $\mathrm{Var}[X] = r(1-\pi)/\pi^2$. Some literature refers to this form of the negative binomial as the *Pascal probability function*.

The form of the function where a noninteger r is allowed, referred to as a negative binomial function, is characterized as

$$\Pr(X = x) = \frac{\Gamma(x+r)}{\Gamma(r)x!} \pi^r (1-\pi)^x$$

where Γ is the gamma function (Abramowitz and Stegun, 1972).

Now reformulate the water-well drilling example. Previously, the probability that the driller has at least one success out of five holes has been investigated. Now, let the probability of success π for a randomly drilled hole be the same (i.e., 0.2) and ask the question: How many dry holes, on average, are drilled prior to the first success? For $\pi = 0.2$ and $r = 1$, $E[X] = 1(1-0.2)/0.2 = 4$. However, $\mathrm{Var}[X] = 1(1-0.2)/0.2^2 = 20$ and the standard deviation is 4.47. Thus, it is not unusual for, say, nine wells to be drilled prior to a success. Specifically, $\Pr(X \le 9) = 0.89$, so there is a 0.11 probability that more than nine wells must be drilled prior to a success.

When $r = 1$, $\Pr(X = x) = \pi(1 - \pi)^x$. This special form is called the *geometric distribution*. It is a discrete waiting-time distribution because it allows an investigator to compute the probability of x failures prior to a success.

8.3 POINT PROCESSES

Point processes are random processes that occur in time or space or both. Point processes are important in almost every earth science discipline. Examples include:

- The location, time of occurrence, and magnitude of earthquakes. This is an example of a space–time process, since earthquake data includes location, time, and magnitude.
- Occurrences (locations) of cases of mercury poisoning in children.
- Shows of gas hydrate in the U.S. Gulf of Mexico.
- Time sequence of deaths in coal mines.

Point processes may occur in four dimensions (x, y, z, t), where x, y, and z are geographic coordinates and t is time. Outcomes are not generated from a designed experiment, such as observations taken at prespecified locations along a transect. The distribution of points in space is interesting because it may yield information about underlying processes that generated such patterns. Our discussion here is restricted to two dimensions. There are three broad classes of point patterns: regular (Figure 8.5a), random (Figure 8.5b), and clustered (Figure 8.5c). Also shown is a highly clustered pattern (Figure 8.5d).

Figure 8.5a shows a regular structure, usually a lattice structure. Think of this result as a pattern that begins with a single point. A second point is repelled a certain distance from the first point. Subsequent points are some minimum distance away from established points. A regular structure does not require points to be exactly regular in the sense of an equal distance between points, but it cannot contain clusters. It is not the result of systematic sampling (Chapter 9), where points can be regularly spaced by design. This pattern is not common in nature. Figure 8.5b shows a random pattern of points. Think of these points as being generated sequentially, where the location of any one point does not influence the location of any other point. Finally, Figure 8.5c and d represent clustered point patterns, with the points in Figure 8.5d being more tightly clustered. This pattern can be constructed as follows. Within a neighborhood of the first point, a second point is attracted to the first point and thus located nearby. The process continues and yields clusters. All of these patterns except the random one (Figure 8.5b) exhibit some type of structure. Modeling and identifying such a structure can yield information about the underlying process.

Consider an example from medical geography, where an investigator notices what appears to be a large number of reported cases of childhood leukemia in a given neighborhood. There is suspicion that this may have been caused by polluted drinking water. An important question is: Can this spatially compact grouping of cases of

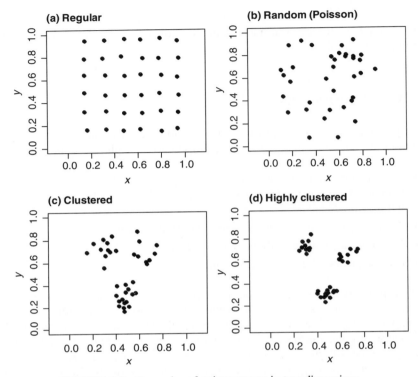

FIGURE 8.5   Examples of point patterns in two dimensions.

leukemia occur by chance? Even if points are distributed spatially at random, there can be a few clusters (Figure 8.5b). But in a clustered pattern there are more points spaced closer together than is indicated by a random generation of points. The way to address this problem is to compute indices and fit models to examine spatial point patterns.

The *index of dispersion* for count data is defined as

$$D = \frac{\sigma^2}{\mu}$$

and is estimated by $\widehat{D} = \widehat{\sigma}^2/\widehat{\mu}$. This index may serve as a guide to the general form of the distribution. When a distribution is Poisson, $D = 1$, since the variance equals the mean. If there are an equal number of points in each cell, the variance is zero and thus $D = 0$. When the distribution is aggregated (clustered), $\sigma^2 > \mu$ and thus $D > 1$.

8.3.1   Testing for a Poisson Process

A common approach to examining spatially distributed points is to begin with a test for randomness. If points are spatially random, further investigation may not be

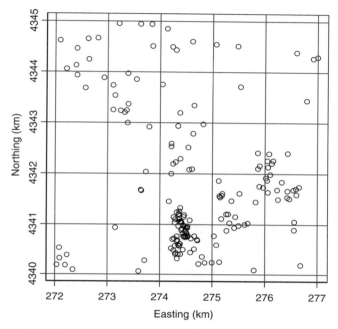

FIGURE 8.6   Subregion of the water-well yield case study.

necessary since there may be no underlying cause for the distribution of points. A traditional approach to investigating spatial randomness is to "toss" a square grid over the region of interest, count the number of points within each cell, fit a Poisson distribution to the count data, and evaluate the fit.

The first example is an investigation of the water-well yield case study data (Appendix I) by overlaying the data with an arbitrarily chosen 1×1 km grid and then counting the number of occurrences in each cell. A subregion is shown in Figure 8.6. The frequency distribution of point counts and Poisson model fit for the complete data set is shown in Table 8.2.

Table 8.2 is constructed as follows. The bin (column 1) is an identification number. The n points (column 2) refer the number of points in a cell (Figure 8.6 shows a subset of the cells). A point is the location of a water well. The frequency of points observed in a cell is given in column 3. For example, there are 34 cells that contain 0 points, 22 cells that contain 1 point, and so on. Note that cells with 0 frequencies do not appear in column 3. For example, no cells contain 17 points. The total number of observations (points) is $T_o = \sum_{i=1}^{21} n_i O_i = 754$ and the total number of cells $T_c = \sum_{i=1}^{21} O_i = 162$. An estimate of the mean of a Poisson distribution is $\widehat{\lambda} = T_o / T_c$. For this example, $\widehat{\lambda} = 4.65$ points per cell or is expressed as an intensity of 4.65 points/km^2. The estimated variance $\widehat{\sigma}^2 = 34.85$ and the dispersion index $D = 7.49$. It is clear that this distribution is unlikely to be Poisson; however, for illustrative purposes the goodness-of fit-procedure is completed. The Poisson probabilities $\Pr(X = n_i \mid X \sim \text{Poisson}(\widehat{\lambda}))$ are given in column 4. Of course, $\sum_{i=0}^{\infty} \Pr(X = n_i \mid X \sim \text{Poisson}(\widehat{\lambda})) = 1$.

TABLE 8.2 Frequency Distribution of Points per Cell and Computation of a Chi-Square Statistic for a Poisson model, Water-Well Yield Case Study Locations

(1) Bin	(2) n Points	(3) O (Obs. Freq.)	(4) $\Pr(X = n_i \mid \text{Poisson})$	(5) E_i (Expected Counts)	(6) $(O_i - E_i)^2/E_i$
1	0	34	0.00952	1.54	683.1
2	1	22	0.04431	7.18	30.6
3	2	13	0.10312	16.71	0.8
4	3	24	0.15998	25.92	0.1
5	4	13	0.18615	30.16	9.8
6	5	9	0.17328	28.07	13.0
7	6	6	0.13442	21.78	11.4
8	7	7	0.08938	14.48	3.9
9	8	6	0.05200	8.42	0.7
10	9	5	0.02689	4.36	
11	10	1	0.01252	2.03	
12	11	6	0.00530	0.86	
13	12	1	0.00205	0.33	
14	13	4	0.00074	0.12	
15	14	3	0.00024	0.04	
16	15	2	7.59×10^{-5}	0.01	
17	16	2	2.21×10^{-5}	0.00	
18	18	1	1.56×10^{-6}	0.00	
19	21	1	1.97×10^{-8}	0.00	
20	28	1	1.56×10^{-13}	0.00	
21	48	1	8.75×10^{-32}	0.00	
Sum bins 10–21:		28	0.04784	7.7	

The expected count (column 5) for the ith bin is $E_i = T_o \times \Pr(X = n_i \mid X \sim \text{Poisson}(\widehat{\lambda}))$, $i = 1, \ldots,$ 21. In bins 10 through 21, the expected counts (column 5) are all less than five. When the count expected is less than five, as frequently occurs in the tail of a distribution, the tail needs to be folded (i.e., categories combined) to make the value expected at least five, so that the test statistic is a reasonable approximation to a chi-square distribution. The column sums for bins 10 through 21 are shown in the bottom row of Table 8.2. To construct a reasonable approximation to the chi-square goodness-of-fit statistic, the data in bins 10 through 21 of Table 8.2 are replaced by their sum (the bottom line in Table 8.2). The chi-square contribution from the sums of bins 10–21 is 52.9.

The null and alternative goodness-of-fit hypotheses are

$$H_0: \text{data are Poisson}$$
$$H_a: \text{data are not Poisson}$$

The test statistic is

$$X^2 = \sum_{i=1}^{k} \frac{(O_i - E_i)^2}{E_i}$$

If H_0 is true, X^2 has an approximate chi-square distribution with $k - 1 - c$ df, where k is the number of bins. In the expression for df, 1 df is lost because $T_o = \sum_{i=1}^{k} E_i$ is required, and c degrees of freedom are lost if c model parameters need to be estimated from the data. In this example, only one parameter λ is estimated; thus, $c = 1$. When X^2 is large, H_0 is rejected. For a significance level α, H_0 is rejected if $X^2 > \chi^2_{\alpha,k-1-c}$. The p-value $= \Pr(Y > X^2 | Y \sim \chi^2_{k-1-c})$.

In the water-well yield case study, $X^2 = 806.27$, $k = 10$ and the corresponding p-value is infinitesimally small. Thus, H_0 is clearly rejected and the conclusion is that the data are not from a Poisson distribution. The search for a better model should begin. The next candidate model can be a negative binomial distribution.

A final concern about the chi-square test is that results can be sensitive to the size and placement of the grid. In addition, there is a related edge effect. An alternative is the Kolmogorov–Smirnov goodness-of-fit test (implemented by the kstest function in the R-project library spatat; Baddeley, 2008) where the null hypothesis is complete spatial randomness. For this test, $D = 0.15$ and the p-value $= 1.2 \times 10^{-15}$. Thus, the null hypothesis is rejected.

A variety of procedures, including spatial graphics, have been implemented in the R-project function spatstat (Baddeley and Turner, 2005).

8.3.2   Clark and Evans Nearest-Neighbor Index

One of many alternative approaches to investigating spatial distributions is the *Clark and Evans nearest-neighbor index* (CEI) (Clark and Evans, 1954). This index is the ratio of the mean nearest-neighbor distance to the expected distance if the data points are completely (spatially) random: namely Poisson distributed. Thus, the CEI is

$$\text{CEI} = \frac{\bar{d}}{1/\left(2\sqrt{\widehat{\lambda}}\right)}$$

where $\bar{d}$ is the mean nearest-neighbor Euclidean distance over all n points and $\widehat{\lambda} = n/A$ is the estimated points per unit area (A being the area of the region) under complete randomness.

If the CEI is approximately 1, the data points are distributed randomly; if it is less than 1, the points are clustered; and if it is greater than 1, the points are regularly spaced. For the water-well yield case study data $\widehat{\lambda} = 4.65$, the mean nearest-neighbor distance $\bar{d} = 0.16$ km, so CEI $= 0.69$, suggesting clustering. Because the CEI considers only the nearest neighbor, it is called a *first-order point pattern method*. It cannot account for distributions of distances beyond the nearest point.

8.3.3   Ripley's K

Ripley's K (Venables and Ripley, 1997) is more powerful than CEI because it considers neighbors up to a distance d from an arbitrary point. Ripley's K is a second-order method, which is defined as

$$\widehat{K}(d) = \frac{\hat{\lambda}^{-1}}{n} \sum_{i=1}^{n} \sum_{j=1}^{n} \delta_{ij}(||x_i - x_j|| \le d)$$

where $||x_i - x_j||$ represents the Euclidean distance between points x_i and x_j and

$$\delta_{ij} = \begin{cases} 1 & ||x_i - x_j|| \le d \\ 0 & ||x_i - x_j|| > d \end{cases}$$

A plot of $\widehat{K}(d)$ versus d for the water-well yield case study data (dashed line, Figure 8.7) shows a departure from the theoretical Poisson point pattern (solid line, Figure 8.7). The former line is smooth because of the large number of distance pairs. In a smaller data set, this graph looks ragged. Ripley's L function is a standardized version of Ripley's K function, defined as

$$L(d) = \sqrt{\frac{K(d)}{\pi}}$$

For a spatially random process, $L(d) = d$.

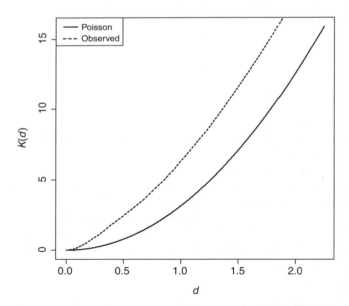

FIGURE 8.7   Ripley's K plot and complete spatial randomness (Poisson) for the water-well yield case study data.

8.3.4   Fitting Cluster Models

Indices computed for the water-well yield case study data suggest the presence of clustering. Many models are available to fit a clustered pattern of points, including the negative binomial, Neyman, and Thomas distributions. The negative binomial appears to be a reasonable fit and will be presented.

When modeling a spatial distribution, the number of failures r is relabeled as k, which is allowed to assume noninteger values. It is interpreted as a dispersion parameter. The fit to the negative binomial is as follows. Let $\widehat{\mu}$ and $\widehat{\sigma}^2$ be the sample mean and variance of the count data. A first approximation to the dispersion parameter estimate for the water-well yields case study data is

$$\widehat{k} = \frac{\widehat{\mu}^2}{\widehat{\sigma}^2 - \widehat{\mu}}$$

yielding $\widehat{k} = 1.154$ for the well-water yield case study data. Reich and Davis (2002) provide the following guidelines for interpreting $\widehat{k}$:

- $\widehat{k}$ *small positive*: highly aggregated (clustered) population
- $\widehat{k}$ *large positive*: slight clumping in the population
- $\widehat{k} \to \infty$: negative binomial approaches a Poisson distribution
- $\widehat{k} < 0$: population is regularly distributed

Since, in this example, $\widehat{k}$ is small positive, a highly aggregated population is indicated and the negative binomial distribution is fit using the function goodfit from the R-project library vcd (Ricci, 2005). The fitting method is to minimize the chi-square goodness-of-fit statistic. The resultant parameter estimates are $\widehat{k} = 0.843$ and $\widehat{\pi} = 0.153$. The data are shown in Table 8.3. Since the points and observed

TABLE 8.3   Frequency Distribution of Counts from Water-Well Yield Case Study Data and Fit to a Negative Binomial Distribution

(1) Bin	(2) n Points	(3) O (Obs. Freq.)	(4) Pr($X=n_i$| Neg. Binomial)	(5) E_i (Expected Counts)	(6) $(O_i - E_i)^2/E_i$
1	0	34	0.20591	33.357	0.01
2	1	22	0.14691	23.800	0.14
3	2	13	0.11461	18.566	1.67
4	3	24	0.09195	14.896	5.56
5	4	13	0.07479	12.116	0.06
6	5	9	0.06133	9.936	0.09
7	6	6	0.05057	8.192	0.59
8	7	7	0.04185	6.780	0.01
9	8	6	0.03474	5.628	0.02
10	9	5	0.02890	4.682	0.02
Sum bins 11–21:		23	0.14843	24.046	0.05
Totals		162	1.00000	162.000	8.22

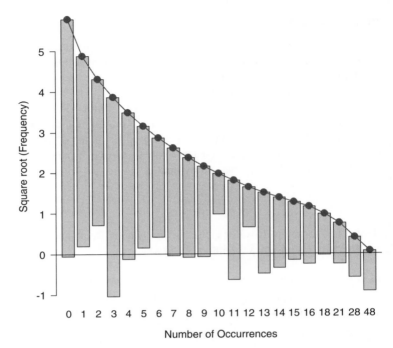

FIGURE 8.8   Histogram of frequency and negative binomial fit to the water-well yield case study data.

frequency are the same as in Table 8.2, only the folded results are shown. The p-value corresponding to $X^2 = 8.22$ with $df = 11 - 2 - 1 = 8$ is 0.412. Thus, the hypothesis that this sample came from a negative binomial distribution cannot be rejected.

The plot of the observed frequency and negative binomial fit (Figure 8.8) is also from the R-project vcd library. Note that this display is somewhat different from the usual histogram and model fit. The tops of the bars are even with the fit, and a lack of fit is denoted by the length of the bar below or above zero. In addition, only bins containing observations are plotted. Additional information on point processes may be found in Cox and Isham (1980).

8.4   LATTICE DATA AND MODELS

A *lattice* is simply an array of values. It can be in one, two, or more dimensions. Typically, the lattices used in the social sciences, agriculture, and economics are two-dimensional. Some are defined on a regular grid. The prediction points discussed in Section 6.7.4 constitute a lattice, where the block kriging value is a block average. However, many lattices are irregularly spaced. Often, agricultural and health statistics are reported by county. The variable of interest can be the proportion of land owned by the federal government, the number of cases of lead poisoning, the number of abandoned mine sites, average income, or farm crop yield. Typically, these are values that cannot be associated with a point estimate. Rather, the data are associated with a county or other spatial unit. Some graphics are defined by nodes, such as transportation and communication systems, where the interest can be in finding an optimal route to handle a given volume of traffic. GIS typically allow users to deal with both types of lattice structures.

Data analysis of lattices may involve smoothing and trend removal along the lines discussed previously. However, results may be weighed by the size of the lattice. Markov processes and other techniques for fitting lattice structures have been described by Cressie (1993).

8.5 PROPORTIONS

Proportions are derived from discrete data. Instead of considering x successes in n trials, a *proportion $p = x/n$* $(0 \leq x \leq n)$ can be defined. The corresponding population proportion is usually denoted as π. Sometimes the notation $\hat{\pi} = x/n$ is used in place of p. Comparing a proportion to a constant is useful to see if there are statistically significant changes over time, such as the success rate of exploratory drilling for oil or proportion of new houses that are equipped with water-saving devices. Comparisons can be made between proportions from two or more populations. For example, an energy company may wish to compare the success rates of exploratory drilling in two basins or between different technologies. A public health official may wish to compare the proportion of a fish catch that has dangerous levels of mercury between reservoirs or over time. A confidence interval can be computed on a proportion as a measure of uncertainty.

Sample size is an important consideration in deciding the best method for estimating and testing proportions. For one population, a guideline for a large sample is an n such that $n\pi \geq 5$ and $n(1 - \pi) \geq 5$. For a confidence limit or comparison of proportions π_1 and π_2, a guideline for large samples is an n_1 and an n_2 such that $\{n_1\pi_1 \geq 5$ and $n_1(1 - \pi_1) \geq 5\}$ and $\{n_2\pi_2 \geq 5$ and $n_2(1 - \pi_2) \geq 5\}$. Of course, since π is rarely known in advance, it must be estimated via expert judgment, analog, or a pilot study.

8.5.1 Large Sample Size

Define a proportion p as

$$p = \sum_{i=1}^{n} \frac{\delta_i}{n}$$

where $\delta_i = 1$ if the ith trial is a success and 0 otherwise. In this sense, p is a mean. For large n, the central limit theorem applies. Therefore, the methodology in Chapter 3 developed to estimate confidence limits and test hypotheses for a normally distributed population can be used.

The approximate $(1 - \alpha)\%$ CI on π is

$$p - z_{\alpha/2}\sqrt{\frac{p(1 - p)}{n}} \leq \pi \leq p + z_{\alpha/2}\sqrt{\frac{p(1 - p)}{n}}$$

where p is an estimate of π. The test statistic for the hypothesis H_0: $\pi = \pi_0$, where π_0 is a constant, against the alternative H_a: $\pi < \pi_0$ or H_a: $\pi > \pi_0$ is

$$Z = \frac{p - \pi_0}{\sqrt{p(1 - p)/n}}$$

The testing procedure is identical to that described in Chapter 3 for the means of continuous data.

The approximate $(1 - \alpha)\%$ CI on $\pi_1 - \pi_2$ is

$$(p_1 - p_2) - z_{\alpha/2}\widehat{\sigma}_{\pi_1 - \pi_2} \leq \pi_1 - \pi_2 \leq (p_1 - p_2) + z_{\alpha/2}\widehat{\sigma}_{\pi_1 - \pi_2}$$

where

$$\widehat{\sigma}_{\pi_1 - \pi_2} = \sqrt{\frac{p_1(1 - p_1)}{n_1} + \frac{p_2(1 - p_2)}{n_2}}$$

Independent sampling between the two populations is assumed. The corresponding test statistic for the hypothesis H_0: $\pi_1 - \pi_2 = 0$ against the alternative H_a: $\pi_1 - \pi_2 < 0$ or H_a: $\pi_1 - \pi_2 > 0$ is

$$Z = \frac{(p_1 - p_2) - (\pi_1 - \pi_2)}{\sqrt{\bar{p}(1 - \bar{p})(1/n_1 + 1/n_2)}}$$

where $\bar{p} = (n_1 p_1 + n_2 p_2)/(n_1 + n_2)$ since $\pi_1 = \pi_2$ by hypothesis.

8.5.2 Small Sample Size

For small samples and indeed even moderate-sized samples, it is computationally possible to perform exact tests. One and two population cases are discussed. Exact tests are also available for the $k > 2$ population comparisons.

Exact Test for One Population   The National Weather Service National Hurricane Center (2006) categorizes hurricanes using the Saffir–Simpson hurricane scale into one of five categories, with category 1 being the least severe and category 5 being the most severe. Categories 3 through 5 hurricanes are considered to be major hurricanes for purposes of this example. Suppose that the number of major hurricanes is believed to increase over time. Let π be the historical probability of a major hurricane and suppose that the test is H_0: $\pi = 0.2$ versus H_a: $\pi > 0.2$ at the 0.05 significant level. Of course, if $\pi < 0.2$, H_0 is accepted. Suppose in a given season and region that four out of 10 hurricanes are classified as severe. Does this evidence support accepting or rejecting the null hypothesis? Since $n\pi = (10) \times (0.2) = 2 < 5$, the normal approximation is not appropriate. Thus, an exact test is used and a p-value is computed.

The exact test is constructed in the following way. If the null is true, the question is: What is the probability of a result at least as extreme as the test statistic result? This means that if X is the number of major hurricanes observed, the "at least as extreme" condition occurs if $X = 4, \ldots, 10$. If H_0 is true, two (10×0.2) or fewer major hurricanes are expected. Four major hurricanes are a more extreme result in the direction of H_a. Observing five through 10 hurricanes is an even more extreme result. If the probability of a hurricane in a given season is constant, a binomial model is reasonable. The test statistic resulting in a p-value is formulated from a binomial model and $\Pr(X \geq 4) = 1 - \Pr(X \leq 3) = 0.12$. Thus, H_0 at the 0.05 significance level cannot be rejected, and therefore it is not possible to conclude that the four major hurricanes constitute a statistically significant departure from the historical probability level of 0.2.

Suppose in the example above that the alternative is H_a: $\pi \neq 0.2$. Four or more successes in the right tail by symmetry correspond in the left tail to zero successes. Obviously, fewer than zero successes are impossible. Thus, the exact test is $\Pr(X = 0) + \Pr(X \geq 4) = 0.23$, and therefore H_0 cannot be rejected at the 0.05 significance level.

Exact Test for Two Populations (Fisher's Exact Test)   *Fisher's exact test* is a statistical test used to determine nonrandom associations between two categorical variables. It is most often applied to the analysis of 2×2 tables (two categorical variables and two levels within each category). Suppose that the localities for the hurricanes observed in the previous example are in region A. In addition, hurricanes are observed in another region, called region B. A question of interest may be to see if the proportion of severe hurricanes is the same in both regions against the alternative that it differs. Data for this example, the observed counts, are shown is Table 8.4. Clearly, the sample sizes are too small for the normal approximation, so Fisher's exact test is used.

This test is constructed as follows. Let E_{ij} be the expected number of observations (counts) in row $i = 1, 2$, column $j = 1, 2$. Let R_i be the row totals (marginals) and C_j be the column totals (marginals). Finally, let G be the grand total. Under the assumption of independence, the expected count is cell (i, j) is $E_{ij} = (G)(R_i/G)(C_j/G)$. Values expected for the hurricane example are shown in Table 8.4. Since row and column totals are fixed, only E_{11} (the number of severe hurricanes in region A) needs to be calculated directly. Because independence is assumed, the probabilities R_i/G and C_j/G are multiplied to find the joint probability of occurrence in cell (i, j). Independence in this example is equivalent to testing that the population proportion of severe hurricanes in region A is the same as in region B. The procedure under the null hypothesis is similar to that in the binomial exact test. That is, the given result and those more extreme are tabulated (Table 8.4). Summing the probabilities for these outcomes yields a one-sided p-value. A number of possibilities exist for determining a

TABLE 8.4   Hurricane data for Fisher's exact test

	Region A	Region B	Row Totals
Observed			
Severe	4	2	6
Not severe	6	10	16
Column totals	10	12	22
Expected (Assuming Independence)			
Severe	2.72	3.27	6
Not severe	7.28	8.73	16
Column Totals	10.00	12.00	22
More Extreme 1			
Severe	5	1	6
Not severe	5	11	16
Column totals	10	12	22
More Extreme 2			
Severe	6	0	6
Not severe	4	12	16
Column totals	10	12	22

two-sided *p*-value, including doubling the one-sided value (Agresti, 2002). The *p*-value for the one-sided alternative is 0.23; thus, it is not possible to reject the null hypothesis of equality between the regions.

8.6   CONTINGENCY TABLES

A *contingency table* consists of count data from two or more factors and two or more levels in each factor. The hurricane data shown in Table 8.4 comprise an example of a 2×2 contingency table. There are two factors, region and severity, and two levels in each factor, regions *A* and *B* and severe and not severe.

The levels of a factor can be ordered or unordered. Typically, questionnaire responses such as poor, fair, and good are considered ordered, as well as geologic periods, severity of earthquakes, and categories of hurricanes. Geographic areas, rock type, and species types typically are unordered. Order can be used to provide more powerful tests. In the hurricane data example, region is unordered whereas severity is ordered.

The usual null hypothesis is that the factors are independent against an alternative that they are not. In the hurricane example above, independence between samples implies that knowing the level of a region (*A* or *B*) will not provide information about severity.

TABLE 8.5 Two-Factor Contingency Table with Row and Column Marginals

Factor II Level	Factor I Level			Totals
	1	2	3	
1	n_{11}	n_{12}	n_{13}	$n_{1.}$
2	n_{21}	n_{22}	n_{23}	$n_{2.}$
Totals	$n_{.1}$	$n_{.2}$	$n_{.3}$	n

Two key assumptions must be met:

1. Each observation is classified in only one cell.
2. Observations are independent. An example of lack of independence is respondents from the same household.

Table 8.5 is a 2×3 contingency table in which the n_{ij}'s are the observed cell counts. An equivalent notation for observed counts in cell (i,j) is O_{ij}. The $n_{i.}$'s are the row totals, the $n_{.j}$'s are the column totals, and n is the grand total.

There are three sampling schemes to consider. In each, an observation is classified into one and only one cell. They are:

1. n, the sample size, is fixed in advance; row and column totals are determined by the experiment. This is multinomial sampling. Observations are cross-classified as being in level 1, 2, or 3 of factor I and level 1 or 2 of factor II. Sampled values are independent.
2. n is fixed and either row or column marginal totals are fixed. Without loss of generality, assume that the marginal row totals are fixed. The samples between rows are independent, and within each sample (row) values are dependent.
3. n is fixed and both row and column marginal totals are fixed in advance. The n observations have an equal chance of being classified in the (i,j) cell or in any other cell. The application is to determine, usually by Fisher's exact test, if the event of an observation being in a given row is the same as it being in a given column.

It is important to understand the sampling scheme so that the proper model can be used. The relationship between the sampling scheme and test are illustrated in subsequent examples.

The question should also be asked: Is there a response variable? A discrete equivalent of regression occurs when one of the factors is considered a response variable and the others are the explanatory variables. In this case, the logistic regression model, discussed in the next section, may be appropriate. In the absence of a response variable, the problem is the discrete equivalent of a multivariate problem, where the interest is to model relationships between two or more factors.

These are called *association models*; a discussion of association models follows. Applications include:

- Examination of the relationship between type of building structure and archaeological period of Puebloan peoples of the southwestern United States
- Understanding the relationship between volcanic vents by age (geologic period) and the composition of erupted products
- Comparing the accuracy between vegetation maps made by satellite with those made by ground observers

It is important to know if the factors are independent. Several models and tests address this question. The appropriate test for a given situation is a function of the assumption and type of table, as described above.

Analytic methods are illustrated with a geological example—an examination of grain size versus color and a biological example—a study of species versus location. They are introduced in the next subsections. Further information on contingency table and other methods of categorical data analysis may be found in the literature (Agresti, 2002 and Thompson, 2006).

Geological Example: Color Versus Grain Size   In the analysis of sediments, an issue is the possible independence of the color of rock and grain size. Suppose that there are three colors, red, green, and white, and two types of grain size, fine and coarse. In this example, the correct classification is assumed. Thus, any finding of independence or dependence is conditioned on the observation being in the correct cell. Let $n = 65$ be the number of independent samples that are classified by color and grain size, resulting in the contingency table in Table 8.6. The expected values are also shown in the table. In this example, only the sample size n is fixed in advance.

TABLE 8.6   Geological Example, Color Versus Grain Size, Observed and Expected Assuming Independence

Observed

| | Color | | | |
Grain Size	Red	Green	White	Totals
Fine	20	10	8	38
Coarse	12	5	10	27
Totals	32	15	18	65

Expected

| | Color | | | |
Grain Size	Red	Green	White	Totals
Fine	18.70	8.80	10.50	38.00
Coarse	13.30	6.20	7.50	27.00
Totals	32.00	15.00	18.00	65.00

TABLE 8.7 Biological Example: Harmful Algal Bloom Presence Counts by Station (Location) and Species

Observed

Station	Ppis	Pshum	Kmic	Csub	Cver	Hakash	Fjap	Row Totals
				Species				
IP2	0	0	4	0	0	5	0	9
LA2	0	2	5	3	4	5	1	20
LA5	2	3	5	5	4	4	0	23
LA6	2	1	5	3	0	2	0	13
Col. totals	4	6	19	11	8	16	1	65

Expected

Station	Ppis	Pshum	Kmic	Csub	Cver	Hakash	Fjap	Row Totals
				Species				
IP2	0.55	0.83	2.63	1.52	1.11	2.22	0.14	9.00
LA2	1.23	1.85	5.85	3.38	2.46	4.92	0.31	20.00
LA5	1.42	2.12	6.72	3.89	2.83	5.66	0.35	23.00
LA6	0.80	1.20	3.80	2.20	1.60	3.20	0.20	13.00
Col. totals	4.00	6.00	19.00	11.00	8.00	16.00	1.00	65.00

Source: Data from Humphries (2003).

Biological Example: Species Versus Location The data in Table 8.7 were collected by the Delaware Department of Environmental and Natural Resources. They consist of two factors, station (location) and species. A major thrust of this study is an examination of the ecology of harmful algal bloom (HAB), the proliferation of a toxic or nuisance algal species that negatively affects natural resources and humans. The stations are locations in the Delaware inland bay. Only stations with at least seven samples are included. The species (Ppis, Pshum, ..., Fjap) are known producers of HAB. The counts are the number of HABs observed for a given species at a given station. The question of interest is: Does the distribution of HAB species counts vary by station? The null hypothesis is that they do not versus the alternative that they do. Note that the table is sparse, with many zero counts. This experiment is run for a given time and 65 samples are observed. An alternative way to run this experiment is to continue sampling until a certain number of observations are observed at each station. Expected values are computed as in Table 8.6.

8.6.1 Methods of Contingency Table Analysis

This section begins with a discussion of three tests: Pearson's chi-square, the Fisher–Freeman–Halton test, and the G-test. These tests may be applied to $K \geq 2$ factor models with independent samples and nominal (unordered) data. They are also appropriate for ordered data; however, other tests may be more powerful (Liu and Agresti, 2005). For the following discussion, assume that the contingency table has two factors, A and B, with r and c levels ($r, c > 1$) respectively. The null hypothesis is that factors A and B are independent against the alternative that they are dependent. Geological and biological examples are used to illustrate methods of testing.

Pearson's Chi-Square   *Pearson's chi-square test statistic* for an $r \times c$ contingency table is the best known of the goodness-of-fit tests. The test statistic is

$$X^2 = \sum_{i=1}^{r} \sum_{j=1}^{c} \frac{(O_{ij} - E_{ij})^2}{E_{ij}}$$

where O_{ij} ($= n_{ij}$) is the number of observations in cell (i, j) and E_{ij} is the expected number of observations (assuming independence) in cell (i, j). A moderate-to-large sample is required for the statistic X^2 to be approximately chi-square. The resulting p-value is an asymptote; that is, it is based on the assumption that X^2 is chi-square. A commonly accepted guideline for judging when the sample size is adequate for the chi-square distribution to hold is that at least 80% of the E_{ij}'s ≥ 5 and none of the E_{ij}'s < 1 (Cochran, 1954). For 2×2 tables and when Cochran's guideline doesn't hold, an exact test is recommended.

For the geological example (Table 8.6), this guideline is satisfied and $X^2 = 2.09$ with $\mathrm{df} = (r - 1)(c - 1) = 2$, using the standard asymptotic computation described above. The corresponding p-value $= 0.35$. Thus, the null hypothesis of independence cannot be rejected. Pearson's chi-square test is inappropriate for the biological example (Table 8.7).

Monte Carlo Approximation to the Pearson Chi-Square Test   Instead of relying on an asymptotic result such as Pearson's chi-square test, a *Monte Carlo approach* is available in most software packages (i.e., function chisq.test in the R-project library stats), which constructs the test statistic by random sampling from the contingency table cells. For the geological example, with 10,000 replications, the p-value computed via Monte Carlo simulation is 0.38, which results in the same conclusion as before.

Two principal advantages of using the Monte Carlo estimate of the p-value are that it is unbiased and that a confidence interval can be generated. Its use is not recommended for small or sparse data sets.

Yates Correction   *Yates correction* for continuity to the chi-square approximation is recommended for a 2×2 table because the chi-square distribution is continuous. It is obtained by replacing $O_{ij} - E_{ij}$ for the four cells with $|O_{ij} - E_{ij}| - 0.5$. The test statistic becomes

$$X^2 = \sum_{i=1}^{2} \sum_{j=1}^{2} \frac{(|O_{ij} - E_{ij}| - 0.5)^2}{E_{ij}}$$

The effect of this correction factor is to reduce X^2, especially for small n, and thus increase the p-value, making the test conservative.

Fisher–Freeman–Halton Test   The *Fisher–Freeman–Halton test* is a generalization of the Fisher's exact test for an $r \times c$ unordered table, where r and/or c is greater than 2. It is also know as the *generalized Fisher's exact test*. This is implicitly a two-sided test. Like the Fisher exact test described above, this test assumes fixed marginals, however, it can be applied to contingency tables that have only one marginal total fixed. In the R-project, a generalization of the Fisher–Freeman–Halton test is in the algorithm aylmer (Aylmer is Sir Ronald Fisher's middle name). This implementation (West and Hankin, 2008) allows for the presence of structural zeros. When an exact test cannot be computed, it is almost always possible to compute a Monte Carlo result via simulation. For the geological example, this test yields a *p*-value of 0.37. For the biological example, the *p*-value $= 0.41$.

G-Test   The *G-test* (also referred to as the G^2 or *likelihood ratio test*) is a goodness-of-fit alternative to Pearson's chi-square test of independence (Sokal and Rohlf, 1995). The *G*-test is a *likelihood ratio test* (LRT), defined as

$$G = 2 \sum_{i=1}^{r} \sum_{j=1}^{c} O_{ij} \ln \frac{O_{ij}}{E_{ij}}$$

It is recommended over Pearson's chi-square when for any cell (i, j), $|O_{ij} - E_{ij}| > E_{ij}$; however, when the sample size is small, the chi-square test is recommended. The *G*-test is formed by the ratio of the maximum likelihood function under the null hypothesis to that with the constraint relaxed. For the *G*-test, the null is that the values observed are equal to the expected values for all i and j to within the sampling error against the alternative that it is not. The condition $|O_{ij} - E_{ij}| > E_{ij}$ does not hold for the geological example; however, it does for the biological example, and the test results yield $G = 24.9$ and *p*-value $= 0.13$. Thus, the null hypothesis cannot be rejected. The relatively small *p*-value may be the result of a small sample and should be interpreted cautiously.

Ordered Categorical Data   The fine and course grain size levels in the geological model represent ordered categorical data. Liu and Agresti (2005) review procedures that capture the ordered categories and are therefore more powerful than those described previously.

Loglinear Model   A *loglinear model* provides more options for examining the effects in a contingency table, including interaction between factors. It is more like regression. An example of a loglinear model for an $r \times c$ table with factors (variables) A and B is

$$\log(E_{ij}) = \mu + \lambda_i^A + \lambda_j^B + \lambda_{ij}^{AB} \qquad i = 1, \ldots, r, \, j = 1, \ldots, c$$

where μ is the overall mean of $\log(E_{ij})$, E_{ij} is the expected number of observations in cell (i, j), λ_i^A is the effect (called the *main effect*) of factor A on cell frequencies, λ_j^B is the effect (the main effect) of factor B on cell frequencies, and λ_{ij}^{AB} is the effect of the interaction between factors A and B on cell frequencies.

There are as many parameters as cells; thus, the loglinear model above will always fit the cell count data exactly. This is called a *saturated model*. If independence is assumed, the interaction term λ_{ij}^{AB} can be eliminated from the model. The power of a loglinear model is that various combinations of interactions can be tested when there are at least three variables. Goodman (1978) provides a classical treatment of the loglinear model. More modern texts include Agresti (2002), Andersen (1996), and Hosmer and Lemeshow (2000). Gotelli (2000) presents analysis methods for species co-occurrence patterns.

8.6.2 Choice of a Method

The choice of a contingency test depends on the hypothesis of interest, the table size, the sparseness of the count data, and assumptions about marginal totals. Chi-square and LRT results are typically asymptotic, which means that they are suspect for small samples. However, most algorithms provide for a Monte Carlo approximation of the chi-square and LRT tests, which is typically unbiased. We discussed earlier the cell size guidelines in which the expected value in 80% of the cells must be at least 5 and all the cells have an expected value of at least 1. We recommend using an exact test or Monte Carlo approximation rather than asymptotic approximation, however, for very large n the differences in p-values are minimal.

8.7 GENERALIZED LINEAR MODELS

The *generalized linear model* (GLM) is a more general formulation of the relationship between a response variable and explanatory variables (McCullagh and Nelder, 1989). It includes models where the response variable is discrete and distributed as binomial, Poisson, or negative binomial. In the treatment of these response variables, normal distribution assumptions are not needed. In addition, homogeneity of variance is not a requirement. Solutions are typically based on the method of maximum likelihood and are achieved iteratively. Two important models, logistic regression and Poisson regression, are presented. To simplify the notation, two predictor variables, X_1 and X_2, are used in the next examples. The results generalize to $K > 2$ variables.

8.7.1 Logistic Regression

Logistic regression is similar in form and purpose to linear regression (discussed in Chapter 4) except that the response variable is binary (Menard, 1995). The form of the logistic regression model with two predictor variables X_1 and X_2 (also known as *covariates*) is

$$\log \frac{\pi}{1 - \pi} = \beta_0 + \beta_1 X_1 + \beta_2 X_2$$

where π is the response probability defined on the open interval $(0,1)$, which can be considered the probability of success. The quantity $\pi/(1-\pi)$ is called the *odds ratio*. It is the ratio of the probability of success to the probability of failure and, not surprisingly, $\log(\pi/(1-\pi))$ is called the *log odds ratio* or *logit*(π). By simple algebra, the probability of a success is

$$\pi = \frac{\exp(\beta_0 + \beta_1 X_1 + \beta_2 X_2)}{1 + \exp(\beta_0 + \beta_1 X_1 + \beta_2 X_2)}$$

A goal of logistic regression is to correctly predict the outcome of an experiment as successful or unsuccessful. Because the equation is nonlinear, it is solved iteratively to obtain parameter estimates. Logit(π) is also know as a *link function* in GLM terminology. It connects the explanatory variables to the response variable. The logit (π) is also referred to as a *linear predictor*. The predictor variables can be continuous, ordinal, nominal, or a combination of all three. In addition, interactions between predictor variables may be incorporated into the model.

Applications include predicting the probability of the occurrence of heavy metals in groundwater, creating a hazard map to show the probability of landslides, and predicting the risk of lyme disease. Wildfire researchers use logistic regression to estimate the probability of lightning strikes. As stated in Chapter 7, logistic regression can be an alternative to discriminant analysis. Its requirements are less stringent than classical least squares because it does not require normality assumptions, nor does it assume homoscedasticity. However, error terms are assumed to be independent.

Logistic regression is applied to a subset of the Seattle landslide database (Shannon & Wilson, 2000). The data set consists of 1326 landslides from 1890 through 1999. Many factors influence landslides including topography, geology, vegetation, land use, and climate. This example predicts the size of a shallow colluvial (SC) landslide [small (S) or large (L)] from a slope height, vegetation, topography, and geological unit (described below). Other geological units, event date, presence of groundwater, and damage assessment are among the variables included in the database. A small subset of the data used in the following example is shown in Table 8.8. The purpose of this example is to see if landslide characteristics and stratigraphy can predict the size of the landslide. Landslides are restricted to those from a natural cause (as opposed to

TABLE 8.8 Subset of the 565 Observations for Shallow Colluvial Landslides from the Seattle Landslide Database Used to Fit the Logistic Model

Size	Slope Height	Vegetation	Topography	Geological Unit 1
S	10	G	MS	HC
S	10	B	MS	HC
L	70	B	SS	HC
S	60	S	SS	HC
L	65	T	SS	HC
L	25	T	MS	HF

a human triggered mechanism) and those where ground or surface water may have been involved in the trigger mechanism. The variable definitions are:

- Size is the response variable. It has two levels, small (S) and large (L), where $S \leq 10,000$ ft^2 of ground displacement and large $L > 10,000$ ft^2 of ground displacement.
- Slope height (SH) is the elevation difference (in feet) between the headscarp and the toe of the slide. It is a continuous variable, although usually estimated to within 5-f intervals in this study.
- Vegetation (V) is a categorical variable with four levels: sparse cover or bare ground (S), grass (G), brush (B), and wooded (T).
- Topography (Top) is the average slope angle. A moderate slope (MS) $\leq 40\%$; a steep slope (SS) is $> 40\%$.
- Geological unit 1 (GU 1) is the youngest geological unit involved in ground displacement. Possible subunits (levels) are: colluvium (HC), fill (HF), glacial till (QT), glacial outwash sand (QS), and lacustrine clay/slit (QC). Only two levels, HC (level 1) and HF (level 2), are present in the subset used in this example.

The coding of categorical variables is important because it affects the solution and interpretation of coefficient estimates. If each of the vegetation levels is coded separately as a 0 or 1, indicating absence or presence, respectively, the resulting matrix of predictor variables is singular, and thus the model will not yield a unique solution. Instead, vegetation S is coded as a zero (i.e., the absence of G, B, and T). The other three levels are coded as 1 if present and 0 if absent. So when the G, B, and T are zero, vegetation is S. It is conceivable that vegetation can be considered an ordered categorical variable. In geological unit 1, HC is coded as 1 when present and 0 when absent. Topography can be considered an ordinal variable, however, given that there are only two levels, the parameter estimate is the same. Of course, the size of the landslide (small or large) is a binary variable with two levels.

The candidate model is

$$\text{logit}(\pi) = \beta_0 + \beta_1 \text{SH} + \beta_2 \text{V\_G} + \beta_3 \text{V\_B} + \beta_4 \text{V\_T} + \beta_5 \text{Top} + \beta_6 \text{GU\_1}$$

where π is the probability of a landslide, V_G is grass vegetation, V_B is brush vegetation, V_T is wooded vegetation, Top is topography, and GU_1 is the youngest geological unit. Parameter estimates are obtained using R-project function glm (general linear model) with the response variable defined as binomial or, equivalently, the link function defined as logit. The output is presented in Table 8.9.

The output (Table 8.9) is similar to that of linear regression. The slope height is highly significant, as is geological unit 1. The positive sign of this estimated coefficient implies that fill (HF) is more likely than colluvium (HC) to be associated with a large landslide. It is important to recall that significance of coefficients does not

TABLE 8.9 Logistic Regression Output for the Seattle Landslide Example

Coefficient	Estimate	Std. Error	z-Value	$\Pr(>|z|)$	Signif.[a]
(Intercept)	− 3.8556	0.6739	− 5.7220	0.0000	***
Slope height	0.0329	0.0043	7.6500	0.0000	***
Vegetation G	1.0448	0.6938	1.5060	0.1321	
Vegetation B	0.1598	0.6355	0.2510	0.8015	
Vegetation T	0.9787	0.6294	1.5550	0.1199	
Topography	− 0.9833	0.3085	− 3.1870	0.0014	**
Geological unit 1	0.9489	0.2794	3.3970	0.0007	***

[a] ** 0.01, *** 0.001.

Deviance Residuals

Min.	Q_1	Median	Q_3	Max.
− 1.713	− 0.609	− 0.389	− 0.222	2.482

Null deviance: 527 on 570 degrees of freedom.
Residual deviance: 425 on 564 degrees of freedom.
AIC = 439.

imply a causal relationship with the response variable. According to Rex Baum of the U.S. Geological Survey (personal communication, July 28, 2008) the fill is not causative but, rather, just reflects an association. Topography is the next most significance variable. The negative sign of the coefficient implies that larger landslides are more likely to be associated with moderate slopes (MSs). The next two variables in terms of decreasing levels of significance are wooded vegetation (T) and grassy vegetation (G). The positive slopes imply that these are associated with the larger landslides. In addition to these predictor variables, its interaction effects can be incorporated into the model.

A new term used in logistic regression is *deviance*. To understand it, the likelihood function is defined. Suppose that $y_1, \ldots, y_n$ represent the observed values of independent random variables $Y_1, \ldots, Y_n$ from a binomial process, where $y_i \sim \text{binomial}(m_i, \pi_i)$ with sample size m_i and probability of success π_i. So y_i is 1 or 0 from a binomial distribution with probability of success π_i and m_i trials. If $f(y_i)$ is the density function for the ith random variable, the likelihood function L is

$$L = \prod_{i=1}^{n} f(y_i)$$

It is usually more convenient to write the log-likelihood function $\ln(L)$. Some algebra shows that it is equal to

$$\ln(L(\pi_1, \ldots, \pi_n \mid y_1, \ldots y_n)) = \sum_{i=1}^{n} \left[y_i \ln\left(\frac{\pi_i}{1 - \pi_i}\right) + m_i \ln(1 - \pi_i) \right]$$

Since each observation (landslide) is considered separately, $m_i = 1$ for all i. The maximized likelihood function (MLE) of $\ln(L)$ finds parameter estimates so that $\ln(L)$ is as large as possible given the data. The coefficients are a function of π_i,

$$g(\pi_i) = \log\left(\frac{\pi_i}{1 - \pi_i}\right) = \sum_{j=0}^{k} \beta_j X_{ij} \qquad i = 1, \ldots, n$$

The deviance up to a constant is $-2\ln(L)$. There are two deviance terms, null deviance and residual deviance. The *null deviance* (also called the *reduced model*) contains an offset (equivalent to the mean) and an intercept only if one is present. The *residual deviance* results from a comparison of two models. When the null model is true, under certain assumptions of asymptotic convergence,

$$\text{likelihood ratio (LR)} = X^2 = -2\ln\left(\frac{L_N}{L_F}\right)$$

is approximately chi-squared, where L_N and L_F are the MLEs of the null model and full model, respectively.

For the Seattle landslide data, $-2\ln(L_N) = 527$ and $-2\ln(L_F) = 425$, which implies that $X^2 = 527 - 425 = 102$ with $570 - 564 = 6$ df. This implies that the full model is highly significant. This result must be interpreted cautiously. A more appropriate use of the residual deviance is in comparing different models: for example, seeing if the addition of geologic unit 2 significantly reduces unexplained variability. The analysis above can be performed with any two models, say M_1 and M_2, where the parameters in M_2 are a subset of those in M_1.

Akaike's Information Criterion (AIC) is

$$\text{AIC} = 2k - 2\ln(L_F)$$

where k is the number of coefficients in the model. The $2k$ serves as a penalty function on the number of coefficients. In the landslide example, there are seven coefficients, including the intercept, and therefore $\text{AIC} = (2)(7) + 425 = 439$. Lower values of AIC indicate a preferred model. As with the residual deviance, AIC is useful when comparing models. Other forms of model evaluation, such as residual plots used in linear regression, do not translate to logistic regression because the response variable is either 0 or 1.

One way to judge performance is to construct a classification table (Table 8.10). The cut value in this context is a probability arbitrarily chosen at 0.5 to classify a landslide as small or large. If the prediction probability is 0.5 or greater, a landslide is classified as large, and if it is less than 0.5 it is classified as small.

Table 8.10 shows the correct and incorrect classifications that occur at a cutoff of 0.5. Of the 99 large landslides, 25 are classified correctly at this cutoff level, as are 455 of the 472 small landslides. Overall, 84% ($100 \times 480/571$) are classified correctly; however, the fact that only 25% of the large slides are classified correctly is a concern,

TABLE 8.10   Landslide Classification with Cut Value 0.50

	Model Classification		
Size of Landslide	Small (Cut Value < 0.5)	Large (Cut Value ≥ 0.5)	Actual Totals
Small	455	17	472
Large	74	25	99
Totals	529	42	571

since predicting large landslides is the top priority. Before returning to this subject, a graphical view of the results is shown in Figure 8.9. The cut value of 0.5 is represented by the horizontal line at 0.5 on the vertical axis. The vertical line at a linear predictor value of 0 classifies landslide size according to the results in Table 8.10. The S-shaped curve shows the linear predictors versus the probability of a large slide. This graph, like Table 8.10, shows that the model does a reasonable job of predicting small landslides but not large ones. To increase the probability of classifying large land-slides correctly, the model may need to be reformulated. Given the existing model and data, a way to increase the probability of predicting a large landslide correctly is to lower the cut value, but the penalty is to increase misclassification among small landslides. For example, suppose that the cut value is lowered to 0.25. The resulting classification is shown in Table 8.11. Clearly the probability of correctly classifying a landslide as large has increased from 25% to 46%, but this is accompanied by an

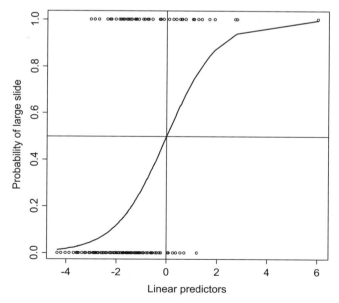

FIGURE 8.9 Probability of a large slide versus linear predictors (Table 8.10). The circles at the top (probability = 1.0) are the large slides plotted against their linear predictor; the circles at the bottom (probability = 0) are the small slides plotted against their linear predictor.

TABLE 8.11   Landslide Classification with Cut Value 0.25

	Model Classification		
Size of Landslide	Small (Cut Value < 0.25)	Large (Cut Value ≥ 0.25)	Actual Totals
Small	396	76	472
Large	53	46	99
Totals	449	122	571

increase in the misclassification of small landslides. Overall, 77% of the slides are classified correctly. Costs of misclassification errors can be used as weights when they can be determined. The ideal model is one that classifies every landslide in the data set correctly and, more important, can be used to predict the occurrence of a new landslide.

Typical regression diagnostics used in Chapter 4 are not appropriate here because the response variable is binomial rather than continuous. However, various residuals, including deviance, studentized, and Pearson, can be computed. Figure 8.10 shows the distribution of the deviance residuals. As noted, deviance is a measure of the discrepancy between the fitted and null models; thus, each observation contributes a unit to that measure. Since the values observed are either 0 (small) or 1 (large), a bias results from underestimating the expected fits. This is illustrated by the large number of slightly negative residuals (Figure 8.10).

The Seattle landslide data contains other variables that allow further insightful modeling and analysis. These include the date of the landslide and amount of damage.

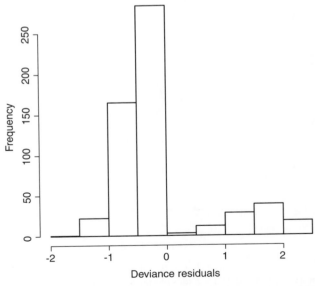

FIGURE 8.10   Histogram of deviance residuals from Seattle landslide data.

Several types of landslides are included in these data. Chleborad et al. (2006) and Coe et al. (2004) use Poisson and binomial probability models to estimate future occurrences of precipitation triggered landslides. Baum et al. (2005) performed a hazard assessment for the Seattle, Washington area.

Another tool for internal validation is the bootstrap. The procedure for implementing the bootstrap is similar to the model-based resampling in linear regression, where for each bootstrap replication a new draw is made from the distribution of errors and added to the estimated model to generate a new dependent variable. In logistic regression, the following steps are performed to generate a new set of n response variables:

1. Predict the response from the initial estimated model. The response is a number between 0 and 1. Call the ith predicted value π_i.
2. Generate n continuous $(0,1)$ uniform random numbers. Call the ith number u_i.
3. Let y_i^* be the ith bootstrap response variable, $i = 1, \ldots, n$. If $u_i < \pi_i$, $y_i^* = 0$; otherwise, $y_i^* = 1$.

After completing these steps, perform a bootstrap estimate. Examples of bootstrap estimates for logistic regression models have been given by Davison and Hinkley (1997).

Multinomial logistic regression is an extension of logistic regression to the situation where the response consists of more than two classes. When the response can be ranked, a model called *ordinal logistic regression* can be used. There is also a stepwise version of logistic regression, which is similar to classical stepwise regression.

8.7.2   Poisson Regression

Poisson regression is similar to logistic regression except that the response variable Y is Poisson distributed with mean λ and is conditioned upon predictor variables X_1 and X_2 for the two-explanatory-variable case. Thus, the probability function

$$\Pr(Y = k | \lambda, X_1, X_2) = \frac{e^{-\lambda}\lambda^k}{k!} \qquad k = 0, 1, \ldots$$

The form of the Poisson regression model is

$$\log(\lambda) = \beta_0 + \beta_1 X_1 + \beta_2 X_2$$

An application is to model the number of tropical cyclones in the southwestern Pacific Ocean during the season as functions of water current and temperature. Other applications include a model of the number of occurrences of a virus or other disease as a function of climate and/or spatial location.

Sometimes it is necessary to model a rate, such as rate of incidence of toxic poisoning per square kilometer, which may be adjusted by the number N of toxic waste sites by county. This model may be formulated as

$$\log(\lambda) = \log(N) + \beta_0 + \beta_1 X_1 + \beta_2 X_2$$

An example is derived from a subset of data from the U.S. Geological Survey World Energy Assessment Team (2000). The data consist of the number of oil fields discovered (N) and their corresponding total size V in millions of barrels of oil (mmbo) for 56 accumulation units (Table 8.12) less than 900 mmbo. The reason for this cutoff is that there appear to be separate processes at work, one for small fields (under 900 mmbo) and another for fields of at least 900 mmbo. Accumulation units are homogeneous regions where oil and/or gas may be found. For example, a sedimentary basin may be an accumulation unit. The size distribution of V (not shown) is highly right-skewed, and thus $U = \ln(V)$ is used as the explanatory variable. The Poisson regression model is

$$\ln(E[N]) = \beta_0 + \beta_1 U$$

The output from this model (Table 8.13) is similar to that of the logit model.

TABLE 8.12 Number of Fields Discovered Versus Total Size of Fields (Millions of Barrels of Oil

N	V (mmbo)	N	V (mmbo)	N	V (mmbo)
5	7	17	124	16	322
3	12	7	134	7	332
7	14	2	156	14	406
2	20	7	169	9	426
4	46	20	176	3	454
4	46	5	179	14	474
7	53	5	190	9	484
3	54	6	210	18	562
13	57	16	231	7	590
3	77	12	233	19	617
5	77	18	245	12	633
4	78	8	246	11	637
3	80	19	263	34	672
10	94	23	276	24	677
11	101	7	278	9	679
10	103	7	285	34	812
12	106	27	292	13	860
7	110	16	295	28	891
7	120	16	302		

Source: Data from U.S. Geological Survey World Energy Assessment Team (2000).

TABLE 8.13   Poisson Regression Output Estimating the Number of Fields Found Versus Field Size

Coefficients	
(Intercept)	U
0.1741	0.4165

df: total (null) = 55; residual = 54.
Null deviance = 272.2.
Residual deviance = 167.7.
AIC = 399.2

The deviance residuals are similar to the residuals in least-squares regression. As with logistic regression, the AIC is an adjusted log likelihood ratio. A goodness-of-fit test on $U = \ln(V)$, the slope parameter estimate, is

$$\frac{\text{null deviance } - \text{ residual deviance}}{\text{total (null) df } - \text{ residual df}} = \frac{272.2 - 167.7}{55 - 54} = 104.5.$$

If the null is true, this statistic is approximately $\chi^2(1)$; however, it is clearly highly significant, indicating a nonzero slope and thus an association between U and $\ln(N)$. A graphical procedure to assess goodness of fit is to examine the scatter of data points against the fitted model (Figure 8.11).

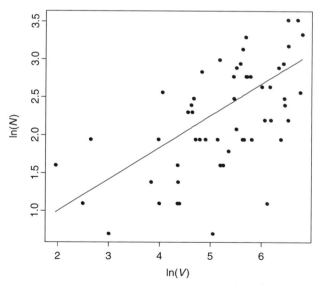

FIGURE 8.11   Plot of $\ln(V)$ versus $\ln(N)$ and a line representing the Poisson regression fit.

Another useful link function is the *inverse normal* or *probit*, defined as $\eta = F^{-1}(\pi)$, where F^{-1} is the inverse of the cumulative normal distribution and π is the probability of success in a binomial process. A probit transformation reflects the belief that the underlying process is normal even though only two categories are observed. For information on these and other forms of the general linear model, see the book by McCullagh and Nelder (1989).

8.7.3   Weight-of-Evidence Models

Weight-of-evidence was formally developed as a statistical procedure by I. J. Good in 1950 based on Bayes theorem (Bernardo et al., 1985). In the context of American jurisprudence, often it is the responsibility of a jury to determine if sufficient evidence has been presented in court by the prosecution to convict a person accused of a crime. The judge and/or jurors assign weight to evidence based on its believability. Ultimately, they make a decision based on a standard of proof and if in their judgment the weight of evidence is in support of or in opposition to the proposition.

Weight-of-evidence models for mineral resource assessment use conditional probabilities to associate attributes on a series of geological maps and the locations of known mineral deposits (Agterberg et al., 1990, 1993). As a simple example, let M be the event of mineralization, having two outcomes, yes and no, and E the event of evidence in support of mineralization being present. The investigator is interested in $Pr(M\,|\,E)$. Using Bayes' theorem, this can be expressed as

$$Pr(M\,|\,E) = \frac{Pr(E\,|\,M)Pr(M)}{Pr(E)}$$

Similarly, define $\overline{M}$ as the event of no mineralization and $\overline{E}$ as the event of no evidence in support of mineralization. Thus,

$$Pr(\overline{M}\,|\,E) = \frac{Pr(E\,|\,\overline{M})Pr(\overline{M})}{Pr(E)}$$

Now define the odds ratio as

$$O(M|E) = \frac{Pr(M\,|\,E)}{Pr(\overline{M}\,|\,E)} = \frac{Pr(E\,|\,M)Pr(M)}{Pr(E\,|\,\overline{M})Pr(\overline{M})}$$

The log-odds ratio is

$$\ln(O(M|E)) = \ln\left(\frac{Pr(E\,|\,M)}{Pr(E\,|\,\overline{M})}\right) + \ln\left(\frac{Pr(M)}{Pr(\overline{M})}\right)$$

where

$$W^+ = \ln\left(\frac{\Pr(E\,|\,M)}{\Pr(E\,|\,\overline{M})}\right)$$

is a positive weight in support of E if $\Pr(E\,|\,M) > \Pr(E\,|\,\overline{M})$. The log-odds ratio for mineralization is

$$\ln(O(M)) = \ln\left(\frac{\Pr(M)}{\Pr(\overline{M})}\right)$$

The products from this analysis are maps that show the probabilities of mineral deposit occurrences and the associated uncertainties. Weight-of-evidence methods often use binary data (Bonham-Carter, 1994). That is, a 0 is used to signify the absence of, say, a rock type and a 1 is used to signify the presence of the rock type. Real numbers can be used to represent geochemical and other variables.

An important element to note with regard to constructing weights of evidence for a map and/or map unit is that no weight can be established unless a deposit occurs in the region of interest. In the broad sense of resource assessment, this is an important point to consider. The analyst has to be diligent in searching for analog areas in which the weights can be established.

With the advent of geographic information systems (GISs), the application of weight-of-evidence analysis expanded rapidly in mineral resource assessment (Bonham-Carter, 1994; Porwal et al., 2001) and to many other fields, such as landslide hazards (Neuhauser and Terhorst, 2007).

The combining of data from multiple maps in the weight-of-evidence approach is achieved using Bayesian analysis and therefore requires the assumption of conditional independence between the information taken from maps. The structure of the data must be considered carefully to avoid incorrect estimation of the probabilities of occurrence of undiscovered mineral deposits (Bonham-Carter, 1994). Singer and Kouda (1999) provide a rather thorough exposition of the problem of using Bayes' rule and multiple maps and warn of the proneness, particularly when the prior probabilities are low, for the weight of evidence to overestimate the probabilities of occurrence of undiscovered mineral deposits.

8.8   SUMMARY

Discrete distributions and processes commonly observed in the earth sciences include the binomial, the Poisson, and the negative binomial. The binomial process involves a fixed number of trials, whereas in a negative binomial, the interest is in number of trials needed to achieve a given number of successes. A negative binomial process stops when a given number of successes are achieved. Events are assumed to be independent. If the outcome of a given trial is influenced by the outcome of previous trials, the trials are not independent, and neither the binomial nor the negative binomial model is appropriate. The Poisson process is important because it characterizes a completely random process where the number of arrivals (outcomes) in a bounded interval after time t is independent of number of arrivals before time t. In investigations of time between earthquakes, or locations of toxic poisoning, a starting point is to see if these events occurred at random. If not, the investigator should look for patterns which often involve the search for clusters, or regularity. Distance, area, and volume can be substituted for time in the examination of random processes.

The investigation of point processes involves point counting, often after placing a grid on the spatial region of interest. A chi-square goodness-of-fit test can be applied to tests for conformity with a Poisson or other discrete distribution. Other tools to investigate spatial patterns are the Clark and Evans nearest-neighbor index (Clark and Evans, 1954) and Ripley's K.

Lattice data, an array of values, is another component of spatial modeling. Lattice data are often defined by geopolitical boundary areas such as counties.

Binomial data are often transformed into proportions and where the sample size is large; the normal approximation can be used for confidence interval estimation and hypothesis testing. For small samples, exact tests need to be applied. For two populations, Fisher's exact test may be used, especially when row and column marginals are fixed.

Contingency tables are used to categorize count data into levels within factors. In the analysis of contingency data, the usual initial null hypothesis is independence between factors. If this is rejected, further investigation may be warranted to understand the nature of the dependency. Numerous tests are available for contingency tables, including Pearson's chi-square, likelihood ratio, exact, and Monte Carlo approximations. The choice of an appropriate test is a function of the size and sparseness of the table, number of observations, assumption about marginal row and column totals, and levels of factors.

Two important special cases of the general linear model are logistic regression and Poisson regression. In logistic regression, the dependent variable is a log odds ratio. In Poisson regression, the dependent variable is assumed to be Poisson distributed.

EXERCISES

8.1 Suppose that a data set consists of 500 observations. One variable is sand grain size, measured in continuous units, and the other is distance from shoreline at low tide, again in continuous units. The question to be addressed is: Is there a relationship between grain size and distance? Is it better to model and analyze the raw data or construct a contingency table using fine, medium, and course grains and some distance intervals? Compare and contrast the use of these two approaches.

8.2 Pine beetle infestation is a serious problem in many regions of the United States. Suppose that an initial estimate in southwestern Colorado suggests that 50% of pine trees are infected. To investigate this conjecture, a forester decides to take a random sample of 100 trees.

(a) What type of process results from this infestation? Explain.

(b) Describe how a forester might construct a sampling plan.

(c) If the forester finds that 40% of the trees are infected, is the conjecture supported? Explain.

The following data are for Exercises 8.3 to 8.5. A small city is partitioned into 50 blocks. Table 8.14 shows the distribution of reported cases of flu by city block. There are four blocks in which no occurrences are reported, 10 blocks where one occurrence is reported, and so on.

TABLE 8.14

Number of Blocks	Number of Occurrences
4	0
10	1
12	2
9	3
7	4
5	5
3	6

8.3 What model might one use to fit these data? What assumption is required?

8.4 Fit the data to the model selected, compute a goodness-of-fit measure, and draw conclusions.

8.5 What additional information is desirable, and how can it help?

8.6 A question raised is to see if there is any correlation between the number of energy-saving devices in a household and the viewing of public television 8 or more hours per week. Two possible ways to conduct this study are to decide in advance to sample 15 public TV watchers and 15 who do not watch 8 or more hours, and then record the response by zero, one, two, three, or more energy-saving devices. Another way is to sample 30 people who watch TV and record the number of public TV and nonpublic TV watchers and the number of energy-saving devices according to the classes above. Compare and contrast the two approaches. Should the same method of analysis be used in each? Explain.

8.7 Refer to Exercise 8.6. Suppose that the data are collected according to the second of the two approaches, with the results shown in Table 8.15. Is there

TABLE 8.15

	Number Energy Saving Devices			
	0	1	2	>=3
Public TV >= 8 hr	5	8	3	1
Public TV < 8 hr	7	6	0	0

evidence that public TV watchers have more energy devices in their homes? (*Careful!* What model should be used? What conclusions can be drawn?)

8.8 Is it reasonable to assume that the number of airplanes arriving at London's Heathrow Airport in a given time period is Poisson distributed? Explain.

8.9 The area around an abandoned gold mine is suspected of being contaminated by high levels of mercury in the soil. Suppose, by analogy, the probability of any given soil sample being above a threshold is estimated to be 0.15. How many samples need to taken to find five contaminated samples? What assumptions need to be made to answer this question? How should one sample?

For Exercises 8.10 and 8.11, use the data given in Table 8.16. The results of a soil sampling program in the state of Georgia found that at the sampling locations, Al_2O_3 is greater than 13%.

TABLE 8.16

Lat.	Long.	Lat.	Long.	Lat.	Long.
31.500	− 84.383	34.483	− 84.567	34.933	− 84.367
32.383	− 83.233	34.483	− 84.567	34.933	− 84.367
34.200	− 84.600	34.517	− 84.550	34.933	− 84.367
34.367	− 84.400	34.483	− 84.567	34.850	− 84.367
34.367	− 84.400	34.483	− 84.583	34.883	− 84.150
34.133	− 84.517	34.633	− 84.517	34.783	− 83.917
34.133	− 84.517	34.583	− 84.667	34.917	− 83.900
34.350	− 84.400	34.667	− 84.600	34.917	− 83.900
34.150	− 84.133	34.767	− 84.583	34.867	− 83.933
34.150	− 84.133	34.667	− 84.500	34.867	− 83.667
34.267	− 83.900	34.750	− 84.717	34.917	− 83.883
34.267	− 83.900	34.750	− 84.717	34.917	− 83.650
34.167	− 83.883	34.750	− 84.717	34.167	− 83.883
34.467	− 84.417	32.750	− 84.017	34.633	− 84.517
33.817	− 84.550				

8.10 Do these data suggest that high values (above 13%) of Al_2O_3 are clustered? Discuss and use two methods to answer this question.

8.11 What assumptions need to be made for the conclusion to be valid?

For Exercises 8.12 and 8.13, a researcher is interested in studying trace elements of arsenic in drinking water in the city of Denver, Colorado. A sample of 20 people is taken, and through hair follicle analysis it is found that eight have trace amounts of arsenic.

8.12 Does this suggest that 50% or more of the population in the city of Denver has traces of arsenic levels at the 5% significance level? State reasons for use of the test statistic chosen.

8.13 Discuss issues that may affect the results of this study.

8.14 Consider the landslide data discussed in this chapter. (A subset is given in Table 8.8.) What is the effect of using the location of the landslide as an explanatory variable? Is there a way to use the spatial data? Explain.

8.15 Assume that an earthquake of 6.7 or greater on the Richter scale occurs in the San Francisco Bay area every 65 years. If a Poisson process is assumed, what is the probability of an earthquake of magnitude 6.7 or greater occurring in the next 30 years? Is a Poisson process reasonable? Comment.

8.16 A small portion of historical harricane data is shown in Table 8.17. A concern is the possible increase in the number and/or intensity of hurricanes hitting the southeastern United States. Is this a reasonable concern? Explain.

TABLE 8.17

Year	Month	States Affected	Highest Saffir–Simpson Category	Central Pressure (mb)	Max. Winds (kt)	Name
2003	Sept.	NC, VA	2	957	90	Isabel
2004	Aug.	NC	1	972	70	Alex
	Aug.	FL, SC, NC	4	941	130	Charley
	Aug.	SC	1	985	65	Gaston
	Sept.	FL	2	960	90	Frances
	Sept.	AL, FL	3	946	105	Ivan
	Sept.	FL	3	950	105	Jeanne
2005	July	LA	1	991	65	Cindy
	July	FL	3	946	105	Dennis
	Aug.	FL, LA, MS, AL	3	920	110	Katrina
	Sept.	NC	1	982	65	Ophelia
	Sept.	FL, LA, TX	3	937	100	Rita
2005	Oct.	FL	3	950	105	Wilma

Source: Data from the U.S. National Oceanographic and Atmospheric Administration, http://www.aoml.noaa.gov/hrd/hurdat/ushurrlist.htm.

8.17 In the biological example (Table 8.7), the null hypothesis cannot be rejected. What can one infer from this?

9 Design of Experiments

9.1 INTRODUCTION

There are two broad categories of design: sampling design and design of experiments. In the earth sciences, design of experiments is often overlooked. There have been many large, costly projects for which, with minimal planning, crews have been sent into the field to collect data. Too often these efforts result in data that are less valuable than if proper planning and design had been done beforehand. Design involves selecting samples to satisfy the requirements of the problem under consideration. A principal benefit of design is to make the results of a study more generalizable and thus more useful than if samples are collected helter-skelter. Design criteria may include achieving a needed level of precision, increasing the efficiency of estimation, minimizing the cost of sampling, and/or making the results robust to departures in assumptions.

Sampling design is typically used in surveys and field studies (also called *observational studies*), where control of the factors that may influence the outcome of the experiment is not always possible. Field studies include sampling people, animals, airborne contaminants, soils, and permafrost thickness. Conversely, design of experiments usually occurs in a laboratory or pilot plant under controlled conditions. An example is the study of the effect of waves on shorelines in a wave laboratory, where amplitude, frequency, depth, and other attributes are controlled and set to fairly precise values. Many of the traditional design of experiments have been developed to study agricultural, genetic, and industrial processes.

Data collection in most earth science applications is costly and time consuming. Often there is only one opportunity to collect the data. Thus, proper design is paramount.

9.2 SAMPLING DESIGNS

Most sampling designs in the earth sciences involve a spatial component. Indeed, this is also true in human geography, where sampling of individuals or households may be of interest. In most earth science disciplines, sampling is driven by the need to estimate some quantity over a spatial region. For example, a petroleum geologist may

Statistics for Earth and Environmental Scientists, By John H. Schuenemeyer and Lawrence J. Drew
Copyright © 2011 John Wiley & Sons, Inc.

wish to develop a design to site wells or seismic lines to find oil. A geochemist may wish to create a map of the regional deposition of arsenic or mercury. A climatologist may need to understand patterns of precipitation in the Amazon.

9.2.1 Basic Principles

We will illustrate basic principles of sampling design through study of the deposition of arsenic. Arsenic occurs in nature and as a by-product of mining and other chemical processes. In high levels it can cause serious health problems. Health and earth scientists have conducted numerous studies to identify levels of arsenic. Geographic areas of interest are as small as a few city blocks or as large as a country. For example, see the USGS report by Hinkle and Polette (1999) for a study of arsenic in groundwater.

Key questions that must be answered prior to conducting a survey are:

- What is the population to be sampled? In an arsenic study, is it streams, stream beds, an entire area? A clear definition of the population, stated as precisely as possible, is important.

- What is the sampling frame? Often, one does not have access to the entire population—some of it may be off-limits because it is a cemetery, a shopping center, a national park, or people who are institutionalized. The *sampling frame* is that part of the population that is eligible to be sampled. It may be a list of people or places, or most often in the earth sciences, a map.

- What is the elementary or sampling unit? In the arsenic example, it may be 50 grams of soil or a vial of water. In the case of an energy use survey conducted by a government agency, it can be a business, a household, or an individual. In other studies, it may be a tree, a branch, or a leaf. It is clearly important to identify the elementary unit prior to conducting the survey. An analysis or measurement of the sampling unit yields the variable(s) of interest.

- What attributes are to be measured or analyzed? In a groundwater study of arsenic, does the investigator only care about arsenic? She probably also wants to know about the presence and strength of other elements. In most earth science applications, the major cost in obtaining the sample is collection, not the analysis of a few additional attributes.

- How large a sample is needed? Clearly, this is a function of the desired level of precision and variability within the system being studied.

- Where samples are to be taken? This is a key question in survey design. Samples can be taken at random locations, stratified, in clusters, or in numerous other ways. Considerations, addressed in the next section, include cost and efficiency.

- Are samples to be taken with or without replacement? With replacement means that the population size remains constant throughout the sampling process.

Once a sampling frame has been decided upon, an important question to be answered is: How should the sample be collected? The initial temptation is to obtain a convenience sample by asking colleagues, friends, or relatives. Another example of a convenience sample is one collected near roads or in interesting places. Unfortunately, the resulting sample is rarely representative of the population of interest. It is difficult to obtain a representative sample without benefit of a sampling plan, even when one sets out conscientiously to do so. Investigators tend to be attracted to things that they consider interesting and therefore sample them more frequently than items that have attributes of lesser interest. Indeed, even if a representative sample is achievable via convenience sampling, it is difficult to convince anyone that it is so. The best way to achieve a representative sample is by taking a *random sample*.

A variety of sampling schemes incorporate randomness. The choice is based on such factors as cost, spatial distribution of the data, variability of the data, and timing (U.S. Environmental Protection Agency, 2002). We begin our discussion with a simple random sample and then discuss systematic, stratified, and cluster sampling. To keep the notation simple, it is assumed that the sampling frame is equivalent to the population.

9.2.2 Simple Random Sample

A *simple random sample* is one that gives each element in the population an equal chance of being selected. This is conceptually the easiest sampling plan to design and explain, but it may not be the easiest or least costly to implement. Three key issues are sampling with or without replacement, deciding on the sample size n, and implementing the sampling design.

A detailed discussion of sampling with and without replacement is deferred until the next section; however, *sampling with replacement* implies that a sample of size n begins by drawing a single element of the population. Before drawing the second element of the population, the first is put back and the population is randomized. When $n \ll N$ ($\ll$ means "is small compared to"), the simpler assumption of sampling with replacement is used because the chance of selecting the sample element multiple times is quite small. Unless stated otherwise, assume sampling with replacement, even though in practice it is seldom done.

How does one decide on n? Recall from Chapter 3 that a $(1 - \alpha)\%$ CI on μ_y is

$$\bar{y} \pm \frac{z_\alpha \sigma_y}{\sqrt{n}}$$

assuming that the sample size is sufficiently large to justify the use of the z (standard normal) distribution. Let B be defined as the error of estimation

$$B = \frac{z_\alpha \sigma_y}{\sqrt{n}}$$

Then

$$n \geq \left[\frac{z_\alpha^2 \sigma_y^2}{B^2} \right]$$

where n is the smallest integer not less than the quantity in brackets. Note that z_α is sometimes referred to as a *reliability coefficient*; $z_\alpha = 3$ assures virtual certainty, while $z_\alpha = 2$ corresponds approximately to a 95% CI. Clearly, the choice of z_α and B is problem-specific. In practice and when possible, the investigator should increase the size of n by 10% or 20% because σ_y^2 is usually estimated from historical information, analogy, or a pilot study. Often, a few observations will be missing from the resultant sample, for various reasons. The formula above illustrates the following points:

- A greater degree of confidence (higher z_α^2) requires a larger sample size.
- More variability (i.e., a greater σ_y^2) requires a larger sample size.
- A smaller B requires a larger sample size.

If the goal is to find a sample size n necessary to estimate a proportion π, the corresponding expression for sample size is

$$n \geq \frac{z_\alpha^2 (1 - \pi)\pi}{B^2}$$

where an estimate of π is required.

All of this is simple and straightforward. When sampling from a list, a simple random sample is easy to achieve. If the list consists of N items and one wants to take a sample of size n, one can randomly permute the list and use the first n items. Suppose, however, that an investigator wants to determine arsenic levels in soil in New Hampshire and decides that a sample of 36 is reasonable. How is such a sampling scheme designed? To obtain a simple random sample of size 36 in two dimensions, x and y, generate 36 random numbers for x in the range (x_{min} to x_{max}). Do the same for y and plot the (x,y) pairs on a map. See, for example, Figure 9.1. This, however, is the easy part. Someone has to locate these points physically, a job made slightly easier by a global positioning system (GPS) but not trivial. Following this, someone needs to take the samples. In addition, there are issues of what to do if the sample point is in the middle of a lake or a shopping center. These issues need to be considered in the design phase. There may be other factors that make simple random sampling a poor choice. Suppose, for example, that prior information suggests that arsenic variability in the lower part of the state is less than in the upper part. It hardly makes sense to sample both parts at the same density.

9.2.3 Systematic Sampling

A *systematic sampling design* begins by selecting an element at random but then is followed by sampling at even intervals. Systematic sampling may involve sampling from a list, along a line or transect, or on a grid. Several examples are discussed briefly.

- *Transect.* A transect is basically a line segment. An example is to collect vegetation data at 15-m intervals perpendicular to a shoreline (see U.S. Geological Survey Upper Midwest Environmental Sciences Center, 2005).

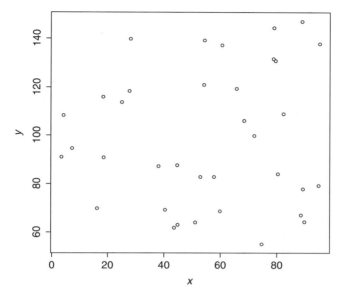

FIGURE 9.1   Random sample of size 36 in two dimensions.

- *Seismic grid.* Seismic grids used in geophysics are geomagnetic lines across a region. The U.S. Geological Survey's oil and gas resource estimate of the National Petroleum Reserves Alaska is based partially on information obtained from a seismic grid (Houseknecht, 2002). The patterns of geophysical lines allow an investigator to determine the size of the largest object, say an oil field, which the grid may contain. Kaufman (1994) discusses how to determine the probability of an object being missed by a seismic grid.
- *Grid drilling and sampling.* Drilling on a grid has long been used in earth science disciplines. For example, see Drew (1967) for a discussion of grid drilling in the search of oil and gas deposits. Grids have been square, rectangular, and hexagonal. Grid patterns are often used in soil sampling.

To further illustrate systematic sampling, consider the transect example. Suppose that a transect is 150 m long (N) and a decision is made to sample every 15 m(k). Thus, the sample size $n = N/k$, or 10. The initial starting point on the transect p_0 is determined by selecting a uniformly distributed random variable in the range (0,15). Every subsequent point is located at $p_0 + 15i$, where $i = 1,\ldots, n - 1$. This is a statistical design because the first sampled observation along the transect between 0 and 15 m is selected at random.

Subsequent sample points are determined by adding 15 m to the starting value. So if the initial random number (the starting point) is 6 m, sampling occurs at 6, 21, . . ., 141 m. The only point that is randomly chosen is the first one. How does systematic sampling compare with simple random sampling? The variance of the mean of a systematic random sample is

$$\text{Var}(\bar{y}_{sy}) = \frac{\sigma_y^2}{n}[1 + (n-1)\rho]$$

where ρ is a measure of correlation between pairs of elements within the same systematic sample. If ρ is close to 1, $\text{Var}(\bar{y}_{sy})$ is approximately σ_y^2 and thus greater than that of a simple random sample. For n large and ρ close to zero, the variance is close to that of a simple random sample; however, systematic sampling will usually be simpler and less costly to implement. A difficulty is estimating ρ. It can be estimated, but this requires knowledge of all of the k samples or prior knowledge (see, e.g., Scheaffer et al., 1996). If there is prior knowledge of spatial correlation, the sampling plan may be altered to reflect this phenomenon (see Chapter 6). After data have been obtained from a stratified sampling design, a semivariogram should be computed to investigate spatial correlation (Chapter 6).

There are two main reasons for choosing systematic sampling. It is easy to implement, and when used as a search vehicle for regularly shaped objects, such as an oil pool that may be represented by an ellipse, probabilistic statements can be made. A major potential downside to systematic sampling occurs when a cyclic effect in the data and the sampling interval coincide.

9.2.4   Stratified Sampling

As the name suggests, *stratified sampling* is sampling within strata. A *geologic stratum* is a layer of sedimentary rock having approximately the same composition throughout. Typically, strata are arranged one layer on top of another. However, there are numerous ways to define strata. In geography, a climate zone may be a stratum. Soil types may constitute strata. In sampling people, classifications (strata) such as occupation, education, income, and gender are often used.

Why sample by stratum? There are three main reasons. One is if the variability of the unit of interest differs significantly by stratum. The second is if one or more of the strata is small and may be insufficiently sampled by simple random sampling. The third is ease of sampling.

As with systematic sampling, one needs to compute the variance of the estimator to make comparisons with other sampling plans. The following notations are used:

- L is the number of strata.
- N_h is the total number of units in stratum h.
- $N = \sum_{h=1}^{L} N_h$.
- n_h is the number of units sampled in stratum h.
- $W_h = N_h/N$, the stratum weight.
- w_h is an estimate of the stratum weight.
- $\bar{y}_h$ is the sample mean of the variable of interest in the hth stratum.
- σ_h^2 is the variance of the variable of interest in the hth stratum.

The overall sample mean is

$$\bar{y}_{st} = \sum_{h=1}^{L} W_h \bar{y}_h$$

Of course, in earth science field studies, N_h is not countable, so W_h is typically estimated as the proportion of area. For example, if stratification is by four soil types and they occupy 25, 50, 10, and 15 acres, the weights w_h may be 0.25, 0.50, 0.10, and 0.15, respectively. The variance of the mean of the stratified sample is

$$V(\bar{y}_{st}) = \sum_{h=1}^{L} w_h^2 \left(1 - \frac{n_h}{N_h}\right) \frac{\sigma_h^2}{n_h}$$

and for $n_h \ll N_h$,

$$V(\bar{y}_{st}) \doteq \sum_{h=1}^{L} \frac{w_h^2 \sigma_h^2}{n_h}$$

So how is n_h determined? There are several ways. If prior information suggests that variances differ significantly among strata, the *Neyman allocation*

$$n_h = \frac{n W_h \sigma_h}{\sum_{h=1}^{L} W_h \sigma_h}$$

is preferred; W_h and σ_h are replaced by their estimates. Two other methods are *equal allocation* $n_h = n/L$, where n is the total sample size and *proportional allocation*, $n_h = n W_h$. Equal allocation may be used when it is desired to achieve similar levels of precision in each stratum, assuming, of course, similar variances. This sampling procedure is useful when one or more of the strata are small and may be missed by other sampling procedures. Proportional allocation may be appropriate when variances are unknown.

Georgia Soil Example   Stratified sampling is illustrated with data from four formations in the state of Georgia (http://csat.er.usgs.gov/statewide/downloads. html, accessed May 12, 2005). The response variable is K_2O (potassium oxide in micrograms per gram, μg/g). Assume that the variances by formation are as shown in Table 9.1. For stratified sampling (and some other more complex designs), it is necessary to iterate between the choice of sample size and desired precision. Suppose that, initially, $n = 200$. If the estimates in the Neyman allocation equation are substituted,

$$n_1 = (200) \left(\frac{0.2 \times 7.07}{0.2 \times 7.07 + 0.1 \times 10 + 0.5 \times 1.41 + 0.2 \times 10} \right)$$
$$= 55$$

Similarly, $n_2 = 39$, $n_3 = 28$, and $n_4 = 78$. Note that the smallest sample size is allocated to the largest estimated relative size, the Norfolk–Ruston formation. Why?

Within a stratum it is possible to take a simple random sample, a systematic sample, or use other sampling schemes, depending on the situation. In addition, it is

342DESIGN OF EXPERIMENTS

TABLE 9.1   Georgia Soil: Estimates of Variability by Formation

Formation	Stratum Number h^a	Estimated Variance	Estimated Relative Size of Stratum
Cecil–Appling	1	50	0.2
Decatur–Dewey–Clarksville	2	100	0.1
Norfolk–Ruston	3	2	0.5
Porters–Ashe	4	100	0.2

http://sat.er.usgs.gov/statewide/downloads.html.

possible to incorporate the cost of sampling into the Neyman allocation; however, this makes little sense from a statistical standpoint because cost has nothing to do with precision.

Stratified sampling is not always better than simple random sampling. It can be worse (less efficient) if, for example, the estimates of strata variances are biased.

9.2.5   Cluster Sampling

Objects within a cluster have one or more similar attributes. Across clusters, attributes may be heterogeneous. Typically, in earth science applications, clusters are objects that are spatially close together. In a study of energy use, a set of clusters may be formed by people living in rural areas, small towns, or large metropolitan areas. A climate application is to cluster land cells based on characteristics such as temperature, precipitation, and cloud cover. Reasons to use cluster sampling are cost savings and variance reduction. When within-cluster variation is small, it is often more efficient to sample within clusters than across a larger population. Cluster samples may be defined according to exogenous methods or by statistical procedures discussed in Chapter 7.

A major impetus to cluster sampling is cost reduction. For example, to assess exposure of arsenic for people who live in major metropolitan areas is time consuming and prohibitively expensive. However, if one is able to assume a reasonable degree of similarity between metropolitan areas, it may be possible to take a simple random sample of metropolitan areas and within the selected areas test for arsenic exposure rather than sample all metropolitan areas. The following notation is used:

- N is the number of clusters in the population.
- n is the number of clusters selected in a simple random sample.
- m_i is the number of elements in cluster i, $i = 1, \ldots, N$.
- $\bar{m} = (1/n) \sum_{i=1}^{n} m_i$, the sample mean cluster size.
- $M = \sum_{i=1}^{N} m_i$, the number of elements in the population.
- $\bar{M} = M/N$, the population mean cluster size.
- y_{ij} is the jth observation in the ith cluster.
- $y_i = \sum_{j=1}^{m_i} y_{ij}$ is the total of all observations in the ith cluster.

An estimator of the population mean μ is

$$\bar{y}_{cl} = \frac{\sum_{i=1}^{n} y_i}{\sum_{i=1}^{n} m_i}$$

and the estimate of the variance of $\bar{y}_{cl}$ is

$$V(\bar{y}_{cl}) = \frac{N-n}{Nn\bar{M}^2} s_c^2$$

where $s_c^2 = \sum_{i=1}^{n} (y_i - \bar{y}_{cl} m_i)^2 / (n-1)$. Typically, $\bar{M}$ will not be known but can be estimated by $\bar{m}$. This process is called *single-stage cluster sampling*. In it, one randomly samples from a population of clusters and then examines all of the elements within each cluster.

9.2.6   Multistage or Hierarchical Cluster Sampling

Often, there are too many elements in a single cluster to query or analyze each one. Thus, one needs to sample within clusters. There are many ways to do this. It may be determined that within each cluster there are clusters, and thus another stage of cluster sampling is added. An investigator may find strata within clusters or may decide that a simple random sample or a systematic sample is best. For estimates of means and variances for various multistage sampling schemes, see the book by Levy and Lemeshow (1991).

9.2.7   Sampling Proportional to Size

Sampling proportional to size (SPS) is a form of unequal probability sampling. Rather than taking a random sample, it may make more sense to sample proportional to size. For example, in sampling toxic waste sites, it is likely that an investigator will want to assign more weight to the larger sites rather than equal weight to all sites to be sampled.

Another area where SPS is important is the search for hidden targets. Consider the circular objects located in a rectangle (Figure 9.2) but hidden below the surface. A randomly located point (a thrown dart or drill hole) on average will find the largest target first. Consider the following example from oil and gas exploration.

Assume that N deposits are labeled $1, \ldots, N$ in a region and their sizes are $\{y_1, \ldots, y_N\}$. Finite population *sampling without replacement* and proportional to size implies that the probability that all N fields are discovered in the order $1, 2, 3, \ldots, N$ is

$$\prod_{k=1}^{N} \left(\frac{y_k}{y_k + \cdots + y_N} \right)$$

This construction can be used to estimate the number of undiscovered deposits $N - n$ given that n deposits have been discovered. For additional details, see the article by Andreatta and Kaufman (1986).

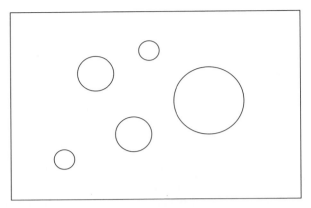

FIGURE 9.2   Sampling proportional to size.

If one assumes that targets are found by simple random sampling, as opposed to sampling proportional to size, parameter estimates can be seriously biased. A critical component of estimation is knowledge of the sampling plan.

9.2.8   Quadrat Sampling

A quadrat is a measured and marked rectangle or circle used in ecology for counting abundance and/or determining density. *Quadrat sampling* is the act of sampling from a quadrat. Typically, quadrat sampling is used in a large area where enumeration is not possible.

Quadrat sampling can be systematic, say along a transect, stratified, or random. Random sampling is illustrated using locations of yields from the water-well yield case study. A subregion of the data is displayed in Figure 8.6. For this exercise a circle of radius 0.5 m is "tossed" in the region of interest and the number of wells counted. This corresponds most closely to simple random sampling with replacement.

Instead of water-well yield locations, think of the points in Figure 9.3 as peat mounds, where the object of a study is to examine the effect of cattle grazing on the mounds. Ten tosses (random samples) are taken. Since this sampling scheme is completely random, circles can overlap. The center of the circles are uniform random numbers (x,y) in the region of interest. The experiment proceeds by having team members identify the mounds in each circle and measure response variables of interest, such as the amount of new growth. The information on mounds obtained from quadrat samples is as follows:

- Sample size $n = 10$; number of circles
- Radius of circle $r = 0.5$ km
- Total points in region $= 754$
- Population density $= 5.812$; points/unit area (km^2)
- Average quadrat count $= 5.528$; average number of points/circle (circles partly outside the region are weighted)

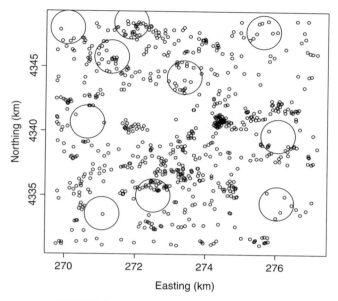

FIGURE 9.3 Quadrat sampling with circles.

- Estimated density $= 7.040$; average quadrat count/circle area (km^2)
- Standard error of the mean $= 2.846$ (points/km^2), which is (standard error of count)/(circle area, km^2).
- 95% confidence interval for estimated density: (0.601, 13.479)
- Index of dispersion $= 9.036$ (variance count/mean count)
- Critical region for index under CSR: (0.300, 2.114)

The algorithm used in this sampling is the SPlus/R function quadrat (Reich and Davis, 2002). As in the investigation of spatial point patterns discussed in Chapter 8, complete spatial randomness (CSR) is typically of interest. CSR is also known as a uniform Poisson process. The conclusion is to reject CSR, since the index of dispersion is outside the critical region. Since CSR is rejected, clustering distributions can be investigated. As noted, other sampling schemes can involve placing a grid over the entire region of interest and then randomly selecting cells for investigation.

9.2.9 Sampling With or Without Replacement and the Finite Population Correction Factor

As stated previously, there are two ways to sample: with replacement and without replacement. A sample of size n with replacement means that a sample of size 1 is taken from the population, the information recorded, the sample put back, and the population mixed. Next, a second sample of size 1 is taken. This process is repeated until a sample of size n is obtained. Sampling with replacement is rarely done in earth science applications, often because it is impossible, unnecessary, and/or ridiculous to do so. Field study examples where sampling with replacement is not done include soil, water, and air sampling. As noted, when $n \ll N$, the distinction between sampling with and without replacement can be ignored because the probability of selecting the same element more than once is extremely (in some cases, infinitesimally) small. When the sample constitutes a relatively large percentage of the population, say 10% or more, the correct assumption is required: namely, sampling without replacement. Examples include sampling of toxic waste sites and studying the discovery of oil deposits where the population size is relatively small.

One may ask: What is the difference between these two approaches? In sampling with replacement, samples are independent because the population remains constant. In sampling without replacement, the population changes and thus samples are dependent. Suppose that the goal of a study is to estimate the mean of $y_1, \ldots, y_n$ and its associated standard error. For simple random sampling with replacement (W), the standard error of the mean is

$$\sigma_{\bar{y}(W)} = \sqrt{\frac{\sigma^2}{n}}$$

For simple random sampling without replacement (WO), $\text{Cov}(y_i, y_j) = -\sigma^2/(N-1)$, $i \neq j$. Some algebra will show that

$$\sigma_{\bar{y}(WO)} = \sqrt{\frac{\sigma^2}{n}\left(1 - \frac{n-1}{N-1}\right)}$$

If $n \ll N$, the two estimators are approximately equal.

A specific example is used for illustrative purposes. The U.S. Environmental Protection Agency (2005) listed 29 toxic waste sites in Denver, Colorado. Suppose that an investigator wishes to sample from five of the sites to estimate a mean toxicity. One cannot assume that sampling with replacement is a reasonable approximation for sampling without replacement, since 5 is a large percentage of 29. Therefore,

$$\sigma_{\bar{y}(WO)} = \sqrt{\frac{\sigma^2}{5}\left(1 - \frac{5-1}{29-1}\right)} = 0.41\sigma$$

Conversely, if one incorrectly assumes sampling with replacement, $\sigma_{\bar{y}(W)} = \sqrt{\sigma^2/5} = 0.45\sigma$. The quantity $\sqrt{(N-n)/(N-1)}$ is called the *finite population correction factor* (FPC). For the toxic waste example, FPC $= \sqrt{(29-5)/(29-1)} = 0.93$ and FPC should be applied. However, if one takes a sample of size 5 from the approximately 200 sites in Colorado, the FPC $= 0.99$ and for most applications it may not be worth worrying about. The FPC, where applicable, is important because its use yields an estimate variance that is lower than if the FPC is not used. This lower estimate will provide a tighter confidence interval on the mean.

9.3   DESIGN OF EXPERIMENTS

When the spatial distribution and concentration of arsenic in a region is studied, a decision should be made on a sampling strategy, based on pilot studies, analogy, and other information. However, in field studies most of the influential factors, such as terrain, type of soil, and precipitation, cannot be controlled.

Conversely, in a formal designed experiment, the ability exists to control many of the factors that may affect the outcome of an experiment. Thus, they can be changed in a systematic way to obtain generalizable results. Design allows the efficient allocation of experimental units. A detailed treatment of design of experiments is beyond the scope of this book; however, we illustrate the power of this important method using several examples.

Much of the seminal applied work in experimental design and the terminology are derived from agriculture and biology. A typical application in plant biology is to increase corn yield. When this occurs in a greenhouse, factors that may affect yield can be changed in a systematic way under controlled conditions. Among the factors that a plant biologist may wish to consider controlling are variety of corn, soil type, watering times and amount, fertilizer, temperature, humidity, and sunlight. A field study may follow the completion of a designed experiment to obtain insight into the response when many of the factors cannot be controlled.

The models discussed in the following sections assume a univariate continuous response variable that is normally distributed or can be transformed into a normally distributed variable. Multivariate procedures and those that permit a relaxation in normality assumptions are available but are beyond the scope of the book.

There are many different designs. The choice may be a function of the availability of data, the number of factors and levels, the ability to replicate the experiment, cost, time, and other considerations. Whatever the design, it must be executed as planned; otherwise, erroneous inferences can be made during analysis. The personnel implementing the design must receive proper training to ensure that it is executed according to plan.

9.3.1   Two-Factor Catapult Example

Initially, a designed example using a catapult (Figure 9.4) is presented, and then the assumptions and conditions are varied to illustrate other types of models. An earth science example is also presented. The catapult is one used in design of experiment classes and can be made or purchased from a scientific education store. The authors recommend its use as a way to have students do a quick and illustrative hands-on experiment.

The goal of the catapult experiment is to determine the level of each factor to make the ball travel the maximum distance. Among the factors that can be set are the number of rubber bands, the length that the catapult arm is extended (arm length), the start angle, the angle where the catapult stops (stop angle), and the height of the pivot point for the rubber bands. In addition, there may be an operator effect. For simplicity, only two factors are varied in this example, the number of rubber bands and the arm

FIGURE 9.4   Catapult.

length. The other factors are held constant. A more complex example where all of these factors are varied is given in the fractional factorial example in *NIST/SEMA-TECH e-Handbook of Statistical Methods* (2006).

9.3.2   Experimental Unit

A critical initial step is to determine the *experimental unit*. It may be a plant, a core sample, a rain gauge, or a person. It is something that is going to be treated and the response to the treatment measured. In the case of a plant, the amount of water it receives, type of fertilizer, and amount of sunlight can be varied. Each of these classes is called a *factor*. Factors may be continuous, such as amount of water, or discrete, such as type of fertilizer. Levels of factors may be chosen at random or selected by some criterion. For example, the investigator decides the amount of water that plants receive per day—some receive a $\frac{1}{4}$ liter, others $\frac{1}{2}$ liter, and the remainder, 1 liter. In this case, there are three levels for the factor "water".

Sometimes the number of experimental units available is less than the number needed to run a proposed design. When this occurs, either additional experimental units must be found or the design changed to make do with the number available. Occasionally, the number exceeds what is required to run the experiment. When this occurs, the subset needed to run the experiment should be selected at random to avoid a potential source of bias that may occur when judgment is used to select the experimental units. It is difficult to impossible to select a random sample using expert judgment.

9.3.3   Completely Randomized Designs

The two-way *completely randomized design model* is

$$Y_{ijk} = \mu + \alpha_i + \beta_j + (\alpha\beta)_{ij} + \varepsilon_{ijk} \qquad i = 1, \ldots, a, \quad j = 1, \ldots, b, \quad k = 1, \ldots, m$$

"Two-way" refers to the two factors, which are labeled A and B. In this model μ is the grand mean, α_i is the effect due to the ith level of factor A, β_j is the effect due to the jth level of factor B, and $(\alpha\beta)_{ij}$ is the interaction between levels i and j of factors A and B, respectively. Factors A and B are called main effects. The usual classical assumption about the error is that the ε_{ijk} are iid $\sim N(0,\sigma^2)$. The general form of this model can easily be extended to the situation where there are more than two factors and any desired interactions are specified. In the authors' experience, interactions that exceed three factors are difficult to interpret.

What constitutes a completely randomized design experiment? The key component is that every experimental unit is assigned randomly to each level of all factors. For example, in an experiment with two factors where there are two levels in the first factor and three levels in the second factor, six experimental units (2×3) are required for one complete replication of the experiment.

9.3.4   Randomization and Replication

Randomization, replication, and blocking are essential components of a successful experimental design. Randomization and replication are discussed in this subsection; blocking is discussed later.

Randomization has two components. One is the random assignment of experimental units to treatment combinations. The other is the run order of factor–level combinations. Consider an experiment based on two factors, A and B, with two and three levels, respectively (Table 9.2). The design is to run two replications of each factor–level combination. The 12 experimental units are numbered 1 through 12. One way to run the experiment is have the investigator select the experimental units and assign them to factor–level combinations. For example, she might select experimental unit 4 to be assigned to replication 2, factor A, level 1, factor B, level 1. A second

TABLE 9.2   Randomization of Experimental Units and Run Order

Experimental Unit (EU) ID Number	Replication	Factor–Level	Assignment Number for EU	Run-Order Randomization
1	1	A_1B_1	5	4
2	1	A_1B_2	1	8
3	1	A_1B_3	6	11
4	1	A_2B_1	11	1
5	1	A_2B_2	10	7
6	1	A_2B_3	4	2
7	2	A_1B_1	7	9
8	2	A_1B_2	2	5
9	2	A_1B_3	3	3
10	2	A_2B_1	12	12
11	2	A_2B_2	9	10
12	2	A_2B_3	8	6

approach is to assign them on the bases of their numbering (i.e., experimental unit 1 is assigned to replication 1, factor A, level 1, factor B, level 1; experimental unit 2 is assigned to replication 1, factor A, level 1, factor B, level 2; and so on). The third and best option is to randomize the experimental units and then assign them to replication, factor–level combinations (Table 9.2).

Why is randomization of experimental units to factor–level combinations important? Say that an experiment occurs in a greenhouse in which the temperature is uneven, but that this is unknown. Specifically, suppose that it is cooler near the door and that the temperature rises toward the back. Further, suppose that all the seeds (experimental units) of one variety of grass are planted by the door, and all of another variety are planted at the back of the greenhouse. At the conclusion of the experiment, any difference in grass quality may be due to unanticipated temperature differences rather than to factors such as the amount of water and the type of soil that is controlled. Randomization in the location of seed variety will help to minimize this unforeseen temperature effect, since seeds of different varieties are planted in all areas of the greenhouse.

The second important component of randomization is run order. The investigator can easily run this experiment in the order that the replication–factor–level is given in Table 9.2: that is, A_1B_1 run first, followed by A_1B_2, and so on. A bias similar to that just described can occur. For example, some unknown factor, perhaps an unanticipated and unmeasured change in humidity, can be present during the runs of A_1 and then change for the runs of A_2. Randomization of run order helps protect against this. The run-order randomization is shown (Table 9.2) as a random permutation of the numbers $(1,\ldots,12)$. So A_2B_1, replication 1 is run first, followed by A_2B_3, replication 1. The experimental units assigned to these two runs are EU 11 and EU 4, respectively.

Replication is important for several reasons. It can ensure more powerful tests by making additional degrees of freedom available to estimate the error. Replication can allow an investigator to test interactions between factors that may be impossible without replication. In addition, replication offers some protection against the loss of data. In the example above, two replications are selected and no distinction is made between replications in the assignments of experimental units or run order. If the experiments are conducted at different times or in different greenhouses, it may be preferable to run all of replication 1 followed by replication 2. This is called blocking.

9.3.5   Fixed-Effects Model

A *fixed-effects model* is one where the level of each factor is chosen purposefully by the investigator from the populations of interest. Often, the levels constitute the population of interest. In the catapult example, for the rubber band factor, one level is one rubber band; the other level is two rubber bands. Three levels of arm length are chosen as 0, 2, and 4 cm. Both factors are fixed effects—chosen purposefully. They are not chosen at random from some larger population of the number of possible rubber bands or arm-length settings. Catapult settings and resulting data for the catapult experiment are shown in Table 9.3. Note that 12 runs are performed, not 6. The reason is that a second replication is needed to estimate a possible interaction effect between

TABLE 9.3   Catapult Experiment Design Variables and Response

(1) Run Order	(2) Y: Distance (cm)	(3) $\ln(Y)$: Distance (cm)	(4) Number of Rubber Bands (NB)	(5) Arm Length (AL) (cm)	(6) Number of Bands Coded	(7) Arm Length Coded
1	28	3.33	1	0	-1	-1
2	99	4.60	2	2	1	0
3	105	4.65	2	4	1	1
4	119	4.78	2	4	1	1
5	35	3.56	1	0	-1	-1
6	45	3.81	1	2	-1	0
7	45	3.81	1	4	-1	1
8	84.5	4.44	2	2	1	0
9	28.5	3.35	2	0	1	-1
10	52	3.95	2	0	1	-1
11	60	4.09	1	4	-1	1
12	37.5	3.62	1	2	-1	0

the number of rubber bands and arm length. An *interaction effect* occurs when the mean response for levels of one factor differ over the mean response for levels of a second factor. For example, suppose that with one rubber band, the means of all three levels of arm length are the same but differ when two rubber bands are used. Subsequently, an interaction plot is shown in the interaction effects section.

Each experimental unit (the ball) is subject to every level of both factors twice. This experiment is unusual in that the same experimental unit, a single ball, can be used for all runs. The actual settings for the number of rubber bands and arm length are shown in columns 4 and 5 of Table 9.3. The corresponding coded values, which are input to a computer algorithm, are shown in columns 6 and 7, respectively. Why code? A main reason is interpretability of results, which are illustrated in a subsequent discussion. In addition, many algorithms will interpret number of rubber bands and arm length as continuous variables, which is not the intent in this design.

Run order (Table 9.3, column 1) is another important consideration. The run order is obtained by randomizing the 12 runs to protect against any unanticipated effects that may influence results. How is the catapult experiment conducted? For run 1, one rubber band is used and the arm length is set at 0. The ball is inserted in the catapult and the result, the distance traveled, is observed. Next, the rubber band is removed and the arm length is reset. For run 2, two rubber bands are used and the arm length is set at 2. Note that by chance, the settings are the same for run orders 3 and 4; however, after run 3 is concluded, the setting should not be left in place but, rather, the rubber bands removed and the arm length reset to 0. A reasonable question to ask is: Why reset a level if its value remains the same from one run to the next? If the level is set exactly, there is no need to reset it. However, suppose that in run 3 the arm length is erroneously set to 2.1 cm instead of 2.0. If this same setting is used in run 4, the error is compounded. Resetting the level provides an opportunity to set it correctly or at least that any error in setting the length is random.

TABLE 9.4 ANOVA Results for Catapult Example: Fixed-Effects Model

Source	Factor	df	Sum of Squares	Mean Square	F-Value	p-Value
Number of bands (NB)	A	1	1.048	1.048	22.138	0.003
Arm length (AL)	B	2	1.319	0.659	13.926	0.006
NB × AL	AB	2	0.222	0.111	2.346	0.177
Residuals	—	6	0.284	0.047		

The result, the distances, in centimeters, that the ball travels is shown in column 2 of Table 9.3. The distribution of distances appears to be right-skewed. Therefore, the ln(distance) is chosen as the response variable. The ANOVA results, which are similar to those in Chapters 3 and 4, are shown in Table 9.4. The null hypotheses are:

- Number of bands (factor A) is not statistically significant.
- Arm length (factor B) is not statistically significant.
- Number of bands × arm length (factor AB), the interaction effect, is not statistically significant.

The respective three alternatives are that they are statistically significant. If the significance level chosen is 0.05, number of bands and arm length are highly significant, however, the (number of bands) × (arm length) interaction is not statistically significant (Table 9.4).

Each F-value is formed by dividing its mean square (A, B, or AB) by the mean square due to residuals, which is the mean-square error (MSE). Why is the MSE the appropriate divisor? To answer this question requires an investigation of the expected mean squares, which for this model and factors A, B, and AB (interaction) are shown in Table 9.5. The expected mean squares for factors A and B and the interaction AB differ from σ^2 only by the quantity of interest. Suppose, for example, that the main effect for factor A does not exist. This is equivalent to a null hypothesis of

$$H_0: \alpha_1 = \cdots = \alpha_a = 0$$

TABLE 9.5 Expected Mean Squares for Two-Factor Completely Randomized Fixed-Effects Model

Source	df	Expected Mean Square
A	$a - 1$	$\sigma^2 + \dfrac{bm\sum_{i=1}^{a}\alpha_i^2}{a-1}$
B	$b - 1$	$\sigma^2 + \dfrac{am\sum_{j=1}^{b}\beta_j^2}{b-1}$
AB	$(a-1)(b-1)$	$\sigma^2 + \dfrac{m\sum_{i=1}^{a}\sum_{j=1}^{b}(\alpha\beta)_{ij}^2}{(a-1)(b-1)}$
Error	$ab(m-1)$	σ^2
Total	$abm - 1$	

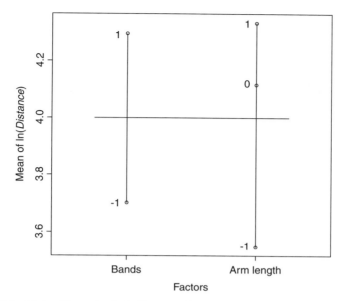

FIGURE 9.5   Main effects plot in ln(distance) for number of rubber bands and arm length settings.

in which case the expected mean square for A, EMS(A), is σ^2. Thus, the ratio of expected mean squares, EMS(A)/EMS(error) is 1, so the corresponding statistic MS(A)/MSE has an f- distribution with $a - 1$ and $ab(m - 1)$ degrees of freedom, where m is the number of replications of the experiment. (see Chapter 3 for a computational formula and discussion.) The importance of the expected mean squares in a fixed-effects model is that they tell the user how to form proper F-statistics to test hypotheses.

Expected mean squares for most standard designs can be found in books on experimental design (see, e.g., Montgomery, 2000). They can also be derived by conceptually simple but tedious algebra.

The df for the main effects is just 1 minus the number of levels. The df for the interaction effect is the product of df from the corresponding main effects. The error term is estimated from the replications at each level of the factors, so its df is $ab(m - 1)$. If $m = 1$, the error cannot be estimated from the data unless no interaction effect is assumed. The main effects plot is shown in Figure 9.5. This plot illustrates a change in mean of ln(distance) from one rubber band (code $= -1$) to two rubber bands (code $= 1$). These two means are averaged over all settings of arm length. It also illustrates the change in mean response for arm length from zero distance (code $= -1$) to length 2 (code $= 0$) to length 4 (code $= 1$). The horizontal line is the estimated grand mean $\hat{\mu} = 4.00$.

Interaction Effects   Interaction between factors is one of the more difficult aspects of experimental design to grasp. An illustration from the catapult example is

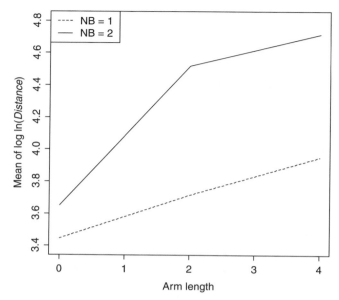

FIGURE 9.6   Weak interaction effect between the number of rubber bands and arm length for the catapult example. NB is the number of rubber bands.

presented. Although the interaction effect is statistically significant only at the 0.177 level, an interaction plot (Figure 9.6) is shown to illustrate the phenomenon. The response variable, ln(distance), is plotted on the vertical axis. Arm length is plotted on the horizontal axis. The dashed line corresponds to one rubber band; the solid line corresponds to two rubber bands. When there is no interaction, these lines should be parallel except for variation attributable to sampling. However, the number of rubber bands makes a small difference in the distance the ball travels at 0 arm length, but it makes a much larger difference at arm lengths of 2 and 4. If knowledge of interaction is important, additional replications should be run.

The catapult data are modified to show a more pronounced example of an interaction effect (Figure 9.7). The illustration indicates that there is minimal change in ln(distance) using two rubber bands as the arm length increases; however, with one rubber band the distance increases as the arm length increases. This is a pronounced interaction effect. However, if one observes this graph in a real catapult experiment, one should be most suspicious of the results, as it seems counterintuitive to the physics of the experiment. In Figure 9.7 the lines cross, but this is not necessary for an interaction effect to be present.

No interaction in the population is characterized by an even spacing (to within sampling error) in ln(distance) between one and two rubber bands for the three arm lengths (Figure 9.8).

The possible presence of interaction illustrates one of the strengths of a designed experiment. Suppose for simplicity that A and B each have two levels ($+$ and $-$) and B is chosen to be held constant at a low level B^- and both high and low levels of A (A^+

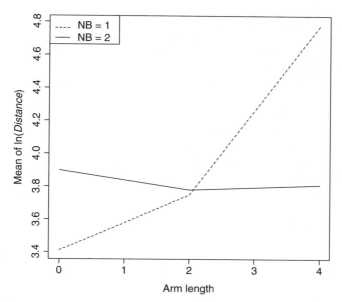

FIGURE 9.7   Strong interaction effect between the number of rubber bands and arm length for the catapult example. NB is the number of rubber bands.

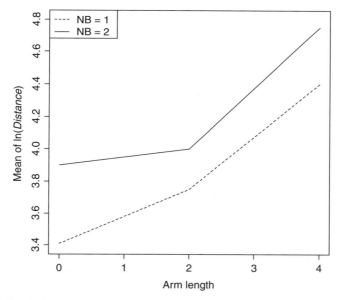

FIGURE 9.8   No interaction effect between the number of rubber bands and arm length for the catapult example. NB is the number of rubber bands.

and A^-) are run. The experiment is repeated at B^+. Suppose that A^+B^- and A^-B^+ yield higher responses than A^-B^-. What does this suggest about A^+B^+? With no interaction, one may reasonably conclude that A^+B^+ will yield an even higher response. However, in the presence of interaction, one cannot assume an additive effect, and the conclusion that A^+B^+ will yield a higher response may be erroneous.

Generally, it is possible to interpret a two-way interaction, that is, interactions between factors A and B. It is more difficult to interpret a three-factor interaction and almost impossible to understand higher-order interactions. Thus, higher-order interactions are often lumped into the error term. Clearly, the problem with doing so is that if an interaction is present, lumping it into error will inflate the error mean square, making it more difficult to reject null hypotheses correctly on main effects and lower-order interactions.

Constraints: Fixed-Effects Model   The two-factor with interaction fixed-effects model just discussed can be written

$$X_{ijk} = \mu + \alpha_i + \beta_j + (\alpha\beta)_{ij} + \varepsilon_{ijk} \qquad i = 1, \ldots, a, \quad j = 1, \ldots, b, \quad k = 1, \ldots, m$$

where $X_{ijk} = \ln(Y_{ijk})$ in the catapult example. This model has $a + b + ab + 1$ parameters. However, the system of equations yielding least-squares estimates of the parameters, called *normal equations*, has $a + b + 1$ dependencies. Thus, it is necessary to impose $a + b + 1$ constraints to obtain a solution to the normal equations. One system of $a + b + 1$ independent constraints is

$$\sum_{i=1}^{a}\hat{\alpha}_i = 0$$

$$\sum_{j=1}^{b}\hat{\beta}_j = 0$$

$$\sum_{i=1}^{a}\widehat{(\alpha\beta)}_{ij} = 0 \qquad j = 1, \ldots, b$$

$$\sum_{j=1}^{b}\widehat{(\alpha\beta)}_{ij} = 0 \qquad i = 1, \ldots, a$$

Therefore, a solution can be obtained and parameters estimated. These are the usual, but not the only constraints that can be chosen, and individual parameter estimates depend on the system of constraints.

Model Evaluation: Fixed-Effects Model   Evaluation is a critical step in the modeling process to ensure its usefulness. This includes viewing residuals. A qq (quantile–quantile)-plot of residuals from the catapult experiment is shown in Figure 9.9. It is used to investigate the normality assumption of the error. The theoretical quantiles are normally distributed and have the same number of points as the residuals. The qq plot of residuals shows four distinct groupings; however, the

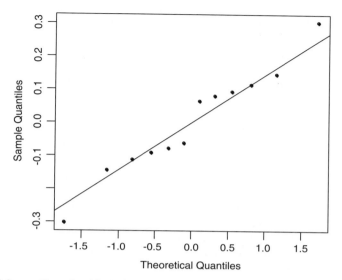

FIGURE 9.9   qq Plot of residuals from the catapult experiment. The line indicates perfect agreement between sample and theoretical quantiles.

sample is small and there is no major departure from the diagonal line, which suggests that the normal assumption is satisfied. A "time series" run order plot of the residuals (Figure 9.10) is constructed to see if there is some effect present that was not considered previously. No pattern appears to exist.

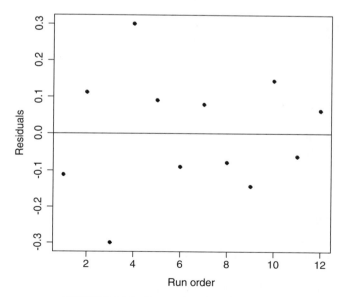

FIGURE 9.10   Run-order plot of residuals.

The conclusion that can be drawn from the model results and graphical evaluation of the catapult experiment is that increasing the number of rubber bands and arm length increase the distance the ball travels for the settings examined in this study.

Time Trends and Drift   A concern in any design model is the effect that time or a time-related variable may have on the results of an experiment. Ideally, it is desirable to make all measurements in as short a time frame as possible; however, this is not always possible. In the catapult example one can think of factors not accounted for in the experiment. These may include wear on the rubber bands (decrease in elasticity) and operator fatigue. As noted previously, randomization of run order can often mitigate these effects. Plotting data by run order or by time may help detect some previously unrecognized problem. There are also blocking designs that can be used if the experimenter has reason to believe that conditions that cannot be controlled may change over time. For example, if an experiment or a replication of an experiment has to be run over multiple days or in different plots of soil. Finally, some experiments may, unbeknownst to the investigator, be sensitive to changes in temperature or humidity that can occur over time.

Response Surface for a Fixed-Effects Model   When one or more of the explanatory variables are continuous, an important question is: Can anything be said about a possible response between the levels? For example, in the catapult example, can an inference be made about distance traveled if the arm length is 3.5 cm? The answer is yes if the investigator is willing to assume a model, such as a low (second or third)-order polynomial. Such a model can be fit to the data and distance estimated at points not included in the design. This procedure, called *response surface fitting*, has been described by Montgomery and Myers (2002). One advantage to estimating response surfaces is that the investigator may find that the optimal setting of level of factors is difficult to implement or costly to achieve. To achieve a less costly solution, it is sometimes necessary to move away from optimality, that is, to change the level of one or more factors slightly away from the optimal setting.

Factors and Levels in a Fixed-Effects Model   Selecting the number of factors and levels is done purposefully based on expert judgment, prior studies, practical considerations of implementation, and cost. In studies where prior information is minimal, many factors are included with only two or three levels per factor. These are sometimes called *screening designs*. In follow up designs, insignificant factors are dropped and the number of levels may be increased.

9.3.6   Random-Effects Model

In the *random effects model*, it is assumed that the chosen levels of all factors are selected at random from a larger population. The inference is on the variability of the population rather than on the specific levels of factors chosen in the sample.

In the catapult fixed-effects model, one and two rubber bands are purposely chosen as the level on the rubber band factor. Suppose, however, that it is possible to use from

1 to 30 rubber bands. If the maximum extent of the arm length is 5 cm, it is possible to randomly select three settings between 0 and 5 cm. Instead of only being able to make inferences at the chosen number of levels, as in the fixed-effects model, with a random effects model it is possible to make inferences about the population. Almost certainly, it is inappropriate to do so for the catapult problem, but in certain situations the goal is not just to make an inference about preselected levels, but to generalize to a larger population. This is illustrated in the following geological example (Griffiths, 1967).

The purpose of the investigation reported by Griffiths is to understand bulk-density variation in a Pocono sandstone. In the design of this experiment:

- Five operators are randomly chosen to analyze sandstone samples. This is the operator factor and the number of levels is five.
- Specimens are randomly chosen from a layer of sandstone 10 ft thick. This is specimen factor and the number of levels is six.
- Each sample is run four times. Thus, there are four replicates.

The resulting model,

$$Y_{ijk} = \mu + \alpha_i + \beta_j + (\alpha\beta)_{ij} + (\delta_{k(ij)} + \varepsilon_{ijk})$$
$$i = 1, \ldots, a, \quad j = 1, \ldots, b, \quad k = 1, \ldots, m$$

appears similar to the fixed-effects model with the exception of the error term. In this model a is the number of operators, b is the number of specimens, and m is the number of replications. For the sandstone example, α_i is the effect of the ith operator, β_j is the effect of the jth sandstone sample, and $(\alpha\beta)_{ij}$ is the interaction effect term between operator and sandstone sample. The error term ε_{ijk} represents the usual error from unassigned and usually unknown sources. The additional error term $\delta_{k(ij)}$ results from the replicates within operator and specimen. The subscript $k(ij)$ is the usual way to represent the error associated with variation that exists within a given operator and specimen for multiple replicates. This term is not present in designs when subsamples are absent.

What makes this experiment different from the catapult experiment? In that experiment, the desire is to investigate results for a few selected settings to provide information on an optimal or nearly optimal result. In the sandstone example, the interest is not the performance of the five operators selected for this experiment but how equally qualified operators from a larger population will perform. Also, the assumption is that the six specimens are from a larger population and the interest is in seeing how much variability exists among specimens. Another interest is to see if certain operators perform better on certain specimens. Hopefully, this is not the case, but it needs to be investigated. In a random effects model, the interest is in examining the *components of variants* because the major concern is to examine and understand variation.

Another major difference between this model and that of the fixed-effects model is the way that hypotheses are tested. The sources of variation and expected mean squares are given in Table 9.6. For the purpose of illustration, a few data values have been changed from those reported by Griffiths (1967). For testing purposes a p-value of 0.05 or less is assumed to be significant.

TABLE 9.6   Sources of Variation, Degrees of Freedom, Expected Mean Squares, and Mean Squares for Sandstone Study

	df	Expected Mean Squares	Mean Squares	F-Stastistic	p-Value
A (operator)	$a - 1 = 4$	$\sigma^2 + m\sigma_{\alpha\beta}^2 + bm\sigma_\alpha^2$	17,765	1.48	0.25
B (specimen)	$b - 1 = 5$	$\sigma^2 + m\sigma_{\alpha\beta}^2 + am\sigma_\beta^2$	33,745	2.81	0.04
AB (interaction)	$(a - 1)(b - 1) = 20$	$\sigma^2 + m\sigma_{\alpha\beta}^2$	12,027	1.46	0.12
Error (replicate)	$ab(m - 1) = 90$	σ^2	8,248		
Total	$abm - 1$				

Source: After Griffiths (1967, Table 19.7).

In a random effects model, the order of testing is important. The interaction effect (AB) is tested first. The F-statistic is based on a comparison (ratio) of the mean square associated with the interaction to that of the error, namely, the test statistic is $F = \text{MS}(AB)/\text{MSE} = 1.46$. If the null hypothesis of no interaction is true, F will have an f-distribution with 20 and 90 df. Since the significance level of an F-statistic $= 1.46$ is 0.12, the null is not rejected at 0.05. The next step is to test the significance of the main effects. If the interaction effect is significant, the main (operator and specimen) effects are difficult to interpret. The F-statistic for the operator effect (component of variance) is $F = \text{MS}(A)/\text{MS}(AB)$. Note that the denominator is not MSE. Why? The operator effect is not significant at the 0.05 level, which means that equally trained operators perform equally well. This inference extends to the population of operators, not just the five used in this experiment because they are selected at random from a larger population. If the investigator is only interested in the five operators employed in a given laboratory, "operator" will be a fixed effect and an inference cannot be made to a larger population. The F-statistic to test specimen significance is formed in a similar manner. Since the p-value $= 0.04$, the specimen effect is significant at the 0.05 level. Thus, there is a specimen effect. To estimate the specific components of variance, such as specimen, view the expected mean squares to isolate the specimen variance σ_β^2 using algebra. The EMS associate with σ_β^2 is found by subtracting the EMS due to interaction from the EMS due to the B effect: namely, $(\sigma^2 + m\sigma_{\alpha\beta}^2 + am\sigma_\beta^2) - (\sigma^2 + m\sigma_{\alpha\beta}^2) = am\sigma_\beta^2$. This quantity is divided by am. The estimate of the component of variance associated with the specimen is obtained by substituting the sample values. Thus,

$$\hat{\sigma}_\beta^2 = \frac{\text{EMS}(B) - \text{EMS}(AB)}{am}$$

$$= \frac{33,745 - 12,027}{(5)(4)}$$

$$= 1086$$

Generally, the component of variance can only be found by the foregoing procedure when the design is balanced. Balanced designs will be discussed subsequently. When it is not possible to find an estimate of the component of variance using simple algebra, approximations are available. The issues of interpreting interactions, and possibly lumping them into the error term, are similar to those for the fixed-effect models.

9.3.7   Mixed-Effects Model

In a *mixed-effects model*, there is a least one fixed effect and at least one random effect. Suppose that in the experiment above, there are only five operators, but the specimens represent a random sample from a larger population of specimens. This is a mixed-effects model. The expected mean squares for a two-factor model with the A effect fixed (operators) and B random (specimens) are

$$\text{EMS}(A) = \sigma^2 + m\sigma_{\alpha\beta}^2 + \frac{bm\sum_{i=1}^{a}\alpha_i}{a-1}$$
$$\text{EMS}(B) = \sigma^2 + am\sigma_{\beta}^2$$
$$\text{EMS}(AB) = \sigma^2 + m\sigma_{\alpha\beta}^2$$
$$\text{EMS}(\text{error}) = \sigma^2$$

These are used to determine the appropriate F-statistics as follows:

- For the A effect, it is MS(A)/MS(AB).
- For the B effect, it is MS(B)/MSE.
- For the AB effect, it is MS(AB)/MSE.

9.3.8   Other Designs

The completely randomized design where experimental units are applied to all combinations of factors is easy to formulate and analyze; however, it may be costly and inefficient or impossible to implement. For example, in a wave laboratory, where the problem is to understand the effects of a tsunami on structures near shore, factors may include type of wave, frequency, amplitude, direction, water bottom surface angle, nearshore sand/soil composition, and type of structure. If there are only two levels for each of these seven factors, $2^7 = 128$ runs are required for one complete replication of the experiment. In a laboratory where setup, run time, and expense are considerations, this may not be possible or desirable. There are numerous designs that overcome some of the practical difficulties posed by a completely randomized factorial design. These include fractional factorial, blocking, analysis of covariance, subsampling, and repeated measures. Each of these designs is described briefly. Many experiments involve combinations of these designs. Details are available in a book by Montgomery (2000) and in other experimental design books.

Fractional Factorial Designs   If it can be assumed that higher-order interactions are negligible (often a reasonable assumption), a *fractional factorial design* can be used.

This design allows the investigator to reduce the number of runs by perhaps one-fourth or one-half at the expense of confounding some of the higher-order interaction effects. Confounding means that certain higher-order interactions will not be able to be estimated. For example, a 2^{7-2} design is one-fourth of a two-level, seven-factorial design. Rather than 128 runs, this design requires only 32 runs for one complete replication. There are specific ways to implement this design so that interactions not deemed to be significant can be confounded (Montgomery, 2000).

Blocking   *Blocking* is a way to negate the presence of nuisance variables that may bias the results of an experiment. In many agricultural experiments, a given plot of land is too small for more than one replication of an experiment to be conducted. To run a second replication, the investigator selects a second plot that matches the first plot as closely as possible. However, it is almost impossible to match the plots exactly in terms of soil characteristics, exposure to sun or water, and other factors that may affect the growth or health of plants of interest. When this occurs, the investigator will designate the plots as blocks and include a blocking term in the model. In a laboratory experiment, there may be only enough test equipment to perform one complete replication of an experiment on a given day, or it may be that replications of an experiment need to be done at different laboratories. In the latter case, one blocks on laboratories. In some experiments it may be possible to complete only a single replication in a given year; thus, year may be a block.

A one-factor completely randomized fixed-effects model with a blocking term τ is

$$Y_{ik} = \mu + \alpha_i + \tau_h + \varepsilon_{ih} \qquad i = 1, \ldots, a, \quad h = 1, \ldots, \nu$$

where there are ν blocks (plots of ground, years, etc.). The factor main effect is α_i and τ_h is the block effect. In the grass example, they are grass and plot, respectively. An important assumption is that blocks and factors do not interact.

The penalty for blocking is a loss of degrees of freedom. If, for example, there are ν blocks, then $\nu - 1$ degrees of freedom are lost to account for a possible block effect. Of course, the penalty for not blocking may be to determine incorrectly that a treatment such as fertilizer accounts for the difference in plant grow when the actual determinant may have been unknown differences in soil type.

A two-factor completely randomized fixed-effects model with a blocking term τ is

$$Y_{ijk} = \mu + \alpha_i + \beta_j + (\alpha\beta)_{ij} + \tau_h + \varepsilon_{ijh}$$

$$i = 1, \ldots, a, \quad j = 1, \ldots, b, \quad h = 1, \ldots, \nu$$

where there are ν blocks (plots of ground, years, etc.). The factor main effects are α_i and β_j. An interaction between the two factors is allowed; however, as noted, interaction between factors and blocks is not. To repeat, blocks are not of primary interest but are used to account for influential unanticipated effects, should they exist.

More complex designs exist. For example, if there are six grasses to test and only four can be planted in a 1-acre plot, an incomplete block design is required.

Analysis of Covariance   When possible, blocking is used to account for nuisance factors that cannot be controlled. Sometimes, however, there is a continuous variable X which cannot be controlled but is linearly related to the response variable Y. The variable X is called a covariate or *concomitant variable*, and its introduction into the design changes the model to an *analysis of covariance* (ANCOVA). For a single-factor fixed-effect experiment, the form of this model is

$$Y_{ij} = \mu + \alpha_i + \gamma(X_{ij} - \bar{X}) + \varepsilon_{ij} \qquad i = 1, \ldots, a, \quad j = 1, \ldots, n$$

where α_i is the effect due to factor A, X_{ij} is the covariate, $\bar{X}$ is its mean, γ is the coefficient associated with the covariate, and there are n observations. A basic assumption is that the treatments (levels of factors) do not influence the covariate. An example of a covariate in the study of wildfires is humidity. Multiple covariates are allowed in the model.

Subsampling   A *subsample* is a sample of a sample. Some examples include:

- To ascertain business energy conservation, a random sample of businesses is selected. A subsample selects five people to be interviewed within each business. The experimental unit is the business. The people within the businesses are subsamples. The resulting information will provide information about variability in responses within a business as well as between businesses.
- To compare operators or laboratories to see if they produce equivalent results. Suppose that the problem is monitoring of mercury in water. Sampling may involve collecting water at several locations in a reservoir. A subsample is the division of the sample into two, and each subsample is assigned randomly to a different laboratory. If 10 vials of water are collected and then split, the sample size is 10, not 20. Two subsamples are associated with each sample.
- A biologist is studying treatment for beetle infestation at an experimental station tree farm. The biologist decides to take a random sample of 20 trees, selects six branches from each tree at random, and then collects three leaves from each branch for analysis. The experimental unit is the tree. Branches and leaves constitute subsamples.

Subsampling is also referred to as nested or hierarchical sampling.

Repeated Measures   As the name suggests, a *repeated-measures design* results from sampling the same experimental unit multiple times. The experimental unit or subject in a repeated-measure design is usually a person or animal. Applications in medicine and animal science are common where the same subject may be given multiple treatments over time, such as different drugs or different doses of the same drug. This reduces the number of subjects required. The important consideration in a repeated-measures design is that measurements made on the same experimental unit at different points in time usually are not independent. Thus, the variability of the response is decreased over an experiment where observations (samples) are independent. A disadvantage of the repeated-measure design is a lack of randomization when contrasted to a completely randomized design where each experimental unit receives each level of each treatment (factor). A *longitudinal study* is the name often given when repeated measurements are taken over long periods of time.

In a repeated-measures design, the total sum of squares is partitioned into within-subject and between-subject sum of squares. The within-subject sum of square is then partitioned into a treatment effect and error. Again, it is important to remember that the sample size n is the number of subjects. The number of treatments applied to a subject is used to estimate the within-subject variability.

Balanced Designs   A *balanced design* is one in which there are an equal number of observations in each factor–level combination. In the catapult example, there are two observations in every arm length–rubber band combination. Thus, it is a balanced design. Balanced designs are typically less sensitive to small departures from the equal-variance assumption and are more powerful. In addition, tables exist for the EMS and appropriate hypothesis-testing procedures.

Because of space, time, cost, and/or variability considerations or the need to obtain more information on particular factor combinations, it may be necessary or desirable to allocate more observations to some factor combinations and fewer to others. In addition, an unbalanced experiment may result from a loss of data or an unanticipated inability to obtain a data value.

For a two-factor experiment, proportional data are the number of observations in cell (i,j), $n_{ij} = n_i.n._j / n$, where $n_i.$ is the row total for the ith level of (say) factor A and $n._j$ is the column total for the jth level of (say) factor B and n is the sample size. It is easy to see that all balanced designs have proportional data. The converse is not true; however, if the data is proportional, the analysis is straightforward. If the experiment is just slightly unbalanced, approximations exist (Montgomery, 2000). For more highly unbalanced experiments, a general linear model approach may be appropriate.

9.4   COMMENTS ON FIELD STUDIES AND DESIGN

The process of formulating a field study or design can be challenging, but failure to use appropriate statistical sampling and/or design methods is costly in time, energy, and money. Failure to design and implement field studies and/or designs properly often results in biased estimates and the inability to make useful inferences. Some comments are offered about these concerns.

9.4.1   Designed Experiments and Field Studies

Designed experiments usually take place in a laboratory, greenhouse, pilot plant, or other venue where factors can be carefully controlled. Field studies, however, involve aspects of design. For example, in weed control various treatments can be compared by identifying similar patches of weeds, treating them differently, and measuring treatment effectiveness. Wildfire control is another example (Gebert et al., 2008). The object of this five-year U.S. Department of Agriculture–U.S. Department of the Interior study (2006) is to learn how to manage fire consistent with ecological and economic components. Variables of interest are vegetation, fuel and fire behavior, soils and forest floor, wildlife, entomology, pathology, and costs. Proposed treatments include untreated controlled burns, prescribed burns only, and cutting followed by prescribed burns. There are multiple response variables resulting from the effects of the controlled burns. Another example is a classical field/longitudinal study in medicine, the Framingham, Massachusetts Heart Study, where over 5000 adults and their dependents have reported on their medical conditions since 1948.

One of the most widely used methods in scientific investigation is stochastic computer modeling. Many of the principles of experimental design should be applied to computer modeling. For computer modeling applications in geology, see the book by Merriam and Davis (2001).

9.4.2 Design Consideration

In any design there are several important factors that must be considered, including the existence of historical data from the study area, data from a pilot study, or data from an analogous region to estimate sample size and variability. Low variability means that a smaller sample will be sufficient to achieve a given level of precision than if the variability is high. Issues include:

- *Spatial dependency.* Where there is a high degree of spatial dependence with respect to the variable or attribute of interest, fewer samples need to be taken than in a region of large heterogeneity. This issue was addressed in Chapter 6.
- *Time.* Time, as noted earlier, is a lurking variable. Surveys of people and physical phenomena taken over time can affect the interpretability of results if populations change. For example, soil or water samples and samples of fish or other living organisms can be influenced by natural and/or man-made events.
- *Transportation mechanisms.* Transportation of particles through water or by wind can have a significant impact on the results of a study and should be considered carefully in a sampling design.
- *Analytical results.* An estimate of measurement error should be obtained in advance of the implementation of a full-scale design. Also, the measurement devices and analytic processes must be able to achieve the range of values needed for a given study. When it is necessary to use multiple laboratories, analytic procedures must be standardized so that results are comparable.

In an experimental design, replication and randomization are critical. A properly designed, implemented, and analyzed statistical design usually yields results that are more generalizable and thus more useful. Designs allow partitioning and estimation of variability, including the study of possible interactions among factors. A designed experiment can be replicated by other scientists. It uses samples efficiently. Also, and most important, the exercise of formulating a design provides insight to the investigator. In the process of design, he or she is required to consider the important factors that may influence the results of an experiment. A design allows for the control of nuisance factors and often protects against unanticipated effects. We argue strongly that even if the execution of a design does not go totally according to plan, and it rarely does, the results will be more useful than the results of an arbitrary, haphazard data allocation and collection process. Data collected without benefit of a design are often not representative of the population of interest and thus do not provide the necessary information to permit valid inferences to be made.

9.4.3   Formulating a Model and Testing the Hypothesis

It has been noted in our discussion of the models that the F-statistics used to test hypotheses depend on assumptions made about fixed and random effects. They also depend on how experimental units are assigned to factors. It is important to answer the following question: Is there complete randomization, blocking, and/or subsampling? We cannot stress strongly enough the necessity of understanding how an experiment is conducted to analyze it properly.

For complex models, deriving the expected mean square is a formidable task. Fortunately, there are now many computer algorithms to assist with this and textbooks (e.g., Montgomery, 2000) that list designs and present appropriate tests. Many of these same algorithms and texts also provide help with designing an experiment so that with some prior information, the reader will be able to determine sample size and power of the tests.

9.5   MISSING DATA

Missing data is a concern in any study, because it may cause bias to be introduced. Missing data takes many forms. Sometimes an entire multivariate observation is missing. Other times, one or more variables are missing. It is important to know why data are missing. Missing data can be unobserved or lost. Rubin (1976) classifies data as follows:

- Missing completely at random
- Missing at random
- Not missing at random

Missing completely at random (MCAR) occurs when the event of an observation being missing is independent of observable data and unobservable parameters. If in a survey of water utilization, the probability that income missing is the same for everyone regardless of income or age, the data are MCAR. *Missing at random* (MAR) missingness may depend on values of observed variables but not on values of missing or unobserved variables. Examples include an observation lost in transit, instrument failure, or if a respondent forgot to answer a question on a survey. MCAR is a stronger assumption about data loss than MAR; however, in practice it is difficult to recognize MCAR. Both types of missing data require that the probability of being missing is not related to any other variable. The converse to MAR is *not missing at random* (NMAR). If, for example, in a household survey of water usage, older people are less likely to respond to a question on income than younger users, the data are not considered to be missing at random. If all the samples from a given geologic stratum are lost in shipment, the data are NMAR. If an analytical procedure fails to detect small values, such as trace elements, these data are NMAR.

With historical data, it is often difficult to determine why data are missing. Sometimes it occurs because a sample location is inaccessible. Sometimes data are lost and sometimes a given analytic procedure cannot analyze certain types of data. The obvious question is: What should we do about missing data? The first option should be to try to determine why the data are missing and, if possible, locate the missing data. Should this be impossible, other options are available. One is to delete observations with missing variables. This may not result in significant bias if the data sets are large and data are MCAR or MAR. However, for smaller data sets and when data are NMAR, this approach is not recommended. One option is to use methods that do not depend on having all data present. An example of such a method is *classification and regression trees* (CART) (Hastie et al., 2003). A second option is to estimate the missing data. Two of the most popular methods to fill in missing data are hot deck imputation and the expectation maximization algorithm (Little and Rubin, 2002). *Hot deck imputation* fills in missing values on incomplete records using values from similar but complete records of the same data set. The *expectation maximization algorithm* is a maximum likelihood procedure that estimates the distributions of all missing variables. Statistical methods such as regression and analysis of variance do not necessarily account automatically for the additional uncertainty associated with replacing missing data, and the user may need to account for missing data by adjusting degrees of freedom.

A final point concerns the amount of missing data that is tolerable. If a significant proportion of the data is missing, say as a guideline greater than 15%, any statistical inference made from such data should be interpreted with great caution. If only a small proportion of the data is missing, say less than 5%, any reasonable method of imputation will probably be satisfactory. It is in this middle area between 5 and 15% where the choice of imputation methods is more critical. Again, it is important to attempt to determine why data are missing.

9.6   SUMMARY

Experimental designs are modeling approaches that provide structure to the investigator to plan an experiment, sample, interpret, and make inferences. A field study, as the name suggests, is conducted outside a controlled environment. Examples include sampling designs to find energy resources, water, archeological sites, and rare species of plants. Design of experiments refers to studies conducted under controlled conditions, typically in a laboratory. They allow closer control of important factors and interactions that may affect the variable of interest. Sometimes the levels of factors can be set to achieve an optimal result. Design models can include fixed and/or random effects. A fixed-effects model is one where levels of factors are set purposefully and inferences are applicable only to those settings. A random effects model is used to study variation in the populations of the factors. In a random effects model, the levels are selected randomly from the population of interest.

Sampling designs used in field studies include simple random, systematic, stratified, cluster multistage, proportional to size, and quadrat. The choice of a sampling design is a function of variability, stratification, cost, and other problem-specific factors. Another consideration in sampling is possible spatial association and clustering. Sampling can be with or without replacement. In most earth science studies, sampling is without replacement; however, the sample size is usually quite small compared with the population size.

A critical factor in the analysis of a field study or designed experiment is to ensure that the analysis corresponds to the way the experiment is conducted. Specifically, the team doing the analysis needs to understand the sampling design and/or how experimental units are assigned to factors and what, if any, restrictions on randomization are present.

Missing data occur in almost every experiment. Understanding the mechanism that may have caused the missing data is critical to estimating them. Data may be missing at random or may be missing according to a systematic pattern. Missing data, if not treated properly, can seriously bias results.

EXERCISES

9.1   In the catapult example, it is noted that between runs the settings are always reset instead of leaving them in place, even though the next run uses the same number of rubber bands or arm length. Explain why.

9.2   In the catapult example, should one consider making arm length a random effect? Discuss.

9.3   In the discussion of systematic random sampling, mention is made of a problem that exists if the sampling interval corresponds to the period of the data. What is the problem? Explain.

For Exercises 9.4 to 9.9, consider the problem of assessing possible toxic waste sites. In the region surrounding a closed paint plant, the presence of lead in the soil is suspected. Suppose that in what is believed to be an analogous area, the standard deviation of lead in soil is 10 ppm. Remediation is recommended if the lead content exceeds 300 ppm.

9.4   A decision is made to sample five toxic waste sites from a population of 29. What are the practical issues of sampling with or without replacement? Make a recommendation and justify your answer.

9.5   How do you determine if the lead level exceeds 300 ppm? What assumptions do you need to made?

9.6   Suppose that for a simple random sampling scheme, the desire is to have the error of estimation be no larger than 5 ppm. What sample size is needed at a 95% confidence level?

9.7   If the estimated mean is 330 ppm, what is the conclusion?

9.8   Suppose that a project cannot afford to obtain the number of samples determined in Exercise 9.6. What are some alternatives?

9.9   Is simple random sampling the best approach? Why or why not? What are some alternatives? Explain.

9.10   On a 1-km length of stream, you are asked to take samples to evaluate water quality. What sampling scheme is recommended? Justify the answer.

For Exercises 9.11 to 9.13, cores logged from sands are present in three geologic spectra: the Lower, Middle, and Upper Miocene. Their respective thicknesses (measured in millions of years) are 6.4, 7.3, and 6.0. Preliminary sampling suggests that the standard deviations are respectively, 0.4, 0.2, and 0.2, measured in the phi scale φ [$\varphi = \log_2$(grain size in mm)].

9.11   If it is judged that 100 total samples are sufficient, how many samples should be taken in each spectrum? Justify your answer.

9.12   Compare and contrast samples taken using simple random sampling and stratified sampling.

9.13   Suppose that in the Lower Miocene, a prior estimate shows the standard deviation to be 1.6. Does your approach to sampling change? Explain.

9.14   Suppose that in a given region, subterranean water is found in lens (elliptical-shaped) reservoirs. Once a reservoir is located, the hydrologist can reasonably well delineate its boundary. Reservoirs differ in size. What sampling scheme should be proposed to find the reservoirs? What are the issues?

9.15   A study is to be commissioned regarding the use of public transportation by people in the United States. Discuss how such a survey might be conducted. Will the strategy differ if this is a phone survey versus an in-person interview?

9.16   Discuss the use of email to conduct a survey.

For Exercises 9.17 to 9.23, suppose that a laboratory is to be constructed for the study of landslides.

9.17   What factors should be incorporated in the design?

9.18   Which factors may be random and which may be fixed? Explain.

9.19   What considerations need to be made in choosing the number of levels of the factors?

9.20   Discuss any possible interaction effects. Explain.

9.21   Suggest a reasonable response variable. Should there be more than one response variable?

9.22   Might there be factors that cannot be controlled? If so, how are they modeled?

9.23   What is the difference in interpretation between fixed and random effects?

9.24   A laboratory has four technicians analyzing soil samples and there is suspicion of an operator effect on the results. How can an experiment be designed to test this conjecture? Is an operator a fixed or a random effect? Explain.

9.25   Explain why doing an experiment one factor at a time will not detect the presence of an interaction effect.

9.26   Intuitively, why is simple random sampling a better choice for ρ close to 1 when the alternative is systematic sampling?

9.27   Two of the variables collected in the Georgia soil sample study are potassium oxide (K_2O) and magnesium oxide (MgO). The investigator is not sure which to choose as the variable of interest. When does the choice matter in terms of the resulting allocation? Do you have any advice for the investigator if the investigator wishes to retain the influence of both variables? Explain.

10 Directional Data

10.1 INTRODUCTION

Examples of directional data include wind, geologic faults, movement of animals, ocean currents, and paleomagnetism. *Directional data* occur in two or three dimensions. Wind direction is an example of circular data in two dimensions; *paleomagnetic data* occur in three dimensions. Surficial orientation of faults is an example of axial data where the direction θ is indistinguishable from that of θ + 180°. Data in two dimensions is referred to as *circular data*; data in three dimensions is called *spherical data*. Although the mathematics and forms of distribution of circular and spherical data differ from those of data on the real line, the basic forms of investigation are similar, including the use of graphics, computation of measures of location and dispersion, and fitting distributions. Circular plots and rose diagrams are used to display circular data. Among the distributions used to describe data on a circle are the circular uniform and von Mises distributions. Goodness-of-fit tests exist for these distributions. Graphical procedures for spherical data involve projections. Point patterns typically include uniform, clustered, and girded distributions. As in two dimensions, measure of location and spread exist. Spherical distributions include the uniform, von Mises–Fisher, and Kent.

10.2 CIRCULAR DATA

The two basic types of data in two dimensions are circular and axial data. Wind direction is an example of circular data, which can vary over 360°. When viewed from the surface, faults vary over 180°, as θ and θ + 180° are indistinguishable. Angular data may be expressed in degrees or radians. Two major points of reference are used to display circular data. One is to have 0° be due east with counterclockwise rotation, so that 90° is north, 180° west, and 270° south. The second system, called the *geographic template*, has 0° due north with the rotation increasing clockwise, so that 90° is east, 180° south, and 270° west.

The following simple example illustrates why a different approach is needed to analyze directional data. Suppose that the wind direction at a weather station is from 30° at 8:00 A.M. and from 330° at 10:00 A.M. (0° being due east). The arithmetic

Statistics for Earth and Environmental Scientists, By John H. Schuenemeyer and Lawrence J. Drew
Copyright © 2011 John Wiley & Sons, Inc.

estimation of the mean is $(30° + 330°)/2 = 180°$. Clearly, the mean wind does not come from $180°$. To obtain a proper estimate requires a transformation of angular data into orthogonal (x and y) components so that $x = \cos \theta$ and $y = \sin \theta$. In this simple example $x_1 = \cos(30°) = 0.866$, $y_1 = \sin(30°) = 0.5$, $x_2 = \cos(330°) = 0.866$ and $y_2 = \sin(330°) = -0.5$. Now let $\bar{x} = (x_1 + x_2)/2 = 0.866$ and $\bar{y} = (y_1 + y_2)/2 = 0$. Thus, the mean wind is from

$$\bar{\theta} = \tan^{-1}\left(\frac{0}{0.866}\right) = 0°$$

10.2.1   Graphical Procedures for Circular Data

In January 2006, 742 observations of wind direction and speed were taken hourly at Mesa Verde National Park (U.S. National Park Service, Air Resources Division, 2008). A subset is shown in Table 10.1. Wind direction can be represented as a point on a unit circle. Often, it is of interest to see if wind direction varies by speed, time of day, or location. A *circular plot* of these data is shown in Figure 10.1. It is similar to a histogram except that points are plotted on the circumference of a unit circle. Points too close to be plotted individually on the circle are stacked. The circle can be considered to be of unit radius, with each dot representing an observation. The wind direction appears to be relatively uniform with somewhat more wind from the $325°$ to $12°$ range ($0°$ being due north). The line at $5.77°$ is the mean direction. The dot on the line represents the center of mass, which will be explained subsequently.

A second graphical procedure is the *rose diagram*, which is equivalent to a histogram in linear statistics and is shown in Figure 10.2. Twelve bins are specified by the user. This is equivalent to the number of histogram bars in linear statistics. The radii of the sectors (petals) are proportional to the square root of the relative frequencies of observations in each bin. Thus, the area of each sector is proportional to the frequency.

Wind measurements typically include direction and speed, as noted above. A special type of rose diagram that uses a color chart to denote speed is called a *wind rose*

TABLE 10.1   Subset of the January 2006 Wind Data from Mesa Verde National Park

Wind Direction (deg)	Wind Speed (m/s)	Date/Time
190	1.2	1/1/2006 0:00
239	5.4	1/1/2006 1:00
250	4.7	1/1/2006 2:00
260	5.1	1/1/2006 3:00
278	4.2	1/1/2006 4:00
285	4.1	1/1/2006 5:00
288	2.3	1/1/2006 6:00
286	1.1	1/1/2006 7:00
317	2.4	1/1/2006 8:00

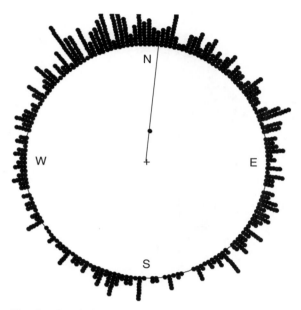

FIGURE 10.1   Circular plot of wind direction data, Mesa Verde National Park, January 2006.

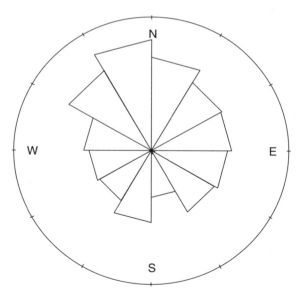

FIGURE 10.2   Rose diagram of wind direction data, Mesa Verde National Park, January 2006.

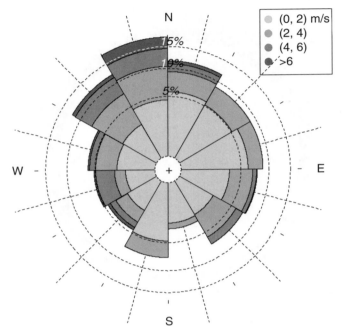

FIGURE 10.3   Wind rose diagram of wind direction and speed (m/s), Mesa Verde National Park, January 2006. (*See insert for color representation of the figure.*)

(Figure 10.3). The colors refer to wind speed in meters per second (m/s). A reverse heat template is used with four colors. The petals of a wind rose indicate the proportion of time that wind comes from a given direction. The bands on the wind rose, denoted by dashed lines, indicate the proportions of time that winds of each magnitude (5%, 10%, and 15%) occurred. This plot is made by windrose in the R-project function circular. It shows clearly that in January 2006 a preponderance of the higher winds came from the north–northwest.

Of course, circular data can be displayed as a traditional histogram that is shown in Figure 10.4. The obvious problem with this display is that the circular (modular) nature of the data is lost.

Axial data are another type of two-dimensional data. An example is geologic faults. Figure 10.5 shows geologic faults taken after Loudoun County Virginia data (Sutphin et al., 2001, Fig. 1). As noted, these data are bidirectional. The representation is constructed by adding a mirror image to the 20 sample values (directions), which indicates the indeterminate nature of bidirectional data.

10.2.2   Measures of Location and Spread

The mean direction is an important measure of location. Let $\mathbf{x}_i = (\cos \theta_i, \sin \theta_i)'$ be the location of the ith point on a unit circle expressed in *Cartesian coordinates* (also called *rectangular coordinates*, where typically, x is the abscissa and y the ordinate).

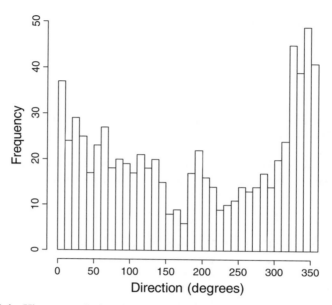

FIGURE 10.4   Histogram of wind direction data, Mesa Verde National Park, January 2006.

Note that because of the restriction that the points lay on the unit circle, data can be represented in polar coordinates by the angle θ. The means of n sines and cosines are

$$\bar{s} = \sum_{i=1}^{n} \frac{\sin \theta_i}{n} \quad \text{and} \quad \bar{c} = \sum_{i=1}^{n} \frac{\cos \theta_i}{n}$$

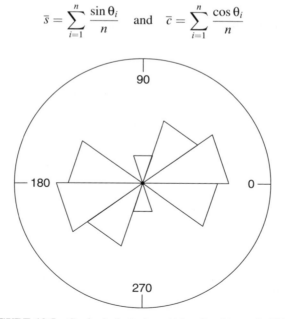

FIGURE 10.5   Geologic fault data. (After Sutphin et al. 2001.)

where θ_i is the angle of the ith observation (direction). The mean angle is then

$$\bar{\theta} = \begin{cases} \tan^{-1}(\bar{s}/\bar{c}) & \bar{c} > 0, \bar{s} > 0 \\ \tan^{-1}(\bar{s}/\bar{c}) + \pi & \bar{c} < 0 \\ \tan^{-1}(\bar{s}/\bar{c}) + 2\pi & \bar{c} > 0, \bar{s} < 0 \end{cases}$$

where $\tan^{-1}$ is the inverse tangent (also called the arctangent) and $\bar{\theta}$ is the mean angle in radians. $\bar{\theta}$ is also the direction of the center of mass $\bar{\mathbf{x}}$ of $\mathbf{x}_1, \ldots, \mathbf{x}_n$. For the Mesa Verde data set, $\bar{\theta} = 0.1007$ rad or $5.77°$ with $n = 742$.

Two measures of dispersion are the *mean resultant length* of the points on the circle and the circular variance. The mean resultant length is given by

$$\bar{R} = (\bar{c}^2 + \bar{s}^2)^{1/2}$$

For the Mesa Verde data, $\bar{R} = 0.258$. Since all points are on a unit circle, $0 \leq \bar{R} \leq 1$. An $\bar{R}$ near 1 implies clustering; an $\bar{R}$ near 0 implies dispersion but not necessarily that points are evenly spaced. Thus, $\bar{R}$ is a measure of concentration. Graphically, it is the length along the mean line (Figure 10.1) from the center of the circle to the dot. The *resultant length* R is the sum of the $\mathbf{x}_i$ or $n\bar{R}$, which for this example is 191.

The circular variance V_c is typically defined as $V_c = 1 - \bar{R}$ and $0 \leq V_c \leq 1$. A circular standard deviation may be defined for the real line (the interval 0 to infinity) as $v_c = (-2\log_e \bar{R})^{1/2}$. See Mardia and Jupp's (2000) book for details.

10.2.3 Circular Distributions and Associated Tests

Two of the most useful distributions are the circular uniform and the von Mises (circular normal distribution).

Uniform Distribution The probability density for the *circular uniform distribution* is

$$f(\theta) = \frac{1}{2\pi} \qquad 0 < \theta \leq 2\pi$$

If θ is an angle between α and β, then

$$\Pr(\alpha < \theta \leq \beta) = \frac{\beta - \alpha}{2\pi}$$

so that $\Pr(\alpha < \theta \leq \beta)$ is proportional to the arc length.

Kuiper's Test *Kuiper's test* is used to determine if a given set of data can be a sample from a specific distribution. It is similar to the Kolmogorov–Smirnov (K-S) test, as both compare cumulative distributions. For the one-sample test, the empirical cumulative distribution is compared to a theoretical cumulative distribution.

Define $F(\theta)$ to be a theoretical cumulative distribution function and

$$S_n(\theta) = \frac{i}{n} \quad \text{if } \theta_{(i)} \leq \theta < \theta_{(i+1)}, \quad i = 0, 1, \ldots, n$$

to be the empirical distribution function where the sample values $\theta_i, i = 1, \ldots, n$ are placed in ascending order as $\theta_{(1)}, \ldots, \theta_{(n)}$; $\theta_{(0)} = 0$ and $\theta_{(n+1)} = 2\pi$ are added. A complicating factor in circular data is that $S_n(\theta)$ and $F(\theta)$ depend on the choice of an initial direction. Kuiper (1960) overcomes this problem by formulating a test statistic that considers

$$D^+ = \max_\theta(S_n(\theta) - F(\theta)) \quad \text{and} \quad D^- = \max_\theta(F(\theta) - S_n(\theta))$$

Recall from Chapter 3 that the K-S test is $D = \max(D^+, D^-)$. Kuiper's test statistic is $V = D^+ + D^-$. In addition to V being invariant under cyclical transformations, the test is as sensitive in the tails as at the median. When applied to the Mesa Verde wind direction data, Kuiper's test statistic $= 5.043$ (Kuiper.test in the R-project library circular). At the 5% significance level the critical value is 1.747, which implies that the null hypothesis of uniformity [represented by $F(\theta)$] is rejected. Other tests of uniformity are discussed by Mardia and Jupp (2000).

Watson's U^2 Test of Uniformity   *Watson's U^2 test of uniformity* is based on a corrected mean-square deviation (Watson, 1961). A computational form of the test statistic is

$$U^2 = \sum_{i=1}^{n} \left(U_i - \overline{U} - \frac{i - 0.5}{n} + 0.5\right)^2 + \frac{1}{12n}$$

where U_i is $\theta_{(i)}$ expressed in radians and $\overline{U}$ is the mean of the U_i's.

Stephens (1969) showed that Kuiper's test may be preferred for small samples. For larger samples, specific alternatives need to be considered in determining the most appropriate test.

Von Mises Distribution and Associated Tests   The probability density function of the *von Mises distribution* is defined as

$$f(\theta \mid \mu, \kappa) = \frac{e^{\kappa \cos(\theta - \mu)}}{2\pi I_0(\kappa)} \quad 0 < \theta \leq 2\pi$$

where κ is a concentration parameter, μ is the mean of the distribution, and $I_0(\kappa)$ is a modified Bessel function of the first kind and order 0. A definition is given by Mardia and Jupp (2000). The von Mises density function is shown in Figure 10.6.

Watson's Test for the von Mises Distribution   The *Watson test* estimates the mean and concentration parameters (μ, κ) from the data using maximum likelihood or other

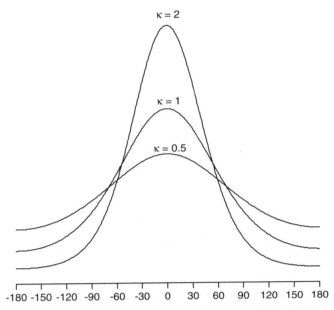

FIGURE 10.6   von Mises density function for $\kappa = 0.5, 1,$ and $2;\ -180° < \theta \leq 180°$.

estimator. Using this estimate, the von Mises cumulative distribution function $F(\theta) = \int_{a=0}^{\theta} f(a|\hat{\mu}, \hat{\kappa})$ is evaluated for the sample data $\theta_i, i = 1, \ldots, n$. If the data are from a von Mises distribution, the resulting transformed distribution should be approximately uniform. A one-sample Watson test is used to test uniformity.

For the Mesa Verde data, the maximum likelihood estimator for the concentration parameter is $\hat{\kappa} = 0.533$ with standard deviation 0.099. (The maximum likelihood is estimated with the function mle.vonmises in the R-project library circular.) For the Mesa Verde directional data, the test statistic is 0.264 for the null hypothesis that the data are von Mises distributed against the alternative that they are not. At the 5% significance level the critical value is 0.066; thus, the null is rejected. (The test is the Watson.test in the R-project function circular.) A von Mises probability plot is shown in Figure 10.7.

Since the uniform and von Mises distribution are rejected at the 5% significance level, the investigator should search for an alternative. Choices for the underlying distribution include a mixture of von Mises distributions or wrapped distribution. As the name implies, a *wrapped distribution* wraps the probability density function around the unit circle so that if y is a value on the real line, $y_w = y(\bmod 2\pi)$, where the w designates a wrapped value. Many standard distributions, including the Poisson and Cauchy, can be wrapped, allowing them to be used to model circular data.

Comments on Circular Data   Many of the tests and estimates used in linear statistics are parallel in construction to circular statistics. These include tests on parameters, two-sample tests to compare parameters from different distributions, and confidence

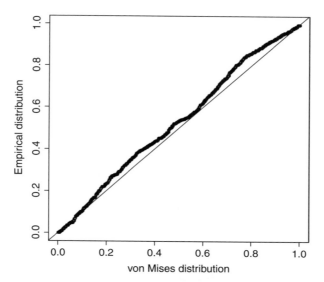

FIGURE 10.7   Probability plot of Mesa Verde directional data versus best fit to a von Mises distribution.

interval estimates. These and other procedures have been described in detail by Fisher (1993) and Mardia and Jupp (2000).

10.3   SPHERICAL DATA

Spherical data include the direction of magnetism in rock, geologic faults, movement of pollutants, wind directions, and ocean currents. Data such as faults are axial data, which means that there is an orientation but not a direction. The latter data are represented on a hemisphere.

Data in three dimensions are typically reported in one of two systems on a unit sphere. One way is a polar coordinate system with a co-latitude θ and a longitude ϕ (see Figure 10.8). In three dimensions the point $\mathbf{x}$ can be represented as $\mathbf{x} = (\cos\theta, \sin\theta\cos\phi, \sin\theta\sin\phi)'$. If for the three dimensions, the endpoint of the vector, $\mathbf{x} = (x_1, x_2, x_3)'$ is specified in Cartesian coordinates, the polar coordinates are $\theta = \tan^{-1}\left(\sqrt{x_2^2 + x_3^2}/x_1\right)$ and $\phi = \tan^{-1}(x_3/x_2)$. Because of the restriction that the point lies on a unit sphere, data can be represented in polar coordinates by θ and ϕ.

Measurements of geological properties are often given as strike and dip (Figure 10.9). The *strike* is the angle that the dipping plane makes with the horizontal (α in Figure 10.9; note that N and Strike are in the horizontal plane). The *dip*, β, is the angle below the horizontal plane of the feature. Often, the strike is given as a compass bearing (i.e., N30°E). Numerically, it is expressed as an *azimuth*, which is clockwise from north, or in this example, 30°. Strike (or azimuth) and dip are also expressed as declination and inclination, respectively.

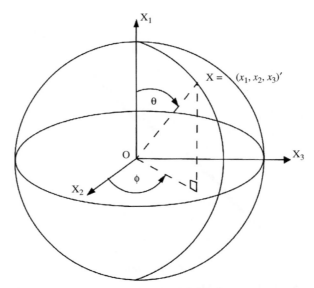

FIGURE 10.8   Spherical polar coordinates. (After Mardia and Jupp, 2000, Fig. 9.1.)

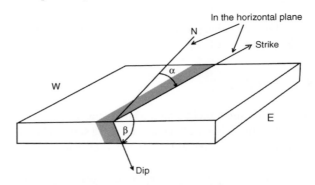

FIGURE 10.9   Strike and dip coordinate system.

In geography and physics, co-latitude is often expressed by its complement, $\delta = 90° - \theta$. It represents a zenith angle originating from the horizontal plane with a domain $-90° \leq \theta \leq 90°$. Computational software for spherical data is less common than for circular data. We have used Pmag Tools, v.4.2a (Hounslow, 2006).

10.3.1   Graphical Procedures for Spherical data

The three patterns that most often occur in earth science applications are shown in Figure 10.10. Note that the girdle distribution (Figure 10.10c) has points distributed around a circumference of the sphere. The spherical data shown in Table 10.2 are Lambert equal area projections. Lambert is a *stereographic projection*, which is a mapping or function that projects a sphere onto a plane. The *Lambert equal-area*

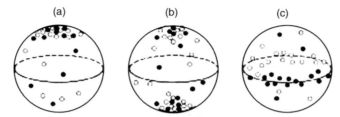

FIGURE 10.10   Spherical data representation: (a) unimodal distribution; (b) bipolar distribution; (c) girdle distribution. The solid circles represent observations on the side of the globe that is viewed. The unshaded circles represent circles on the opposite side of the globe. (From Mardia and Jupp, 2000, Fig. 2.)

projection represents areas accurately but does not preserve angular relationships between curves on the sphere. No projection can do both.

The example data set is from the Integrated Ocean Drilling Program (Acton et al., 2000). This program is designed to understand the Earth through ocean basin exploration. The entire data set for Leg 165 consists of 354 observations. The variables include interval, depth, declination, inclination, and intensity. For the following examples, only declination and inclination are used. Only the 30 observations shallower than 1 m below the seafloor (MBSF) are used (Table 10.2). A projection of the ocean drilling program data (Table 10.2) is shown in Figure 10.11. In Figure 10.11, 0° is north. The declination increases clockwise. The inclination is 90° at the center of the circle (the + symbol) and decreases to 0° at the perimeter. Note the elliptical nature of the point pattern.

TABLE 10.2   Subset of Leg 165 Data from the Integrated Ocean Drilling Program

Declination (deg)	Inclination (deg)	Declination (deg)	Inclination (deg)
277.1	45.6	334.5	62.9
298.2	69.8	342.6	61.2
350.1	77.2	357.5	62.2
281.4	49.0	335.3	61.6
288.1	53.8	339.5	60.9
289.9	60.5	354.0	61.3
276.8	58.4	340.6	65.0
281.0	65.3	345.1	64.0
292.1	70.2	3.2	65.2
301.3	58.5	336.3	73.4
305.6	56.9	337.8	75.6
316.3	61.5	26.5	75.9
329.5	61.0	308.8	72.4
332.9	61.0	298.1	72.5
344.2	62.2	299.0	74.3

Source: Data from Acton et al. (2000).

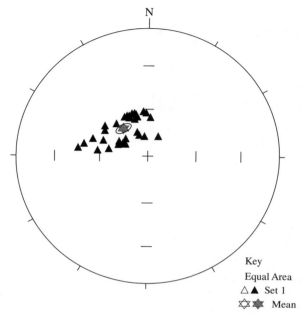

FIGURE 10.11   Projection of Leg 165 data with Kent mean (the six-pointed star) from the Integrated Ocean Drilling Program. The solid triangles and stars represent observations on the side of the globe that is viewed. The unshaded symbols represent observations on the opposite side of the globe and are unseen in this view. (From Acton et al., 2000.)

10.3.2   Measures of Location and Spread

The sample mean vector is

$$\bar{\mathbf{x}} = \frac{1}{n} \sum_{i=1}^{n} \mathbf{x}_i$$

where $\mathbf{x}_i' = (\cos\theta_i, \sin\theta_i \cos\phi_i, \sin\theta_i \sin\phi_i)$ is the ith sample point on a unit sphere. Analogous to the circular case,

$$\bar{x}_1 = \sum_{i=1}^{n} \frac{\cos\theta_i}{n}$$

$$\bar{x}_2 = \sum_{i=1}^{n} \frac{\sin\theta_i \cos\phi_i}{n}$$

$$\bar{x}_3 = \sum_{i=1}^{n} \frac{\sin\theta_i \sin\phi_i}{n}$$

TABLE 10.3   Relationship of Eigenvalues to Distributions

Relative Magnitudes of Eigenvalues	Type of Distribution	Comment
$t_1 \sim t_2 \sim t_3$	Uniform	
t_1 large; t_2, t_3 small		
$\quad t_2 \neq t_3, \overline{R} \sim 1$	Unimodal	Concentrated at one end of t_1
$\quad t_2 \neq t_3, \overline{R} < 1$	Bimodal	Concentrated at both ends of t_1
$\quad t_2 \sim t_3, \overline{R} \sim 1$	Unimodal	Rotational symmetry about t_1
$\quad t_2 \sim t_3, \overline{R} < 1$	Bipolar	
t_3 small; t_1, t_2 large		
$\quad t_1 \neq t_2$	Girdle	About great circle in plane of t_1, t_2
$\quad t_1 \sim t_2$	Symmetric girdle	Rotational symmetry about t_3

Source: After Mardia and Jupp (2000, Table 10.1).

and the mean resultant length is $\overline{R} = (\overline{x}_1^2 + \overline{x}_2^2 + \overline{x}_3^2)^{1/2}$, where $0 \leq \overline{R} \leq 1$. The mean direction of the sample is

$$\overline{\mathbf{x}}_0 = \frac{\overline{\mathbf{x}}}{\overline{R}}$$

As in the circular case, the resultant length is $n\overline{R}$. The total variation is $1 - \overline{R}^2$. Another important diagnostic tool is the scatter of points about the origin. This 3×3 (for $p = 3$) matrix is defined as

$$\overline{\mathbf{T}} = \frac{1}{n}\sum_{i=1}^{n} \mathbf{x}_i \mathbf{x}'_i$$

For the Leg 165 data, the descriptive statistics are

$$\overline{R} = 0.969$$
$$1 - \overline{R}^2 = 0.060$$

The sample mean direction in Cartesian coordinates is $\overline{\mathbf{x}}_0' = (0.295, -0.262, 0.919)$. The declination $= 318.4°$ and the inclination $= 66.8°$ at the mean direction. The eigenvalues of $\overline{\mathbf{T}}$ are $t_1 = 0.012$, $t_2 = 0.048$, and $t_3 = 0.940$. Because of the restriction that points lie on the unit sphere, the eigenvalues sum to 1. Sometimes they are multiplied by n. The interpretation of the eigenvalues is shown in Table 10.3.

10.3.3   Spherical Distributions and Associated Tests

There are many models for spherical distributions including:

- Uniform
- von Mises–Fisher
- Kent

- Fisher–Bingham
- Bingham–Mardia

Most can be considered special cases of the Fisher–Bingham distribution. Two of most commonly used distributions, the von Mises–Fisher and the Kent, are discussed.

von Mises–Fisher Distribution   On the unit circle, this is called the von Mises distribution. On a sphere it is the Fisher distribution. In higher dimensions it is the von Mises–Fisher distribution. Expressed in Cartesian coordinates $\mathbf{x} = (x_1, x_2, x_3)'$ the density function is

$$f(\mathbf{x} \,|\, \mu, \kappa) = C_p(\kappa) e^{\kappa \mu' \mathbf{x}}$$

where $C_p(\kappa) = \kappa^{p/2 - 1}/((2\pi)^{p/2} I_{p/2 - 1}(\kappa))$, $\kappa \geq 0$ is a concentration parameter, and $I_\nu(\kappa)$ is a modified Bessel function of order ν. As in the circular case, the spread of the distribution is inversely proportional to κ. In polar coordinates, $\mathbf{x} = (\cos\theta, \sin\theta\cos\phi, \sin\theta\sin\phi)'$ and $\mu = (\cos\alpha, \sin\alpha\cos\beta, \sin\alpha\sin\beta)'$, where α and β represent the directions at the mean of θ and ϕ, respectively. The form of the distribution in spherical coordinates is

$$f(\theta, \phi \,|\, \alpha, \beta, \kappa) = \frac{\kappa}{2\sinh\kappa}\exp\{\kappa[\cos\theta\cos\alpha + \sin\theta\sin\alpha\cos(\phi - \beta)]\}\sin\theta.$$

If μ is the pole, then θ and ϕ are independent and the density function of θ is

$$f(\theta \,|\, \kappa) = \frac{\kappa}{2\sinh\kappa}\sin\theta e^{\kappa\cos\theta}$$

The density function of ϕ is $g(\phi) = 1/2\pi$. For further details, see the book by Mardia and Jupp (2000). An important property of this class of distributions is that they are rotationally symmetric about their modal directions. Basically *rotational symmetry* means that the object looks the same after rotation. Selected density functions are shown in Figure 10.12. The colatitude distributions (Figure 10.12) become more skewed as κ increases.

A conjecture is made that the Leg 165 data has a Fisher distribution. This hypothesis is rejected at the 0.10 level of significance. To see why, consider a sample size of 30 simulated from the Fisher distribution (Figure 10.13). The point pattern is more circular than that of the Leg 165 data (Figure 10.11). This suggests the Kent as a candidate distribution.

Probability plots for the co-latitude θ and longitude ϕ are useful tools to investigate the goodness of fit. Prior to constructing spherical probability plots, (θ, ϕ) need to be rotated so that the sample mean direction is the pole $\theta = 0$.

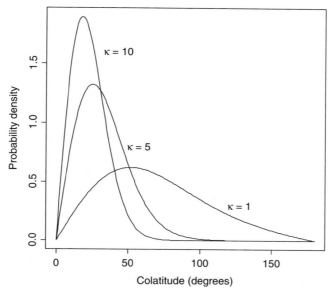

FIGURE 10.12   Probability density of the co-latitude (θ) for $\kappa = 1$, 5, and 10.

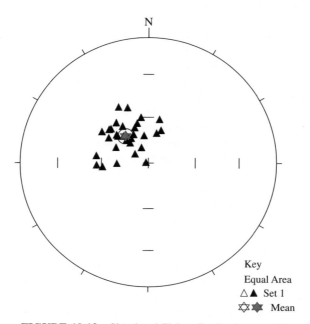

FIGURE 10.13   Simulated Fisher distribution, $n = 30$.

Kent Distribution   The *Kent distribution,* also known as the *five-parameter Fisher–Bingham distribution,* is most useful when the distribution of points is elliptical as opposed to circular. The probability density function expressed in Cartesian coordinates is

$$f(\mathbf{x}|\boldsymbol{\gamma}_1, \boldsymbol{\gamma}_2, \boldsymbol{\gamma}_3, \beta, \kappa) = \frac{1}{c(\beta, \kappa)} \exp\left\{ \kappa\boldsymbol{\gamma}_2'\mathbf{x} + \beta\left[(\boldsymbol{\gamma}_2'\mathbf{x})^2 - (\boldsymbol{\gamma}_3'\mathbf{x})^2 \right] \right\}$$

where $\boldsymbol{\gamma}_1$ is the mean direction, $\boldsymbol{\gamma}_2$ and $\boldsymbol{\gamma}_3$ are the major and minor axes, respectively, $\beta, 0 \leq 2\beta < \kappa$ is the ellipticity or ovalness of the contour, and κ is the concentration parameter. Higher values of β mean more ovalness. $\kappa/\beta \geq 2$ suggests a unimodal distribution; $\kappa/\beta < 2$ suggests a bimodal distribution.

For the Leg 165 data fit, $\overline{R} = 0.969$, $\beta = 14.63$, and $\kappa = 49.73$. Since $\kappa/\beta \geq 2$, a unimodal distribution is implied, which is consistent with the data plot (Figure 10.11).

Axial Distributions   When vectors $\mathbf{x}$ and $-\mathbf{x}$ are indistinguishable, such as for geologic faults, different distributions must be used. These include the Watson, Bingham, and angular central distributions (Mardia and Jupp, 2000)

10.4   SUMMARY

Directional data are circular and spherical. Circular data may be represented in the Cartesian coordinate system or in polar coordinates. The major difference between circular and linear data is that circular data are modular 2π or $360°$. Wind data are an example of circular data. Graphical procedures include plots, which are the circular equivalent of a histogram, and rose diagrams. A special form of the rose diagram, the wind rose, is used to display wind direction and speed. Axial data are circular data, but the directions are unique only to $180°$. Geologic faults projected to a surface are an example of axial data because there is no direction, only an orientation of the fault line. Measures of location and spread require conversion to the Cartesian coordinate system. The mean direction is an important measure of location. The mean resultant length and circular variance are measures of variability. Two circular distributions are the uniform and the von Mises. The uniform is similar to the uniform in linear statistics. Kuiper's test is one of many used to test uniformity. The von Mises or circular normal distribution plays a role equivalent to the normal distribution in linear statistics. It is bell-shaped and the variability decreases as the concentration parameter increases. Watson's test may be used to examine the fit of data to a von Mises distribution.

Spherical data occur in three or more dimensions. As with circular data, it can be either directional or axial. Spherical data can be represented in Cartesian coordinates or polar coordinates; however, in geology it is often represented by strike (azimuth) and dip or by declination and inclination. Patterns of spherical data can often be characterized by unimodal, bipolar, and girdle distributions. Measures of location and spread involve transformation to Cartesian coordinates. The resultant length and total variation are measures of spread. An important diagnostic tool is the relative magnitude of eigenvalues of a variation matrix $\overline{\mathbf{T}}$. Among the distributions that can be used to characterize data on a sphere are the uniform, the von Mises–Fisher, and the Kent.

EXERCISES

Note to the reader. Numerous weather data sets are available online, including those at airports and parks. Consider analyzing data in your region.

10.1   Explain why traditional linear statistics are not appropriate for the analysis of circular data.

Consider the small wind data set shown in Table 10.4, taken on Oct. 19, 2008 from the Chicago O'Hare International Airport. Other days and stations are available.

TABLE 10.4

Hour Min.	Wind Speed	Wind Direction	Hour Min.	Wind Speed	Wind Direction
0051	6	170	1251	22	190
0151	6	190	1351	17	190
0251	6	180	1451	16	180
0351	7	170	1551	17	190
0451	7	170	1651	17	190
0551	7	160	1751	10	190
0651	8	170	1851	10	180
0751	10	180	1951	10	190
0851	11	190	2051	13	200
0951	21	200	2151	17	210
1051	18	190	2251	14	210
1151	17	190	2351	8	240

Source: NOAA National Data Centers, http://www1.ncdc.noaa.gov/pub/.

10.2   Display wind direction and speed from Table 10.4 and comment on the form of the distribution of wind speed and the relationship between wind speed and direction.

10.3   Compute summary statistics on the Chicago wind data.

10.4   Is the wind data uniform? Perform a Kuiper's test of uniformity and comment.

10.5   Is the wind data circular normal? Apply a Watson's test and comment.

10.6   Is there evidence to suggest that wind speed and/or direction is different during the day (say, 0700 to 1900) than during the night?

10.7   What is the difference between directional and bidirectional data?

10.8   In addition to those discussed in this chapter, give other examples of directional data.

The data set given in Table 10.5 is for Exercises 10.9 to 10.12.

TABLE 10.5

Declination (deg)	Inclination (deg)	Declination (deg)	Inclination (deg)
336.3	57.5	356.1	53.6
332.0	32.0	348.6	22.5
325.3	28.7	350.1	22.8
340.9	54.1	359.1	55.2
336.8	36.8	351.4	27.5
328.8	34.9	354.9	28.8
341.1	61.2	0.7	60.2
330.2	40.8	347.7	27.4
319.2	37.4	355.2	28.1
1.3	52.0	358.1	59.6
353.1	28.9	345.0	17.6
355.8	29.4	352.8	18.7
0.9	59.6	350.2	60.7
353.6	31.3	335.3	24.1
356.9	31.0	325.4	22.4

Source: Data from the Integrated Ocean Drilling Program (Acton et al., 2000) in the depth range 10.00 to 10.99 mbsf.

10.9   Show a stereographic projection of the data set in Table 10.5 and comment.

10.10   Compute measures of location and spread for the data set in Table 10.5.

10.11   Compute the eigenvalues and comment on the general type of distribution that is observed (see Table 10.3 for guidelines).

10.12   Either by graphical or statistical procedures, comment on a possible specific distribution (uniform, von Mises–Fisher, Kent, etc.) for the data.

10.13   Either by graphical or statistical procedures, ascertain if there is a difference between the data set in Table 10.5 at 10 mbsf and the one in Table 10.2 for data less than 1 mbsf.

10.14   In addition to those discussed in this chapter, give examples of other types of spherical data.

REFERENCES

Abramowitz M, Stegun IA. *Handbook of Mathematical Functions.* Applied Mathematics Series 55. Washington, DC: U.S. Department of Commerce, National Bureau of Standards, 1972.

Acton GD, Galbrun B, King JW. Paleolatitude of the Caribbean Plate since the Late Cretaceous. *Proceedings of the Ocean Drilling Program: Scientific Results,* 2000;165:149–173.

Agresti A. *Categorical Data Analysis,* 2nd ed. New York: Wiley-Interscience, 2002.

Agterberg FP, Bonham-Carter GF, Wright DF. Statistical pattern integration for mineral exploration. In: Gaal G, Merriam DF, Eds. *Computer Applications in Resource Exploration, Prediction and Assessment for Metals and Petroleum.* Oxford, UK: Pergamon Press, 1990, pp. 1–21.

Agterberg FP, Bonham-Carter GF, Cheng Q, Wright DF. Weights of evidence modeling and weighted logistic regression for mineral potential mapping. In: Davis JC, Herzfeld UC, Eds. *Computers in Geology: 25 Years of Progress.* New York: Oxford University Press, 1993, pp. 13–32.

Aitchison J. *The Statistical Analysis of Compositional Data.* Monographs on Statistics and Applied Probability. Boca Raton, FL: Chapman & Hall, 1986.

Andersen EB. *Introduction to the Statistical Analysis of Categorical Data.* New York: Springer-Verlag, 1996.

Andersen R. *Modern Methods for Robust Regression.* Series on Quantitative Applications in the Social Sciences. Newbury Park, CA: Sage Publications, 2008, p. 07–152.

Andreatta G, Kaufman GM. Estimation of finite population properties when sampling is without replacement and proportional to magnitude. *Journal of the American Statistical Association* 1986;81(395):657–666.

Anscombe FJ. Graphs in statistical analysis. *The American Statistician* 1973;27(1):19.

Arkhipov SM, et al. Soviet glaciological investigations on Austfonna, Nordaustlandet, Svalbard in 1984 to 1985. *Polar Geography and Geology* 1987;11:25–49.

Arnold SF. *Mathematical Statistics.* Englewood Cliffs, NJ: Prentice Hall, 1990.

Baddeley A. 2008, *Analyzing Spatial Point Patterns in R.* Workshop Notes, Version 3. Australia: CSIRO, 2008, p. 199.

Baddeley A, Turner R. Spatstat: an *R* package for analyzing spatial point patterns. *Journal of Statistical Software* 2005;12:1–42.

Banerjee S, Carlin BP, Gelfand AE. *Hierarchical Modeling and Analysis for Spatial Data.* Boca Raton, FL: Chapman & Hall/CRC Press, 2004.

Bates DM, Watts DG. *Nonlinear Regression Analysis and Its Applications*. New York: Wiley, 1988.

Baum RL, Coe JA, Godt JW, Harp EL, Reid ME, Savage WZ, Schulz WH, Brien DL, Chleborad AF, McKenna JP, Michael JA. Regional landslide-hazard assessment for Seattle, Washington. *Landslides* 2005;2(4).

Bayarri MJ, Berger JO. The interplay of Bayesian and frequentist analysis. *Statistical Science*. 2004;19(1):58–80.

Belsley DA. *Conditioning Diagnostics, Collinearity and Weak Data in Regression*. New York: Wiley, 1991.

Belsley DA, Kuh E, Welch RE. *Regression Diagnostics: Identifying Influential Data and Sources of Collinearity*. New York: Wiley, 1980.

Berger JO. Bayesian analysis: a look at today and thoughts of tomorrow. *Journal of the American Statistical Association* 2000;95:1269–1276.

Berger JO. Could Fisher, Jeffreys, and Neyman have agreed on testing? *Statistical Science* 2003;18:1–32.

Bernardo JM, DeGroot MH, Lindley DV, Smith AFM, Eds. Weight of evidence: a brief survey. In: *Bayesian statistics 2: Proceedings of the Second Valencia International Meeting*, Sept. 6–10, 1983. New York: North-Holland, 1985, pp. 49–269. (including discussion).

Bonham-Carter GF. *Geographic Information Systems for Geoscientists: Modeling with GIS*. Kidlington, UK: Pergamon Press/Elsevier Science, 1994.

Box GEP, Hunter WG, Hunter JS. *Statistics for Experimenters: An Introduction to Design, Data Analysis, and Model Building*. New York: Wiley, 1978.

Box GEP, Jenkins GM, Reinsel GC. *Time Series Analysis, Forecasting and Control*, 3rd ed. Englewood Cliffs, NJ: Prentice Hall, 1994.

Bragg LJ, Oman JK, Tewalt SJ, Oman CJ, Rega NH, Washington PM, Finkelman RB. U.S. Geological Survey Coal Quality (COALQUAL) Database: 1998 Version 2.0. U.S. Geological Survey Open-File Report 97-134. CD-ROM. http://energy.er.usgs.gov/products/databases/CoalQual/intro.htm.

Breiman L, Friedman JH, Olshen RA, Stone CJ. *Classification and Regression Trees*. Boca Raton, FL: CRC Press, 1984.

Brillinger DR. 2000. Time series: general. In International Encyclopedia of Social and Behavioral Science. http://www.stat.berkeley.edu/users/brill/Papers/encysbs.pdf. Accessed Sept. 27, 2005.

Brockwell PJ, Davis RA. *Introduction to Time Series and Forecasting*, 2nd ed. New York: Springer-Verlag, 2002.

Burrus CS, Gopinath RA, Guo H. *Introduction to Wavelets and Wavelets Transforms: A Primer*. Upper Saddle River, NJ: Prentice Hall, 1998.

California Environmental Protection Agency. *California Ambient Air Quality Data 1980–2003*. PTSD-05-020-CD. Sacramento, CA: California EPA, Air Resources Board, Planning and Technical Support Division, 2005.

Campbell JA, Franczy KJ, Lupe RD, Peterson F. *National Uranium Resource Evaluation, Cortez Quadrangle, Colorado and Utah*. PGJ/F-051(82). Grand Junction, COl: U.S. Geological Survey, U.S. Department of Energy, 1982.

Carlin BP, Louis TA. *Bayes and Empirical Bayes Methods for Data Analysis*, 2nd ed. Boca Raton, FL: Chapman & Hall/CRC Press, 2000.

Chatfield C. *The Analysis of Time Series*, 6th ed. Boca Raton, FL: Chapman & Hall/CRC Press, 2003.

Chayes F. On correlation between variables of constant sum. *Journal of Geophysical Research* 1960;65:4185–4193.

Chleborad AF, Baum RL, Godt JW. *Rainfall Thresholds for Forecasting Landslides in the Seattle, Washington, Area: Exceedance and Probability.* Open-File Report 2006-1064. Washington, DC: U.S. Geological Survey, 2006.

Clark PJ, Evans FC. Distance to nearest neighbor as a measure of spatial relationship in populations. *Ecology* 1954;35:445–453.

Cleveland WS. *Visualizing Data.* Murray Hill, NJ: Hobart Press, 1993.

Cochran WG. Some methods for strengthening the common tests. *Biometrics* 1954;10:417–451.

Coe JA, Michael JA, Crovelli RA, Savage WZ, Laprade WT, Nashem WD. Probabilistic assessment of precipitation-triggered landslides using historical records of landslide occurrence, Seattle, Washington. *Environmental and Engineering Geoscience* 2004; 10(2):103–122.

Cohen AC. *Truncated and Censored Samples: Theory and Applications.* Statistics: Textbooks and Monographs Series No. 119. New York: Marcel Decker, 1991.

Consumer Reports. Annual auto issue, Apr. 2003, p. 90.

Cooley JW, Tukey JW. An algorithm for the machine calculation of complex Fourier series. *Mathematics of Computation* 1965;19:297–301.

Cox DR, Isham V. *Point Processes.* New York: Chapman & Hall, 1980.

Cox TF, Cox MAA. *Multidimensional Scaling.* 2nd ed. Boca Raton, FL: Chapman & Hall/CRC Press, 2001.

Cressie N. *Statistics for Spatial Data*, rev. ed. New York: Wiley, 1993.

Cressie N, Hawkins DM. Robust estimation of the variogram: I. *Journal of the International Assocation for Mathematical Geology* 1980;12:115–125.

Cutler A. 1997.The Little Ice Age: When Global Cooling Gripped the World. http://www-earth.usc.edu/geol150/evolution/images/Littleiceage/LittleIceAge.htm. Accessed Dec. 3, 2004.

Davis JC. *Statistics and Data Analysis in Geology*, 2nd ed. New York: Wiley, 1986.

Davison AC, Hinkley DV. *Bootstrap Methods and Their Application.* New York: Cambridge University Press, 1997.

Draper NR, Smith H. *Applied Regression Analysis.* 3rd ed. New York: Wiley, 1998.

Drew LJ. Grid-drilling exploration and its application to the search for petroleum. *Economic Geology* 1967;62:698–710.

Drew LJ, Grunsky EC, Schuenemeyer JH. Investigation of the structure of geological process through multivariate statistical analysis—the creation of a coal. *Mathematical Geosciences* 2008;40(7):789–811.

Drew LJ, Schuenemeyer JH, Mast RF. 1995. Application of the modified Arps–Roberts discovery process model to the 1995 U.S. National Oil and Gas Assessment. *Nonrenewable Resources* 1995;4:242–252.

Drew LJ, Karlinger MR, Schuenemeyer JH, Armstrong TR. Hydrogeologic significance of the association between well-yield variography and bedrock geologic structures, Pinardville Quadrangle, New Hampshire. *Natural Resources Research* 2003;12(1):79–91.

Durbin J, Watson GS. Testing for serial correlation in least squares regression: I. *Biometrika* 1950;37:409–428.

Dzwinel W, Yen DA, Boryczko K, Ben-Zion Y, Yoshioka S, Ito T. Nonlinear multidimensional scaling and visualization of earthquake clusters over space, time and feature space. *Nonlinear Process in Geophysics* 2005;12:117–128.

Efron B. Why isn't everyone a Bayesian? *American Statistician* 1986;40:6–11.

Egozcue JJ, Pawlowsky-Glahn V, Mateu-Figueras G, Barceló-Vidal C. Isometric logratio transformations for compositional data analysis. *Mathematical Geology* 2003;35(3): 279–300.

Evans M, Hastings N, Peacock B. *Statistical Distributions*, 3rd ed. New York: Wiley, 2000.

Everitt BS, Landau S, Leese M. *Cluster Analysis*, 4th ed. New York: Oxford University Press, 2001.

Fisher N. *Statistical Analysis of Circular Data*. New York: Cambridge University Press, 1993.

Flury B. *Common Principal Component and Related Multivariate Models*. New York: Wiley, 1988.

Fort Union Coal Assessment Team. *Resource Assessment of Selected Tertiary Coal Beds and Zones in the Northern Rocky Mountains and Great Plains Region*. Professional Paper 1625-A. Washington, DC: U.S. Geological Survey 1999.

Franses P-H. Modeling seasonality in economic time series. In: Ullah A, Gile DEA, Eds. *Applied Economics Statistics: Textbooks and Monographs*. New York: Marcel Dekker, 1998, pp. 553–577.

Fuller WA. *Measurement Error Model*. New York: Wiley, 1987.

Gebert KM, Calkin DE, Huggett RJ Jr, Abt KL. *Economic Analysis of Federal Wildfire Management Programs in the Economics of Forest Disturbances*. New York: Springer-Verlag 2008, pp. 295–322.

Gelman A, Carlin JB, Stern HS, Rubin DB. *Bayesian Data Analysis*, 2nd ed. Boca Raton, FL: Chapman & Hall/CRC Press, 2004.

Gentle JE. *Matrix Algebra: Theory, Computations, and Applications in Statistics*. Springer Texts in Statistics. New York: Springer-Verlag, 2007.

Gneiting T. 2001. *Nonseparable, Stationary Covariance Functions for Space–Time Data*. NRCES-TRS 063. Washington, DC: U.S. Environmental Protection Agency, 2001.

Golub GH, Van Loan CF. *Matrix Computations*, 3rd ed. Johns Hopkins Studies in Mathematical Sciences. Baltimore: The John Hopkins University Press, 1996.

Good IJ. *Probability and the Weighing of Evidence*. London: Charles Griffin; New York: Hafner Publishing, 1950.

Goodman L. *Analyzing Qualitative/Categorial Data*. Reading, MA: Addison-Wesley, 1978.

Goovaerts P. *Geostatistics for Natural Resources Evaluation*. New York: Oxford University Press, 1997.

Gotelli NJ. Null model analysis of species co-occurrence patterns. *Ecology* 2000;81(9): 2606–2621.

Gower JC, Hand DJ. Biplots. *Computational Statistics and Data Analysis* 1996;22(6):651–655.

Griffiths JC. *Scientific Method in Analysis of Sediments*. New York: McGraw-Hill, 1967.

Gros J. Variance inflation factors. *R News* 2003;3(1):13–15.

Hall ME. Pottery styles during Early Jormon Period: geochemical perspectives on the Moroiso and Ukishima pottery styles. *Archaeometry* 2001;43(1):59–75.

Han J, Kamber M. *Data Mining: Concepts and Techniques*. San Diego, CA: Morgan Kaufmann, 2000.

Handcock MS, Stein ML. A Bayesian analysis of kriging. *Technometrics* 1993;35(4):403–410.

Hardy MA. Regression with Dummy Variables, Series on Quantitative Applications in the Social Sciences. Newbury Park, CA: Sage Publications, 1993, p. 07–093.

Harmon HH. *Modern Factor Analysis*, 3rd ed. Chicago: University of Chicago Press, 1976.

Hastie T, Tibshirani R, Friedman JH. *The Elements of Statistical Learning*. New York: Springer-Verlag, 2003.

Hawkins D. *Identification of Outliers*. New York: Chapman & Hall, 1980.

Helsel D. *Nondetects and Data Analysis: Statistics for Censored Environmental Data*. Hoboken, NJ: Wiley, 2005.

Henning C, Kutlukaya M. 2007, Some thoughts about the design of loss functions. *REVSTAT— Statistical Journal*, Mar. 2007;5(1):19–39.

Hinkle SR, Polette DJ. *Arsenic in Ground Water of the Willamette Basin, Oregon*. Water-Resources Investigations Report 98–4205. Washington, DC: U.S. Geological Survey, 1999, p. 28.

Holt C. 1957. Forecasting seasonals and trends by exponentially weighted moving averages. *ONR Research Memorandum, Carnegie Institute*, 1957;52(7).

Hosmer DW, Lemeshow S. *Applied Logistic Regression*, 2nd ed. Wiley Series in Probability and Statistics, Applied Probability and Statistics Section. New York: Wiley, 2000.

Hounslow MW. PMag Tools, v.4.2a. Jan. 2006. http://geography.lancaster.ac.uk/cemp/cemp.htm. Accessed July 1, 2008.

Houseknecht DW, Ed. National Petroleum Reserve–Alaska (NPRA) Core Images and Well Data. 2002. U.S. Geological Survey Digital Data Series DDS-0075 (4 CD-ROMs).

Hsu JC. *Multiple Comparisons: Theory and Methods*. Boca Raton, FL: CRC Press, 1996.

Hu LY, Chugunova T. Multiple-point geostatistics for modeling subsurface heterogeneity: a comprehensive review. *Water Resources Research* 2008;44(W11413):14.

Humphries EM. Pfiesteria piscicida *and* Pfiesteria shumwayae *Associations with Environmental Parameters*. Dover, DE: State of Delaware, Department of Natural Resources and Environmental Control, 2003.

Isaaks EH, Srivastava RM. *An Introduction to Applied Geostatistics*. New York: Oxford University Press, 1989.

Jagger TH, Niu X, Elsner JB. A space–time model for seasonal hurricane prediction. *International Journal of Climatology* 2002;22:451–465.

Johnson DH. 1999. The insignificance of statistical significance testing. *Journal of Wildlife Management* 63(3):763–772. Jamestow, ND: Northern Prairie Wildlife Research Center Online. http://www.npwrc.usgs.gov/resource/methods/statsig/index.htm (version 16SEP99).

Johnson RA, Wichern DW. *Applied Multivariate Statistical Analysis*, 5th ed. Upper Saddle River, NJ: Prentice Hall, 2002.

Joiner BL. Lurking variables: some examples. *American Statistician* 1981;35(4):227–233.

Journel AG, Huijbregts CJ. *Mining Geostatistics*. New York: Academic Press, 1978.

Kaufman G. What has seismic missed? *Nonrenewable Resources Journal* 1994;3(4):304–314.

Kaufman L, Rousseeuw PJ. *Finding Groups in Data: An Introduction to Cluster Analysis*. Hoboken, NJ: Wiley, 2005.

Kimball BA. Geochemistry of water associated with the Navajo Sandstone Aquifer, San Rafael Swell Area, Utah. AWRA Monograph Series 14. In: McLean JS, Johnson AI, Eds. *Regional Aquifer Systems of the United States: Aquifers of the Western Mountain Area*. Bethesda, MD: American Water Resources Association, 1988, pp. 121–134.

Krige DG. A statistical approach to some basic mine valuation problems on the Witwatersrand. *Journal of the Chemical, Metallurgical and Mining Society of South Africa* 1951;52:119–139.

Kriegel HP, Pfeifle M. Clustering moving objects via medoid clusterings. Presented at the 17th International Conference on Scientific and Statistical Database Management (SSDMB'05), Santa Barbara, CA, 2005.

Kuiper NH. Tests concerning random points on a circle. *Proceedings Koninklijke. Nederlandse. Akademic van Wetenschappen* 1960 Series A (63):38–47.

Lau MK. Dunnett–Tukey–Kramer Pairwise Multiple Comparison Test Adjusted for Unequal Variances and Unequal Sample Sizes. http://finzi.psych.upenn.edu/R/library/DTK/html/DTK-package.html. Accessed Aug. 29, 2009.

Lawler GF. *Introduction to Stochastic Processes*. New York: Chapman & Hall, 1996.

Levene H. 1960. Olkin I, et al., Eds. In *Contributions to Probability and Statistics: Essays in Honor of Harold Hotelling*. Stanford, CA: Stanford University Press, 1960, pp. 278–292.

Levy PS, Lemeshow S. *Sampling of Populations: Methods and Applications*. New York: Wiley-Interscience, 1991.

Little RJ. Calibrated Bayes: a Bayes/frequentist roadmap. President's Invited Address, Joint Statistical Meetings, Minneapolis, MN, Aug. 8, 2005. http://sitemaker.umich.edu/rlittle/files/roadmap.pdf. Accessed Sept. 29, 2005.

Little RJ, Rubin DB. *Statistical Analysis with Missing Data*, 2nd ed. Hoboken, NJ: Wiley, 2002.

Liu I, Agresti A. The analysis of ordered categorical data: an overview and a survey of recent developments. *Sociedad de Estadistica e Investigacion Operativa Test* 2005;14(1):1–73.

Mahlman, JD. Science and nonscience concerning human-caused climate warming. *Annual Review of Energy and the Environment* 1998;23:83–105.Available at http://www.gfdl.noaa.gov/~gth/web_page/article/aree_page1.html. Accessed Apr. 24, 2003.

Mann ME, Bradley RS, Hughes MK. Northern Hemisphere Temperature Reconstruction for the Past Millennium. IGBP PAGES/World Data Center-A for Paleoclimatology Data Contribution Series 1999-014. Boulder CO: NOAA/NGDC Paleoclimatology Program, 1999.

Mardia KV, Jupp PE. *Directional Statistics*. New York: Wiley, 2000.

McCullagh P, Nelder JA. *Generalized Linear Models*, 2nd ed. New York: Chapman & Hall, 1989.

McLachlan GJ. *Discriminant Analysis and Statistical Pattern Recognition*. New York: Wiley, 1992.

Menard S. *Applied Logistic Regression Analysis*. Series on Quantitative Applications in the Social Sciences. Thousand Oaks, CA: Sage Publications, 1995, p. 07–106.

Merriam DF, Davis JC, Eds., *Geologic Modeling and Simulations: Sedimentary Systems*. (Computer Applications in the Earth Sciences) New York: Kluwer Academic/Plenum Press, 2001.

Montgomery DC. *Design and Analysis of Experiments*, 5th ed. New York: Wiley, 2000.

Montgomery DC, Myers RH. *Response Surface Methodology: Process and Product Optimization Using Designed Experiments*. 2nd ed. Hoboken, NJ: Wiley, 2002.

Montgomery DC, Peck EA, Vining GG. *Introduction to Linear Regression Analysis*. 3rd ed. New York: Wiley, 2001.

Mosteller F, Tukey JW. *Data Analysis and Regression*. Reading, MA: Addison-Wesley, 1977.

National Geophysical Data Center. http://www.ngdc.noaa.gov/. Accessed Nov. 22, 2004.

National Weather Service National Hurricane Center 2006. http://www.nhc.noaa.gov/aboutsshs.shtml. Accessed Nov. 29, 2006.

Neuhauser B, Terhorst B. Landslide susceptibility assessment using "weights-of evidence" applied to a study area at the Jurassic escarpment (SW-Germany). *Geomorphology* 2007;86:12–24.

Neyman J. Frequentist probability and frequentist statistics. *Synthese* 1977;36:97–131.

NIST/SEMATECH e-Handbook of Statistical Methods. 2006. http://www.itl.nist.gov/div898/handbook/. Accessed Jan. 5, 2007.

Oregon State University O.H. Hinsdale Wave Research Laboratory 2004. http://wave.oregonstate.edu/. Accessed Dec. 29, 2004.

Pawlowsky V, Egozcue JJ. Compostional data and their analysis. In: Buccianti A, Mateu-Figueras G, Pawlowsky V, Eds. *Compositional Data Analysis in the Geosciences from Theory to Practice*. Special Publication 264. London: Geological Society of London, 2006, pp. 1–9.

Porwal A, Carranza EJM, Hale M. Extended weights-of-evidence modeling for predictive mapping of base metal deposit potential in Aravalli province, Western India. *Exploration and Mining Geology* 2001;10(4):273–287.

R Version 2.10.2009-10-23. Accessed Aug. 25, 2008.

Reich RM, Davis R. *Collection of Splus Functions for Analyzing Spatial Data*. Fort Collins, CO: Department of Forest Sciences, Colorado State University, 2002.

Reyment RA, Jöreskog KG. *Applied Factor Analysis in the Natural Sciences*. Cambridge, UK: Cambridge University Press, 1996.

Ribeiro PJ Jr, Diggle PJ. GeoR: a package for geostatistical analysis. *R-News*, June, 2001; 1(2):15–18.

Ricci V. Fitting Distributions with R. 2005. http://cran.r-project.org/doc/contrib/Ricci-distributions-en.pdf. Accessed Feb. 1, 2009.

Ripley BD. *Pattern Recognition and Neural Networks*. Cambridge, UK: Cambridge University Press, 1996.

Rousseeuw PJ, Leroy AM. *Robust Regression and Outlier Detection*. New York: Wiley, 1987.

Rubin DB. Inference and missing data. *Biometrika* 1976;63:581–590.

Ruddiman WF, Raymo ME. *Methane-Based Time Scale for Vostok Ice*. IGBP PAGES/World Data Center for Paleoclimatology Data Contribution Series 2004-016. Boulder CO: NOAA/NGDC Paleoclimatology Program, 2004. ftp://ftp.ncdc.noaa.gov/pub/data/paleo/icecore/antarctica/vostok/vostok_methane_age.txt.

Ruppert LF, Tewait SJ, Wallack RN, Bragg LJ, Brezinski DK, Carlton RW, Butler DT, Calef FJ III, A Digital Resource Model of the Middle Pennsylvania Upper Freeport Coal Bed, Allegheny Group, Northern Appalachian Basin Coal Region, USA. 2001. Chapter D in U.S. Geological Survey Professional Paper 1625-C, Version 1.0, pp. D1–D97 (CD-ROM).

Ryan TP. *Modern Regression Methods*. Hoboken, NJ: Wiley, 2008.

Scheaffer RL, Mendenhall W. III, Ott RL. *Elementary Survey Sampling*, 5th ed. Belmont, CA: Duxbury Press, 1996.

Schuenemeyer JH, Drew LJ. Description of a Discovery Process Modeling Procedure to Forecast Future Oil and Gas Using Field Growth. ARDS 4.01. U.S. Geological Survey Digital Data Series DDS-36. 1996.

Schuenemeyer JH, Power HC. Uncertainty estimation for resource assessment: an application to coal. *Mathematical Geology* 2000;32(5):521–541.

Seber GAF, Lee AJ. *Linear Regression Analysis*, 2nd ed. Hoboken, NJ: Wiley; 2003.

Seber GAF, Wild CJ. *Nonlinear Regression*. Hoboken, NJ: Wiley, 2003.

Shannon & Wilson. 2000, *Seattle Landslide Study*. Seattle, WA: Shannon & Wilson, Inc., Geotechnical and Environmental Consultants, 2000. http://www.seattle.gov/DPD/Landslide/Study. Accessed June 24, 2008.

Singer DA, Kouda R. A comparison of the weights-of-evidence method and probabilistic neural networks. *Natural Resources Research* 1999;8(4):287–298.

Snedecor GW, Cochran WG. *Statistical Methods*. 8th ed. Ames IA: Iowa State University Press, 1989.

Sokal RR, Rohlf FJ. *Biometry: The Principles and Practice of Statistics in Biological Research*, 3rd ed. New York: W.H. Freeman, 1995.

Stephens MA. A goodness-of-fit statistic for the circle, with some comparisons. *Biometrika* 1969;56:169–181.

Sterne JAC, Smith GD. Sifting the evidence: What's wrong with significance tests? *British Medical Journal* 2001;322(7280):226–231.

Stigler SM. *The History of Statistics: The Measurement of Uncertainty Before 1900*. Cambridge, MA: Harvard University Press, 1986.

Strebelle S. 2002. Conditional simulation of complex geological structures using multiple-point statistics. *Mathematical Geology* 2002;34(1):1–21.

Sutphin DM, Drew LJ, Schuenemeyer JH, Burton WC. Characteristics of water-well yields in part of the Blue Ridge Geologic Province in Loudoun County, Virginia. *Natural Resources Research* 2001;10(1):1–19.

Thioulouse J, Dray S. *Journal of Statistical Software* Sept. 2007;22(5). http://www.jstatsoft.org/.

Thompson LA. 2006. *S-Plus (and R) Manual to Accompany Agresti's Categorical Data Analysis* (2002), 2nd ed. 2006. http://home.earthlink.net/~lthompson33/Splusdiscrete2.pdf. Beta version accessed Feb. 21, 2006.

Titterington DM, Smith AF, Makov UE. *Statistical Analysis of Finite Mixture Distributions*. New York: Wiley, 1986.

Tukey JW. *Exploratory Data Analysis*. Reading, MA: Addison-Wesley, 1977.

U.S. Department of Agriculture–U.S. Department of the Interior. Joint Fire Science Program. 2006. http://www.nifc.gov/joint_fire_sci/index.html. Accessed Jan. 2, 2007.

U.S. Environmental Protection Agency. *Guidance for Choosing a Sampling Design for Environmental Data Collection*. EPA QA/G-5S. Washington, DC: U.S. EPA, 2002.

U.S. Environmental Protection Agency. Final National Priority List Site, by Date. 2004. http://www.epa.gov/superfund/sites/query/queryhtm/nplfin.htm. Accessed Nov. 5, 2004.

U.S. Environmental Protection Agency. Techniques for Assessing Water Quality and for Estimating Pollution Loads. 2005. http://www.epa.gov/owow/nps/MMGI/Chapter8/ch8-2.html. Accessed May 16, 2005.

U.S. Geological Survey. The National Geochemical Survey: Database and Documentation. U.S. Geological Survey Open-File Report 2004-1001. http://pubs.usgs.gov/of/2004/1001/. Accessed Jan. 4, 2005.

U.S. Geological Survey National Ice Core Laboratory. 2004. http://nicl.usgs.gov/. Accessed Dec. 3, 2004.

U.S. Geological Survey Upper Midwest Environmental Sciences Center. Pool 8 Transect Data Summary. 2005. http://www.umesc.usgs.gov/data_library/vegetation/transect/pool8/ p8_summary.html. Accessed May 11, 2006

U.S. Geological Survey World Energy Assessment Team. U.S. Geological Survey World Petroleum Assessment 2000: Description and Results. 2000. U.S. Geological Survey Digital Data Series DDS-60.

U.S. National Park Service, Air Resources Division. NPS Gaseous Pollutant and Meterological Data Access. 2008. http://ard-request.air-resource.com/readme.html. Accessed Aug.10, 2008.

van den Boogaarta KG, Tolosana-Delgadob R. Compositions: a unified R package to analyze compositional data. *Computers and Geosciences* 2008;34:320–338.

Venables WN, Ripley BD. *Modern Applied Statistics with S-Plus*, 2nd ed. New York: Springer-Verlag, 1997.

Watson GS. Goodness-of-fit test on a circle. *Biometrika* 1961;47:109–114.

West LJ, Hankin RKS. Exact tests for two-way contingency tables with structural zeros. *Journal of Statistical Software* 2008;28(11). http://www.jstatsoft.org/v28/i11/.

Western Regional Climate Center. Denver and Cheyenne Wells Data. 2005. http://www.wrcc .dri.edu/summary/climsmco.html. Accessed Aug. 25, 2006.

WinBugs Project. Cambridge, UK: MRC Biostatistics Unit, 2005. http://www.mrc-bsu.cam .ac.uk/bugs/. Accessed Jan. 10, 2006.

INDEX

Actual error rate (AER), 283, 284
Adjusted R^2, 112
Agglomerative method, 264, 271, 290
Akaike's Information Criterion (AIC), 171, 176,
    177, 179, 321, 322, 327
Alternative hypothesis, 72
Anisotropy, 214–216, 226, 240, 241
    geometric, 214–216, 239, 240
    zonal, 214, 215, 217, 239
Analysis of covariance (ANCOVA), 363
Analysis of variance (ANOVA), 85, 86, 87, 90, 95,
    96, 97, 110, 111, 125, 127, 149, 340, 352
Apparent error, 132, 133
    rate (APER), 283, 284, 289
Autocorrelation function (ACF), 164, 166, 167,
    168, 170, 171, 174–180, 188, 190, 192
Autocovariance function (ACVF), 162, 164, 166,
    168, 186, 187
Autoregressive
    integrated moving-average model (ARIMA),
        169, 171, 176, 177, 178, 179, 180, 189, 190,
        192
    moving-average model (ARMA), 169, 189
    process (AR), 166–171, 180, 188
Axial data, 374, 379, 386
Azimuth, 376, 386
    tolerance, 215

Backshift operator, 166, 169
Backward selection, 139, 140
Balanced design, 364
Bandwidth, 158
Bar plot, 199, 200
Bayes
    factor K, 80
    theorem, 31, 32, 61, 79, 328
Bayesian
    approach, 61, 62, 71, 78, 95, 96, 97
    confidence interval, 61, 62, 63

Information Criterion (BIC), 171, 177
    kriging, 233, 235, 239
Bernoulli trial, 294, 299
Bias, 18, 19, 20, 22, 23, 34, 35, 50, 70, 71, 83, 106,
    112, 122, 141, 230, 324, 348 350, 362, 366,
    367, 368
Bimodal distribution, 8, 383, 386
Binomial
    probability function, 294, 299, 300
    process, 294, 295, 299, 321, 328, 329
Biplot, 255–257, 261–263
Block kriging (BK), 231, 232, 239, 308
Blocking, 350, 358, 361–363, 366
Bonferroni's method, 88
Bootstrap sample, 70, 84, 140, 141
Bootstrapping, 70, 83, 132, 140, 260
Box-Cox transformation, 27, 69, 286
Boxplot, 14, 15, 16, 33, 34, 68, 69, 87, 88, 97, 117,
    118, 207, 208
Bubble plot, 193–196, 245–247

Cartesian coordinates, 374, 379, 383, 384, 386
Categorical, 2, 137, 286, 311, 320
    data, 2, 234, 314, 317
    explanatory variable, 137
Censored sample, 22, 23
Censoring, 22, 23
Centered
    data, 136
    log ratio transformation (CLR), 252–256
Central limit theorem (CLT), 52, 53, 59, 72
Characteristic values, 249. *See also* Eigenvalues
Characteristic vectors, 250. *See also* Eigenvectors
Circular
    data, 371, 372, 374, 377, 378, 380, 386, 387
    plot, 372, 373
    uniform distribution, 376
Clark and Evans nearest-neighbor index (CEI), 305,
    330

Statistics for Earth and Environmental Scientists, By John H. Schuenemeyer and Lawrence J. Drew
Copyright © 2011 John Wiley & Sons, Inc.